T0141859

Engineering Cyber-Physical Systems and Critical Infrastructures

Volume 6

The aim of this book series is to present state of the art studies, research and best engineering practices, real-world applications and real-world case studies for the risks, security, and reliability of critical infrastructure systems and Cyber-Physical Systems. Volumes of this book series will cover modelling, analysis, frameworks, digital twin simulations of risks, failures and vulnerabilities of cyber critical infrastructures as well as will provide ICT approaches to ensure protection and avoid disruption of vital fields such as economy, utility supplies networks, telecommunications, transports, etc. in the everyday life of citizens. The intertwine of cyber and real nature of critical infrastructures will be analyzed and challenges of risks, security, and reliability of critical infrastructure systems will be revealed. Computational intelligence provided by sensing and processing through the whole spectrum of Cloud-to-thing continuum technologies will be the basis for real-time detection of risks, threats, anomalies, etc. in cyber critical infrastructures and will prompt for human and automated protection actions. Finally, studies and recommendations to policy makers, managers, local and governmental administrations and global international organizations will be sought.

Bharat Bhushan · Arun Kumar Sangaiah ·
Tu N. Nguyen
Editors

AI Models for Blockchain-Based Intelligent Networks in IoT Systems

Concepts, Methodologies, Tools, and Applications

Editors
Bharat Bhushan
Department of Computer Science
and Engineering
School of Engineering and Technology
Sharda University
Greater Noida, India

Arun Kumar Sangaiah
School of Computing Science
and Engineering
Vellore Institute of Technology
Vellore, Tamil Nadu, India

Tu N. Nguyen
Intelligent Systems Laboratory (ISL)
Department of Computer Science
Kennesaw State University
Marietta, GA, USA

ISSN 2731-5002 ISSN 2731-5010 (electronic)
Engineering Cyber-Physical Systems and Critical Infrastructures
ISBN 978-3-031-31954-9 ISBN 978-3-031-31952-5 (eBook)
https://doi.org/10.1007/978-3-031-31952-5

This Springer imprint is published by the registered company Springer Nature Switzerland AG
The registered company address is: Gewerbestrasse 11, 6330 Cham, Switzerland

Preface

The Internet of Things (IoT) is an emerging technology that integrates technologies and aspects coming from varied approaches. In addition to connecting devices, IoT also connects people and other entities thereby making every IoT component vulnerable to a huge range of attacks. Therefore, enabling the IoT devices and WSNs to learn and adapt to various threats dynamically and addressing them proactively need immediate attention. Rapidly growing IoT will soon permeate to each and every aspect of our lives and therefore addressing IoT security threats is of utmost importance. Owing to the massiveness of IoT and inadequate data security, the impact of security breaches may turn out to be humongous leading to severe impacts. In order to analyse the voluminous data, Artificial Intelligence (AI) can serve as a strong analytic tool to deliver accurate and scalable real-time data analysis. However, development of AI-based big data analysis tool have some challenges, such as resource-constrained nature, privacy, security, lack of training data and centralized architecture. To this end, blockchain technology can overcome the existing AI challenges and provide secure sharing of resources and data to various IoT nodes. Blockchain can power decentralized coordination platforms and marketplaces for various AI components. This leads to the adoption of AI to an unprecedented level. The goal of this book is to foster transformative, multidisciplinary and novel Artificial Intelligence (AI)-based approaches that ensures security by taking into consideration the unique security challenges present in the environment. The book also discusses the challenges that need to be addressed in order to implement various security solutions in practical WSNs and IoT systems. The goal of this book is to foster transformative, multidisciplinary and novel Artificial Intelligence (AI)-based approaches that ensures security by taking into consideration the unique security challenges present in the environment.

Greater Noida, India — Bharat Bhushan
Vellore, India — Arun Kumar Sangaiah
Marietta, USA — Tu N. Nguyen

Contents

About the Editors

Dr. Bharat Bhushan is an Assistant Professor in the Department of Computer Science and Engineering (CSE) at the School of Engineering and Technology, Sharda University, Greater Noida, India. He received his Undergraduate Degree (B.Tech. in Computer Science and Engineering) with Distinction in 2012, received his Postgraduate Degree (M.Tech. in Information Security) with Distinction in 2015 and Doctorate Degree (Ph.D. Computer Science and Engineering) in 2021 from Birla Institute of Technology, Mesra, India. In the years 2021 and 2022, Stanford University (USA) listed him in the top 2% of scientists list. He earned numerous international certifications such as CCNA, MCTS, MCITP, RHCE and CCNP. He has published more than 150 research papers in various renowned International Conferences and SCI-indexed journals including *Journal of Network and Computer Applications* (Elsevier), *Wireless Networks* (Springer), *Wireless Personal Communications* (Springer), *Sustainable Cities and Society* (Elsevier) and *Emerging Transactions on Telecommunications* (Wiley). He has contributed with more than 30 book chapters in various books and has edited 20 books from the most famed publishers like Elsevier, Springer, Wiley, IOP Press, IGI Global and CRC Press. He has served as Keynote Speaker (resource person) in numerous reputed faculty development programs and international conferences held in different countries including India, Iraq, Morocco, China, Belgium and Bangladesh. He has served as a Reviewer/Editorial Board Member for several reputed international journals. In the past, he worked as an assistant professor at HMR Institute of Technology and Management, New Delhi, and Network Engineer in HCL Infosystems Ltd., Noida. In addition to being the senior member of IEEE, he is also a member of numerous renowned bodies including IAENG, CSTA, SCIEI, IAE and UACEE.

Dr. Arun Kumar Sangaiah received his Master of Engineering from Anna University and Ph.D. from VIT University, in 2007 and 2014, respectively. He is currently a Professor at the School of Computing Science and Engineering, VIT University, Vellore, India, where he teaches a variety of university courses in Computer Science, such as Software Engineering, Software Project Management, Web technologies and Computer Networks. In 2016, he was a visiting professor at the School of Computer

Engineering at Nanhai Dongruan Information Technology Institute in China for 6 months. In addition, he has been appointed as a visiting professor in Southwest Jiaotong University, Chengdu, Changsha University of Science and Technology, China, Dongguan University of Technology, Guangdong, and Hwa-Hsia University of Technology, Taiwan. Further, he has visited many research centres and universities in China, Japan and South Korea for joint collaboration towards research projects and publications. His areas of research interest include e-Learning, learning engineering, machine learning, software engineering, computational intelligence, IoT, wireless networks, bio-informatics and embedded systems. His outstanding scientific production spans over 230+ contributions published in high-standard ISI journals, such as *IEEE-Communication Magazine*, *IEEE Systems* and *IEEE IoT*. His publications are distributed as follows: 230 papers indexed in ISI-JCR (Q1: 90, Q2: 60, Q3: 40, Q4: 40) and 21 papers indexed in Scopus. In addition, he has authored/edited 8 books (Elsevier, Springer and others) and edited 50 special issues in reputed ISI journals, such as *IEEE-Communication Magazine*, *IEEE-IoT*, *ACM Transaction on Intelligent Systems and Technology*, etc. He has also registered one Indian patent in the area of Computational Intelligence. His Google Scholar Citations reached 1300+ with h-index: 20 and i10-index: 39. Finally, he is a responsible Editorial Board Member and Associate Editor of many reputed ISI journals.

Tu N. Nguyen is an Assistant Professor and Director of the Intelligent Systems Laboratory (ISL) in the Department of Computer Science at Kennesaw State University, Georgia, USA. His research and teaching are hinged on developing fundamental mathematical tools and principles to design and develop smart, secure and self-organizing systems, with applications to network systems, cyber-physical systems and cybersecurity. He has published 1 book and 70+ publications in leading academic journals and conferences. His first academic appointment was at Purdue University Fort Wayne from 2019 to 2021, where he joined the Department of Computer Science as an assistant professor and founding director of the Network Science, Optimization and Security Laboratory. In the summer of 2021, he joined KSU's Computer Science Department and established the ISL. He was honored to receive several awards including the US NSF CRII Award (2021), PFW ETCS Excellence in Research (2021), PFW Chapter Sigma Xi Researcher of the Year (2021) and PFW Student Supervisor of the Year Nominee (2021). He was also honored to receive an Honorary Professorship from Ramco Institute of Technology, India, in 2021. He has engaged in many professional activities, including serving as Editor/Guest Editor for international journals such as an Associate Editor of *IEEE Systems Journal*, Associate Editor of *Journal of Combinatorial Optimization*, Associate Editor of *IEEE Access* and Technical Editor of *Computer Communications* as well as serving as the Guest Editor of many academic journals such as *IEEE Transaction on Computational Social Systems*, *IEEE Journal of Biomedical and Health Informatics*, *IEEE Internet of Things Magazine* and *IEEE Systems, Man, and Cybernetics Magazine* and so on. He is the Editor-in-Chief of the book series: IET Advances in Distributed Computing and Blockchain Technologies. He has been in different organizing committees such as being TPC-chairs and general chairs for several IEEE/ACM/Springer flagship conferences. He has also served as a technical program committee (TPC) member for several international conferences including IEEE INFOCOM, IEEE GLOBECOM, IEEE LCN, IEEE RFID, IEEE ICC, IEEE WCNC and so on. He is a member of ACM and a senior member of IEEE. e-mail: tu.nguyen@kennesaw.edu

From Smart Devices to Smarter Systems: The Evolution of Artificial Intelligence of Things (AIoT) with Characteristics, Architecture, Use Cases and Challenges

Veena Parihar, Ayasha Malik, Bhawna, Bharat Bhushan, and Rajasekhar Chaganti

Abstract An opportunity to continuously collect data about all aspects of a business's operations is provided by the Internet of Things (IoT). Companies globally are quickly leveraging IoT to create novel products and services that are creating new business opportunities and generating new revenue streams. This shift is ushering in a new era of how companies operate and interact with customers. There is a huge potential for businesses that are able to transform the raw IoT data into powerful market perceptions and efficient data analysis is the key to achieving this goal. Organizations must now delve deeper into their data to find innovative procedures to improve efficiency and competitiveness. With recent advancements in science and technology, particularly Artificial Intelligence (AI), organizations are adopting larger, more comprehensive analysis methods. To fully realize the benefits of IoT, companies must integrate it with the rapidly evolving AI technologies, enabling "smart machines" to mimic intelligent behavior and make well-informed decisions with minimal human intervention. This paper is an effort to examine the growth of IoT and explore how the best way to integrate with AI can benefit the business in the future. This paper also discusses the evolution, underlying architecture, applications, and challenges of implementing IoT with the integration of AI.

Keywords IoT · AI · Smart machines · Data analysis · AIoT · Data science · Integration · Cyber-physical system

V. Parihar · Bhawna
KIET Group of Institutions, Delhi-NCR, Ghaziabad, India

A. Malik (✉)
Delhi Technical Campus (DTC), GGSIPU, Greater Noida, India
e-mail: ayasha07.am@gmail.com

B. Bhushan
School of Engineering and Technology (SET), Sharda University, Greater Noida, India

R. Chaganti
University of Texas, San Antonio, USA

B. Bhushan et al. (eds.), *AI Models for Blockchain-Based Intelligent Networks in IoT Systems*, Engineering Cyber-Physical Systems and Critical Infrastructures 6,
https://doi.org/10.1007/978-3-031-31952-5_1

1 Introduction

The internet is constantly evolving, moving from the internet of computers to the Internet of Things (IoT). Furthermore, the development of highly interconnected systems, commonly referred to as Cyber-Physical Systems (CPS), is emerging as a result of the integration of various components including infrastructure, smart objects, people, and physical environments [1]. The combination of IoT and CPS along with Data Science (DS) could arise as the further smart revolution. The challenge then becomes how to manage the large amounts of data generated with limited computational power. The ongoing research related to DS and Artificial Intelligence (AI) is attempting to address this issue, and that's why IoT integrated with AI could become a significant advancement. This encompasses more than just cost savings, advancements in technology, reduced human effort or being on trend. It's about enhancing the quality of life. Nevertheless, there are serious concerns like security and ethical dilemmas that will persist as challenges for the IoT. The ultimate inquiry is not how captivating IoT combined with AI may seem, but rather how it is viewed by the general population—a boon, a burden, or a danger [2].

Advancements in technology have greatly impacted our daily lives, bringing about changes that have made them easier and more convenient. This includes innovations like the IoT, AI, and robotics. As these technologies continue to evolve and become more sophisticated, the opportunities for their use expand. IoT connects physical objects, devices, and sensors to the communication network, allowing for the collection and sharing of data. This network of objects has the ability to transform the way we live and work by providing access to vast amounts of data for better decision-making [3]. The growth of connected devices through IoT is evident in today's world, from household appliances to vehicles and medical devices. However, with the increasing number of devices, there is a growing need for advanced technologies to manage and make sense of the data generated. AI, which simulates human intelligence in machines, can be utilized to analyze and interpret IoT data, providing new insights and leading to improved decision-making [4].

1.1 Importance of AI

AI refers to a computer or computerized system's ability to carry out tasks that are typically linked with human intelligence. This involves computational devices that can replace human knowledge in carrying out specific tasks. AI enables machines to learn, adapt to new sources of information, and perform tasks that resemble human-like activities. As AI becomes more widely adopted and sophisticated, it is likely to take over more jobs from humans. Every sector seeks the capability of AI [5].

- In the medical field, AI is utilized for personalized drug prescription and X-ray analysis, and also as personal health assistants.

- It is also being used in virtual shopping experiences to provide customized recommendations and to help customers make purchasing decisions.
- AI is also being used to analyze production line data to assess demand and detect fraudulent trades.
- One advantage of AI is that it can make decisions logically without emotions and with fewer errors. Additionally, AI does not require rest, or breaks, or can be distracted as it does not tire or become exhausted [6].

1.2 Importance of IoT

The Internet of Things (IoT) has revolutionized the way we live, work, and interact with technology. By connecting physical devices and objects to the internet, IoT enables the exchange of data, leading to increased efficiency and improved quality of life. In industries like healthcare, IoT enables remote patient monitoring, reducing hospital visits, and reducing healthcare costs [6].

- In agriculture, IoT helps to optimize crop yields and reduce water waste. In the smart home industry, IoT enables household devices to be connected, making homes more comfortable and secure.
- Sensors in the automotive industry help detect equipment malfunctions in vehicles while in use and provide the driver with information and repair suggestions.
- The retail industry uses IoT for stock management, to improvise customer knowledge, to enhance the supply chain, and decrease operational expenses [7].
- IoT is also used for asset-monitoring applications, to improve patient care and experience, and to enhance the efficiency of healthcare operations. Furthermore, Table 1 shows the difference or comparison between the two trending technologies named AI and IoT on the basis of some attributes.

AIoT is a brief term used for Artificial Intelligence of Things (AIoT). It is a term that refers to the intersection of AI and the IoT. It is a technology that combines the capabilities of IoT devices and AI algorithms to gather, process, and analyze large amounts of data, allowing for more efficient and intelligent decision-making. AIoT allows IoT devices to be more than just a data collection point by providing the ability to analyze and act on the data in real time. This can lead to more efficient and automated systems, improved predictions and insights, and more effective decision-making. Examples of AIoT include smart home devices that use AI algorithms to learn a person's habits and preferences, self-driving cars that use IoT sensors to gather data on the environment and make decisions, and industrial machines that use AI algorithms to predict maintenance needs and optimize production. Additionally, Fig. 1 presented the functional point of view of AI and IoT.

With this basic knowledge about AI and IoT, defining AIoT gets easier. Essentially, AIoT enables data exchange between numerous devices and improves the machine's learning capabilities due to a consistent network of data exchange. And as these devices keep learning at a tremendous pace, they are now becoming capable of

Table 1 Comparison between AI and IoT

Attributes	AI [8]	IoT [9]
Learning capabilities	AI systems have the ability to learn from data and improve over time	IoT devices primarily collect and transmit data, but lack the ability to learn and improve on their own
Decision making	AI systems can make decisions based on the data it has been trained on	IoT devices primarily rely on pre-programmed instructions and rules to perform their functions
Complexity	AI systems can handle and process complex data, such as natural language and images	IoT devices are designed for specific tasks and typically handle simpler types of data, such as sensor readings and device status
Human interaction	AI systems can interact with humans through natural language and other forms of communication	IoT devices primarily interact with other devices and systems and have limited human interaction
Autonomous	AI systems can operate autonomously, without human intervention	IoT devices are typically controlled by humans or other systems and require human intervention to change their settings or update their software
Applications	AI is used in a wide range of applications which may involve self-driving cars, speech recognition, and image recognition	IoT is used in a variety of applications, including smart homes, connected cars, and industrial internet

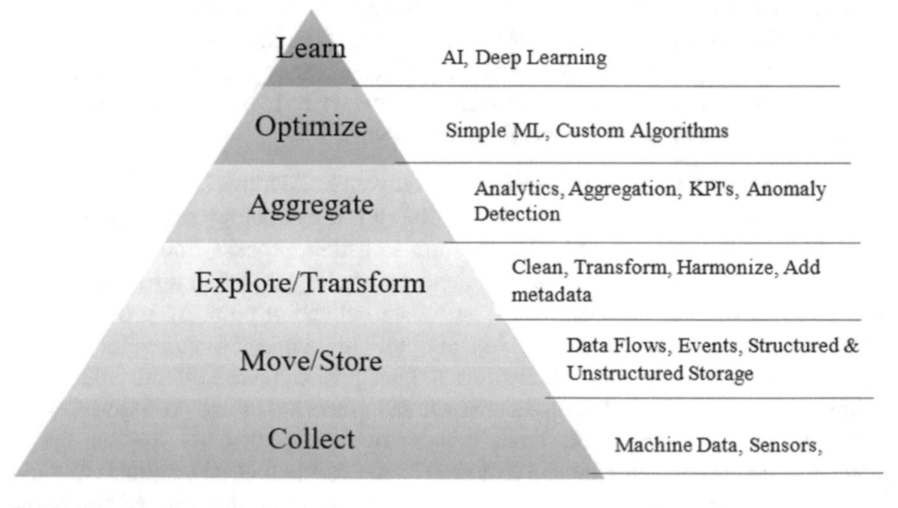

Fig. 1 Functional view—AI and IoT

running programs and analyses by themselves, essentially eliminating their need for Cloud computing. For example, let's take a look at a Smart Car [10]. With features like sensors, cameras, GPS, etc., the car comes with a narrow AI that will be able to see its location and sense other objects in its surroundings, and warn the driver about them accordingly. Due to AIoT, the car can compute all the gathered data on its own, without having to waste internet bandwidth to send the data to a cloud server to receive execution orders. This makes the device perform better and faster, making it increasingly user-friendly as well. Working in tandem, IoT enables information to be processed between machines and servers while AI enables the devices to take actions based on the provided data [11]. Thus, AIoT doesn't just bring in the promise of smart devices, it ushers in a connected network of devices that are able to collect data, store it, share it, and even ACT on it. One question that often comes up while discussing AIoT is how is it any different than what IoT was originally doing. From an outsider's perspective, IoT and AIoT are responsible for data collection and actions. However, there's one key difference between them that should be remembered: IoT can only react to an event whereas AIoT can proactively prepare to act on an event. This proactive behavior means that the devices are now able to self-analyze and correct their processes. So, AIoT will ensure that informed choices are being made based on real-time data and predictive learning patterns, vastly different from the passive nature of IoT [12]. Hence, this paper tries to lighten up the following key points.

- Exploration of the combination of AI and IoT
- Study of the use of AI and IoT to enhance daily life
- Examination of AI and IoT applications in healthcare, transportation, and manufacturing
- Discussion of challenges hindering the full potential of AI and IoT integrations

Further, the paper has been organized in the following manner. Section 2 describes the evolution of AIoT from the beginning. Section 3 discusses the characteristics of AIoT and Sect. 4 elaborates on the architecture thoroughly. Further, Sect. 5 discusses the various use cases of AIoT and Sect. 6 discusses the ethical and societal implications of AIoT. Moreover, Sect. 7 highlights the ethical and societal implications of AIoT, and finally, Sect. 8 concludes the whole discussion of the paper.

2 Evolution of AIoT

The exponential growth of the IoT is driven by declining sensor costs, the widespread use of connected devices, and the advancement of AI. The IoT comprises physical objects with sensors, software, electronics, and network connectivity, enabling them to collect and share data. A recent report from McKinsey estimates that the IoT could have an annual economic impact of up to \$12.6 trillion by 2030 [13]. The AIoT represents the next stage in IoT evolution, where AI transforms data into insights and actions. This integration has the potential to revolutionize industries and society and has already begun to have an impact. The further text is describing the various stages of AIoT evolution over the years [14].

- **Early Stage (2000–2010)**: In the early stages of AIoT, the main focus was on the development of IoT devices and their connection to the internet. These devices were primarily focused on data collection and transmission, using simple protocols such as HTTP and MQTT.
- **Advancement Stage (2010–2015)**: During this stage, there was an increased focus on the development of AI algorithms and machine learning. These technologies were integrated with IoT devices to improve decision-making and automation. The integration of AI algorithms was mainly done in the cloud, where data was transmitted and processed [15, 16].
- **Integration Stage (2015–2020)**: In this stage, AI and IoT technologies were more fully integrated, leading to the development of AIoT. This integration allowed for more efficient and intelligent decision-making and real-time data operation. More advanced communication protocols such as CoAP, LWM2M, and MQTT-SN were introduced to improve data transmission and processing.
- **Maturity Stage (2020–Present)**: In the current maturity stage, AIoT has become more widely adopted across various industries and is being used in a range of applications that involves transportation, medical/healthcare, manufacturing, retail, agriculture, energy, environmental monitoring, home automation, disaster management, smart building, supply chain management, and cybersecurity. Edge computing and fog computing have been introduced to improvise efficiency and reduce the latency of data processing [17].
- **Future Stage**: AIoT is expected to continue to evolve and grow in the future, with the potential for further advancements in AI and IoT technologies, such as 5G communication, LPWAN, and other technologies that can improve communication and data processing [18]. Moreover, the development phases of AIoT are shown in Fig. 2.

3 Characteristics of AIoT

AIoT creates data-driven "learning machines" for enterprise use, automating processes in connected workplaces through real-time data analysis. AI is integrated into components such as programs, chipsets, and edge computing and connected to IoT networks, using APIs to enhance compatibility across devices, software, and platforms [19]. The goal is to improve the system and network operations while extracting value from data, exemplified by reducing road congestion through monitoring and notifications based on real-time traffic flow data. The development of AIoT depends on three phases [20] first, connecting devices to be remotely operated, second, cloud connection for AI inference and, third, device communication through peer-to-peer. Hence, creating AIoT solutions entails data collection, training, and inferencing as its crucial elements. AIoT encompasses several features, a few of which are outlined below:

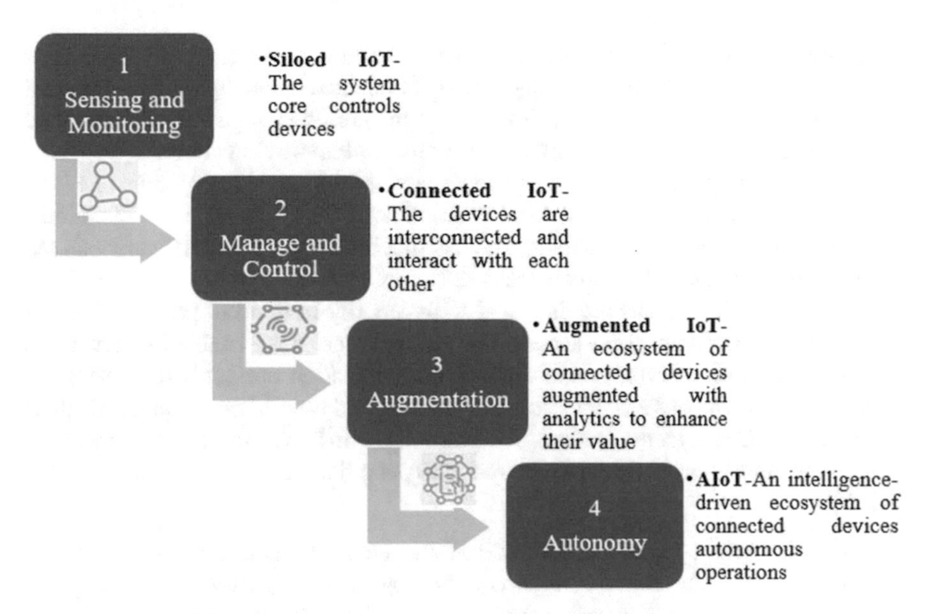

Fig. 2 Evolution of AIoT

- **Combination of IoT and AI**: AIoT is the combination of IoT devices and AI algorithms, which allows for the collection, processing, and analysis of large amounts of data. For example, an AIoT-enabled smart thermostat can collect data on temperature, humidity, and occupancy, and use AI algorithms to learn the occupants' preferences and adjust the temperature accordingly [21].
- **Real-time data operation**: AIoT allows for the real-time operation of data which enables more efficient and intelligent decision-making. For example, an AIoT-enabled manufacturing system can use real-time data from sensors to detect when a machine is about to fail and schedule maintenance before the failure occurs, reducing downtime and increasing production [22].
- **Autonomous**: AIoT systems can operate autonomously, without human intervention, leading to improved efficiency and automation. For example, an AIoT-enabled smart irrigation system can use real-time data on weather and soil conditions to adjust irrigation schedules, reducing water usage and increasing crop yields [23].
- **Predictive and proactive**: AIoT systems can predict future events and take proactive actions, improving overall performance and outcome. For example, an AIoT-enabled predictive maintenance system can predict when a machine is about to fail and schedule maintenance before the failure occurs, reducing downtime and increasing production [24].
- **Scalable**: AIoT systems can be scaled up or down to accommodate changes in demand. For example, an AIoT-enabled transportation system can adjust the number of vehicles in operation to match the demand for transportation, reducing costs and improving service [25].

- **Context-aware**: AIoT devices are aware of their environment and can adjust their behavior accordingly. For example, an AIoT-enabled smart lighting system can adjust the lighting based on the time of day, the presence of people, and the level of natural light, reducing energy consumption and improving comfort [26].
- **Personalization**: AIoT devices can be personalized to meet the specific needs of an individual. For example, an AIoT-enabled personal assistant can learn an individual's preferences and schedule, and use that information to make personalized recommendations and automate tasks [27].
- **Security**: AIoT systems are designed with security in mind to protect data and devices from unauthorized access. For example, an AIoT-enabled smart home system can use encryption and secure communication protocols to protect the data transmitted between devices and the cloud and use biometric authentication to control access to the system. Additionally, AIoT systems can use machine learning algorithms to detect and respond to potential security threats in real-time [28].

The widespread adoption of AI and IoT has been shown in a recent survey by SADA Systems, which found that these technologies are currently the most popular and heavily invested by businesses for increased efficiency and a competitive edge. The results are depicted in the accompanying graph [29].

4 Architecture of AIoT

Contemporary digital society is built on the inescapable foundation of cloud computing, AI, and connected devices. Yet, a greater potential lies in the convergence of these technologies, ready to drive the next phase of digital innovation. The blend of AI and IoT seems very fascinating logically, but the reality of building an AIoT solution is challenging due to architectural and engineering difficulties [30]. This reference architecture addresses these complexities with an event-operated distributed architecture working on multi-tiers. This multi-tier approach works on various constraints related to scalability, security, communication network reliability, and durability. Such architecture logically separates each tier to function independently with its specific workload, privacy, and hardware needs [31]. A closer examination of each tier's features will shed light on how the tiered event-driven architecture addresses the concerns at hand as shown in Fig. 3.

4.1 Things Tier

The purpose of the Things Tier is to sense the environment and collect data from the physical world. The devices in this tier include sensors, cameras, microphones, and other hardware that can digitize physical signals and transmit them to other tiers for

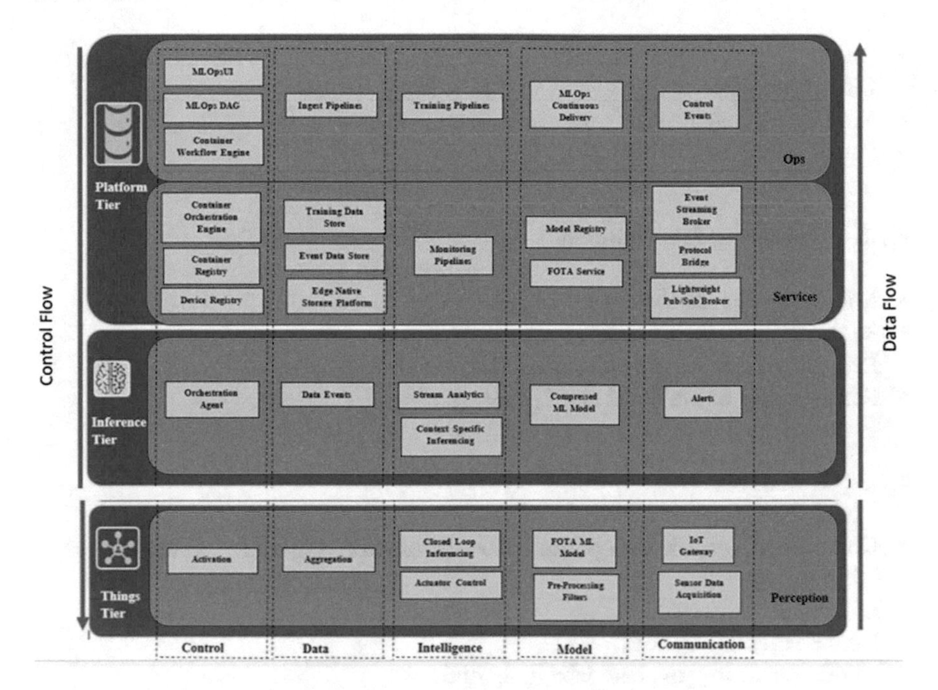

Fig. 3 AIoT reference architecture

further processing [32, 33]. The devices in this tier have limited computing power and storage and are designed to meet specific requirements and operational constraints. The primary responsibility of this tier is to collect and transmit data, which is then processed in other tiers to extract meaningful insights and drive intelligent actions. The Things Tier involves the following responsibilities and operational constraints:

4.1.1 Functions

- The main role of the Things Tier is to collect and transmit data from the physical world to other tiers in the AIoT architecture.
- Sensing the environment using various sensors and digitizing physical signals.
- Performing basic data processing and transmission to other tiers [34].
- Preprocessing data with DSP filters.
- Executing closed-loop inferences.
- Connecting with actuators.
- Offering protocol gateway services for communication between sensor nodes and the gateway.
- Packaging, standardizing, combining, and transmitting data using efficient messaging protocols.

- Responding to command messages and performing actions like initiating model OTA downloads.
- Minimizing data loss.
- Guaranteeing low latency [35, 36].

4.1.2 Operating Environment

- Microcontroller or System on a Chip (SoC)
- 8, 16, or 32-bit architecture
- Real-Time Operating System (RTOS)
- Sensor nodes or mote nodes [37, 38].

4.1.3 Resources

- Constrained networking options—Low bandwidth, high latency, and low reliability
- Limited security options
- Limited upgrade options
- Limited error logging and debugging options
- Difficulty in applying firmware updates and patches
- Cost-sensitive, with a preference for low-cost components [39, 40].

4.1.4 Communication

The Things Tier communicates with other tiers in the AIoT architecture, such as the Gateway Tier and Cloud Tier, using various communication protocols, such as MQTT, CoAP, or HTTP. The communication between the devices in this tier and other tiers must be secure, reliable, and efficient [41].

4.1.5 Security Concerns and Measures

- The Things Tier requires robust security measures to protect the integrity of data and the privacy of sensitive information transmitted over the air.
- To prevent unauthorized access to the sensors and the physical environment, gateway-initiated connections are established with asymmetric key cryptography.
- The identity of the edge devices must be firmly established and strictly verified, to ensure that only authorized devices can communicate with the gateway and other tiers [42, 43].

4.2 *Inference Tier*

The Inference Tier in AIoT refers to the component of the architecture that performs AI-based analysis and decision-making on the data collected by the Things Tier. This tier is responsible for analyzing and processing the data, making predictions and decisions, and transmitting the results to other tiers [44].

4.2.1 Functions

- Receive and process incoming data from the Things Tier.
- Run AI models to generate insights and alerts.
- Scale the number of parallel inferences based on incoming traffic.
- Provide APIs for data retrieval and model management.
- Maintain the models and data stores.
- Perform real-time processing and low-latency inference.
- Ensure data privacy and security [45].

4.2.2 Operating Environment

- Multi-core CPUs or GPUs.
- 64-bit architecture.
- Virtualized or Containerized environments.
- Distributed computing platforms such as Kubernetes [46].

4.2.3 Resources

- Computational resources to run AI models.
- High-speed storage for data and models.
- Scalable computing and storage options.
- Thermal management options.

4.2.4 Network

- High-bandwidth and low-latency network connectivity.
- Load balancing and auto-scaling network components.
- Secure communication using TLS and IPSec protocols [47].

4.2.5 Security

- Encrypted communication between tiers.
- Authentication and authorization of users and devices.

- Secure data stores with access control policies.
- Regular security patches and upgrades [48].

4.3 Platform Tier

The Platform Tier contains MLOps and Platform Services. These services are divided into two categories, with training activities separated from platform services. This tier is responsible for enabling the development and deployment of AIoT solutions, as well as managing the life cycle of these solutions, including updates, maintenance, and security [49].

4.3.1 Functions

- Provide the infrastructure to run machine learning models in production.
- Host the AI/ML pipeline management, orchestration, and deployment.
- Provide API endpoints for the services running in the inference tier.
- Provide secure and scalable storage for the data produced by the sensors.
- Track and store the metadata for each version of the models.
- Provide the ability to roll back to previous versions of the models [50].
- Perform model updates and rollouts.
- Provide the ability to monitor and manage the performance of the deployed models.
- Provide resource management and allocation for the AI/ML workloads.

4.3.2 Operating Environment

- Public cloud, on-premise, or hybrid.
- ×86, ARM, or GPU architecture.
- Virtualized or containerized environment [51].
- Supports distributed computing frameworks such as Kubernetes, Open MPI, and TensorflowOnSpark.

4.3.3 Resources

- High-performance computing workloads.
- Scalable storage solutions.
- Ability to scale out or scale-up the infrastructure on demand [52].
- Thermal management.

4.3.4 Network

- High bandwidth and low latency network connections.
- Connected to the internet.
- Firewall, security groups, and access control policies for the platform services [53].

4.3.5 Security

- Encryption at rest and in transit.
- Authentication and authorization for access to platform services.
- Multi-factor authentication for sensitive operations [54].
- Audit and log for security events and anomalies.

5 Use Cases of AIoT

How do AI and the IoT operate together? Is it not a little strange to see two distinct technologies working in conjunction with one another rather than in competition with one another? IoT and AI, on the other hand, are two distinct technical developments that are each independently transforming sectors all over the globe in their way. On the other hand, the advantages of combining them become much more significant. The main reason why AI skills are used in this situation is for real-time analysis of data acquired by IoT systems. Thus, AI systems include IoT connection and data transfer competence into machine learning models [55]. When AI is combined with IoT, the latter expands beyond its original information-gathering and delivery purposes. In addition, IoT gadgets have a greater capacity to understand and assess the retrieved data. Businesses, sectors, and economies are all susceptible to radical changes brought on by IoT and AI. The combination of AI knowledge with the IoT enables the creation of autonomous decision-making devices. It's important to note that AI and IoT work together, IoT provides the data and AI provides the intelligence to extract insights from that data and make decisions, this is the main concept behind AIoT.

5.1 Increases Productivity in the Workplace

Through the use of AIoT, businesses can maximize their efficiency in all areas of operation. Through the use of machine learning techniques, AIoT-enabled devices can create data, analyze it, and spot trends. Because of this, it can rapidly give operational insights, identify and resolve issues, and further automate labor-intensive procedures [56]. Companies can improve their service while cutting costs by using AI to do monotonous jobs. The use of cameras for quality control in industrial automation is one such example of the automation of vision-based quality inspection. Many programs attempt to monitor and guarantee compliance with rules and regulations.

5.2 *Easy Real-Time Monitoring*

Monitoring of systems in real-time may help in the reduction of costly business disruptions while also saving time. It requires regular monitoring by the system to identify any problems and then either make predictions or choose actions based on those findings. In addition, there is no need for any kind of human interaction, which leads to speedier and more objective findings [57].

5.3 *Risk Management*

It's crucial for businesses of all sizes and in all fields to have a solid risk management plan in place. Distributed, intelligent systems can foresee potential dangers and even mitigate them. Analysis of water levels, employee safety, or public crowds is just a few examples. Organizations may anticipate and respond to future threats with more agility with the aid of AIoT technologies. Insurers have just recently begun to employ such software to oversee the risks associated with machinery and whole plants [58].

5.4 *Reduces Costs*

Intelligent AIoT devices and systems can save expenses. Intelligent systems improve resource efficiency. In smart factories, AIoT devices are used for preventative maintenance and equipment analysis. Here, sensors and cameras monitor machine components to minimize failure and business disruption [59]. In industries such as offshore oil & gas and industrial manufacturing, unplanned breakdown of components can lead to significant losses due to downtime. The predictive maintenance with the help of AI-powered IoT can help in predicting the failure of pieces of equipment and planning maintenance procedures ahead of time, thereby avoiding downtime consequences. Deloitte has found that this approach results in a reduction of 20–50% in time spent on maintenance planning, an increase of 10–20% in equipment availability and uptime, and a reduction of 5–10% in maintenance costs.

5.5 Potential of AIoT in Manufacturing

IoT and AI can also have a major impact on manufacturing. IoT sensors can be placed on manufacturing equipment, allowing for real-time monitoring and data collection. This data can then be analyzed by AI algorithms, identifying patterns and potential issues. For example, an AI algorithm could monitor the vibration patterns of a manufacturing machine and can predict the need for maintenance on a prior basis. Additionally, AI algorithms can be used to optimize production schedules and reduce waste, leading to cost savings for manufacturers. Another area where IoT and AI can be used together in manufacturing is in the area of quality control. IoT sensors can be used to collect data about the quality of products as they are being produced. AI algorithms can then analyze this data, identifying patterns and potential issues. This can help to detect defects early on in the production process, reducing the need for rework and improving overall product quality [60].

5.6 AIoT in Retail

IoT sensors can be used to collect data on things like foot traffic, inventory levels, and customer behavior. AI algorithms can then be used to analyze this data, identifying patterns and making predictions. For example, an AI algorithm could be used to analyze data on foot traffic in a store and predict when it will be the busiest. This could help store managers to schedule staff accordingly, ensuring that customers always receive good service. Additionally, AI algorithms can be used to optimize product placement and pricing, increasing sales and profitability for retailers. Another area where IoT and AI can be used together in retail is in the area of personalization [61]. IoT sensors can be used to collect data on customer preferences and behavior, while AI algorithms can be used to analyze this data and make personalized recommendations. This can help to improve the customer shopping experience, increasing customer satisfaction and loyalty.

5.7 AIoT in Agriculture

One of the most important applications of IoT and AI is in the field of agriculture. IoT sensors can be used to collect data on things like soil moisture, temperature, and crop growth, while AI algorithms can be used to analyze this data and make predictions. This can help farmers to optimize crop yields, reduce water usage, and improve crop quality. For example, an AI algorithm could be used to predict when a field needs to be watered, based on data on soil moisture. This could help farmers to reduce water usage and costs, while still maintaining healthy crops. Additionally, AI methods can also be utilized for the identification of various diseases and pests in crops, allowing farmers to take preventative measures before they cause serious damage [62].

5.8 AIoT in Energy

IoT and AI are also having a major impact on the energy industry. IoT sensors can be used to collect data on things like energy consumption, weather conditions, and equipment performance, while AI will be useful in analyzing this data and making predictions, like, AI could be useful to make a prediction about when equipment is likely to fail, based on data on equipment performance. This could help energy companies to reduce downtime and improve the efficiency of their operations. Additionally, AI algorithms can be used to optimize energy consumption, reducing costs and improving sustainability.

5.9 AIoT in Environmental Monitoring

IoT and AI are also having a major impact on the field of environmental monitoring. IoT sensors can be used to collect data on things like air and water quality, weather conditions, and wildlife populations while AI could be used for the prediction of air quality and suggest preventative measures to mitigate the pollution. This could help cities to reduce pollution and improve public health. Additionally, AI algorithms can be used to monitor wildlife populations and predict population changes, helping conservationists to protect endangered species [63].

5.10 Home Automation with AIoT

IoT devices like smart thermostats, smart lighting, and smart security systems can be controlled by AI algorithms, allowing for automated and personalized control of a home. For example, an AI algorithm could be used to control the temperature of a home based on the occupancy and preferences of the residents. This could help to reduce energy consumption and costs, while still ensuring a comfortable living environment. Additionally, AI algorithms can be used to control lighting and security systems, providing added convenience and security for homeowners [64].

5.11 AIoT in Disaster Management

IoT sensors can be used to collect data on things like weather conditions, ground movement, and water levels, while AI algorithms can be used to analyze this data and make predictions. This can help to predict and prevent natural disasters such as floods, earthquakes, and storms, saving lives and reducing damage. For example, an AI algorithm could be used to predict a flood based on data on water levels and weather conditions, allowing emergency responders to take preventative measures. Additionally, AI algorithms can be used to optimize evacuation routes, reducing the time it takes for residents to reach safety.

5.12 AIoT in Smart Building

IoT and AI are also having a major impact on the field of a smart building. IoT sensors can be used to collect data on things like occupancy, energy consumption, and air quality, while AI algorithms can be used to analyze this data and make predictions. This can help to optimize the consumption of energy, improve indoor air quality, and increase the comfort of building occupants [65]. For example, an AI algorithm could be used to predict the occupancy of a building and adjust the temperature and lighting accordingly, reducing energy consumption. Additionally, AI algorithms can be used to optimize the ventilation system, ensuring that the air inside the building is always fresh and healthy.

5.13 AIoT in Supply Chain

IoT and AI are also having a major impact on supply chain management. IoT sensors can be used to collect data on things like inventory levels, transportation, and logistics, while AI algorithms can be used to analyze this data and make predictions. This can help to optimize supply chain operations, reducing costs and increasing efficiency. For example, an AI algorithm could be used to predict demand for a product and optimize the production schedule accordingly, reducing the risk of stockouts. Additionally, AI algorithms can be used to optimize transportation routes and logistics, reducing transportation costs and reducing the environmental impact of supply chain operations.

5.14 AIoT in Healthcare

IoT and AI are also having a major impact on the healthcare industry. IoT devices like wearable fitness trackers or smart medical equipments can be used for the data collection on things like heart rate, sleep patterns, and medication adherence, while AI algorithms can be used to analyze this data and make predictions. This may be helpful in improving patient outcomes and decreasing healthcare costs. For example, an AI algorithm could be used to predict a patient's likelihood of developing a chronic condition, such as diabetes or heart disease, based on data from wearable devices [66]. This could help healthcare providers to take preventative measures and improve patient outcomes. Additionally, AI algorithms can be used to optimize treatment plans and reduce the risk of medication errors, improving patient safety.

5.15 AIoT for Cybersecurity

As the use of IoT and AI continues to grow, it's important to consider the potential cybersecurity risks. IoT devices and AI algorithms can be vulnerable to hacking and data breaches, putting personal information at risk. To address these risks, it's important to implement robust cybersecurity measures, such as encryption and secure communication protocols. Additionally, AI algorithms can be used for threat detection and to respond to cybersecurity threats, such as malware and phishing attacks. By using AI for cybersecurity, organizations can improve the speed and accuracy of threat detection and response, reducing the risk of data breaches and protecting personal information.

5.16 AIoT in Smart Transportation

IoT and AI are also having a major impact on the transportation industry. IoT devices such as GPS and cameras can be used to collect data on things like traffic conditions, weather, and vehicle performance, while AI algorithms can be used to analyze this data and make predictions. For example, an AI algorithm could be used to predict traffic congestion and optimize traffic flow, reducing travel time and fuel consumption [67]. Additionally, AI algorithms can be used to predict vehicle maintenance needs and optimize vehicle routing, improving fleet management and reducing downtime.

5.17 AIoT in Industrial Automation

IoT and AI are also having a major impact on industrial automation. The IoT components such as sensors and cameras can be useful for the data collection on things like equipment performance, production process, and inventory, while AI algorithms can be used to analyze this data and make predictions. This can help to optimize industrial processes, reducing costs and increasing efficiency. For example, an AI algorithm could be useful in predicting equipment failures, scheduling maintenance, dropping downtime, and growing production. Additionally, AI algorithms can be used to optimize inventory management and production scheduling, improving supply chain efficiency and reducing waste.

5.18 AIoT in Smart City

IoT and AI are also having a major impact on the field of smart cities. IoT devices such as cameras, sensors, and RFID tags can be used to collect data on things like traffic, weather, and public transportation, while AI algorithms can be used to analyze this data and make predictions. This can help to optimize city services and reduce costs. For example, an AI algorithm could be used to predict traffic congestion and optimize traffic flow, reducing travel time and fuel consumption. Additionally, AI algorithms can be used to predict and prevent crime, improving public safety [68].

5.19 AIoT in Predictive Maintenance

AIoT can be used to analyze this data and make predictions about when equipment may fail. By predicting and preventing equipment failure, businesses can reduce downtime and improve efficiency. For example, An AI algorithm can be utilized to anticipate when a machine is prone to malfunction, thus allowing for proactive maintenance scheduling. The use of AI algorithms can also optimize maintenance plans, leading to reduced expenses and increased efficiency [69].

6 Technical Challenges of AIoT

The integration of AI and the IoT presents a number of technical challenges which may include the following.

6.1 Data Processing and Management

With the huge data generation by IoT devices, there arise significant challenges related to data processing and management. This includes the need for efficient algorithms for data preprocessing, feature extraction, and data storage. For example, IoT devices such as sensors and cameras generate large amounts of raw data, which needs to be preprocessed to extract relevant information. Additionally, this data needs to be stored in a way that allows for efficient retrieval and analysis. This can be challenging due to the large volume of data and the need for real-time processing.

6.2 Real-Time Processing

IoT systems often require real-time processing and decision-making, which can be challenging because of the huge volume of data that needs to be analyzed in a lesser period. For example, in industrial IoT applications, a sensor may detect a malfunction in a piece of equipment and an AI algorithm needs to quickly analyze the data and provide a diagnosis to prevent further damage. This requires real-time processing capabilities and low-latency communication to make the decision fast.

6.3 Network Latency

IoT systems often rely on wireless networks, which can be subject to latency and varying network conditions. This can make it difficult to guarantee a certain level of performance for AI algorithms. For example, in a self-driving car, the AI algorithm needs to make decisions quickly, which requires low-latency communication between the car and the network. High latency can lead to delays in decision-making, which can be dangerous in a real-time system.

6.4 Limited Computing Resources

Many IoT devices have limited computing resources, which can make it challenging to run complex AI algorithms on the device. For example, a small IoT device such as a smartwatch may have limited processing power and memory, which makes it difficult to run a complex AI algorithm. This requires finding ways to run AI algorithms on limited resources, such as using edge computing or cloud-based solutions.

6.5 *Security and Privacy*

With the increasing amount of data being collected and transmitted through IoT devices, there is a greater risk of data breaches and hacking. Ensuring the security concerns and privacy of this data is quite challenging. For example, in a smart home, a security camera may capture sensitive information, which needs to be protected from unauthorized access. Additionally, the data transmitted between devices need to be secured to prevent hacking.

6.6 *Scalability*

As the number of IoT devices and data generation increases, the ability to scale the AIoT system to handle the increased load becomes a significant challenge. For example, in a smart city, as the number of IoT devices increases, the system needs to be able to handle the increased data and make sure that the decisions are made in real-time [70].

6.7 *Interoperability*

As the number of IoT devices and platforms increases, there is a need for a common set of standards to ensure interoperability between different devices and systems. For example, in a smart home, different devices may use different communication protocols, which makes it difficult for them to communicate with each other. This makes it difficult to make decisions based on the data from multiple devices.

6.8 *Explainability*

As AI and IoT systems become more complex and autonomous, it may become increasingly difficult to understand how decisions are made and it could be hard to explain the decision to a human. For example, in a medical diagnostic system, an AI algorithm may make a diagnosis based on complex data, which is difficult to explain to a patient or a doctor. This requires finding ways to make the decision-making process more transparent and explainable. The implementation of an application that involves the AIoT as a base may have the above-discussed challenges. However, there are some possible solutions which are discussed below.

- Implementing edge computing solutions, which allow for data processing to be done closer to the source, reduces the volume of data that is to be communicated over the networks.

- Using cloud-based solutions, which allows for data to be stored and processed on powerful servers, which can handle the large volume of data and make real-time decisions.
- Implementing data compression techniques.
- Implementing distributed systems, which allows for data to be processed in parallel, reducing the time it takes to make decisions.
- Using LPWAN like Sigfox and LoRaWAN, which have been specifically designed for IoT and can reduce the latency and increase the range.
- Implementing Quality of Service (QoS) mechanisms, which can be used to ensure that critical data is transmitted with low latency.
- Implementing secure communication protocols, such as Transport Layer Security.
- Using secure data storage solutions, such as cloud-based storage, can be used to store data in a secure and encrypted format [71].
- Implementing firewalls and intrusion recognition systems, which can protect the network from unauthorized access.

7 Ethical and Societal Implications of AIoT

The integration of AI and the IoT has the potential to bring about significant changes in various industries and aspects of daily life. However, along with the benefits, there are also ethical and societal implications that must be considered. Some of the key ethical and societal implications of AIoT include:

- **Privacy and security**: As more data is collected and stored through IoT devices, there is a greater risk of data breaches and hacking. This could lead to the loss of personal information and potentially put individuals at risk of identity theft or other forms of fraud. Additionally, the use of AI algorithms to process and analyze data may lead to the creation of profiles that could be used for targeted advertising or other purposes.
- **Job displacement**: The increased automation brought about by AI and IoT has the potential to displace jobs, particularly in industries such as manufacturing and transportation. This could lead to significant economic and social challenges.
- **Dependence on technology**: IoT and AI systems are becoming more integrated into our lives, and their failures can have a significant impact on daily life. Dependence on these systems could lead to a lack of critical thinking and problem-solving skills and can also lead to a lack of privacy and security.
- **Transparency and accountability**: As AI and IoT systems become more complex and autonomous, it may become increasingly difficult to understand how decisions are made and who is responsible for any negative outcomes.
- **Lack of legal regulations**: Presently, there are no regulations around the development and deployment of AI and IoT systems, which could lead to negative consequences if left unchecked [72].

In summary, the integration of AI and IoT has the potential to bring about significant changes in various industries and aspects of daily life, but it also has ethical and societal implications such as privacy and security, bias and discrimination, job displacement, dependence on technology, transparency, and accountability, and lack of regulation. It's important for society to consider these implications and take steps to mitigate any potential negative consequences.

8 Conclusion and Future Research Directions

The integration of IoT and AI has the potential to improve and revolutionize the way businesses operate and interact with their customers. IoT provides a continuous stream of data about various physical activities, which can be analyzed and transformed into valuable business insights through the use of AI. The combination of these technologies has created new systems of products and services, opened up new business opportunities, and generated new revenue streams. There are also various challenges that must be overcome to fully realize the benefits of this integration, including the need for effective data analysis, the need to integrate AI into IoT systems, and the need to extract value from the combination of these technologies.

The future presents a world where everything is connected and "smart," from the clothes people wear to the food they consume and the homes they live in. This vision is the result of ongoing research and collaboration between various branches of science. However, there is a debate on whether this "smart cyber revolution" will bring about creative destruction or not. Automation has already led to changes in the job market, but with the right policies, it can also lead to re-skilling and up-skilling opportunities. As AI models are deployed in the real world, it's important to consider the potential risks, such as privacy violations, entrenching biases, and reducing accountability. The complex and diverse nature of IoT and CPS networks also presents challenges in monitoring unethical or security breaches. It may even require an AI system to oversee the AI-enabled IoT systems. Ultimately, humans should maintain control over technology to prevent being enslaved by it.

References

1. Bhushan B, Sahoo C, Sinha P, Khamparia A (2020) Unification of blockchain and internet of things (BIoT): requirements, working model, challenges and future directions. Wirel Netw. https://doi.org/10.1007/s11276-020-02445-6
2. Hsiao S-J, Sung W-T (2023) Enhancing cybersecurity using blockchain technology based on IoT data fusion. IEEE Internet Things J 10(1):486–498. https://doi.org/10.1109/JIOT.2022.3199735
3. Malik A, Bhushan B, Kumar A (2022) Association rule-based routing protocol for opportunistic network. In: Saini HS, Singh RK, Tariq Beg M, Mulaveesala R, Mahmood MR (eds) Innovations

in electronics and communication engineering. Lecture notes in networks and systems, vol 355. Springer, Singapore. https://doi.org/10.1007/978-981-16-8512-5_42
4. Hsu CH, Cheng SJ, Chang TJ, Huang YM, Fung CP, Chen SF (2022) Low-cost and high-efficiency electromechanical integration for smart factories of IoT with CNN and FOPID controller design under the impact of COVID-19. Appl Sci 12(7):3231
5. Chakraborty P, Dizon-Paradis RN, Bhunia S (2022) ARTS: a framework for AI-rooted IoT system design automation. IEEE Embed Syst Lett 14(3):151–154. https://doi.org/10.1109/LES.2022.3158565
6. McEnroe P, Wang S, Liyanage M (2022) A survey on the convergence of edge computing and AI for UAVs: opportunities and challenges. IEEE Internet of Things J 9(17):15435–15459. https://doi.org/10.1109/JIOT.2022.3176400
7. Malik A, Bhushan B (2022) Challenges, standards, and solutions for secure and intelligent 5G internet of things (IoT) scenarios, smart and sustainable approaches for optimizing performance of wireless networks: real-time applications. https://doi.org/10.1002/9781119682554.ch7
8. Zia K, Chiumento A, Havinga PJM (2022) AI-enabled reliable QoS in multi-RAT wireless IoT networks: prospects, challenges, and future directions. IEEE Open J Commun Soc 3:1906–1929. https://doi.org/10.1109/OJCOMS.2022.3215731
9. Xu G et al (2021) TT-SVD: an efficient sparse decision-making model with two-way trust recommendation in the AI-enabled IoT systems. IEEE Internet Things J 8(12):9559–9567. https://doi.org/10.1109/JIOT.2020.3006066
10. Nguyen DC et al (2021) Enabling AI in future wireless networks: a data life cycle perspective. IEEE Commun Surv Tutor 23(1):553–595. https://doi.org/10.1109/COMST.2020.3024783
11. Malik A, Kumar A (2022) Assimilation of blockchain with internet of things (IoT) with possible issues and solutions for better connectivity and proper security. In: Sharma R, Sharma D (eds) New trends and applications in internet of things (IoT) and big data analytics. intelligent systems reference library, vol 221. Springer, Cham. https://doi.org/10.1007/978-3-030-99329-0_13
12. Saxena S, Bhushan B, Ahad MA (2021) Blockchain based solutions to secure Iot: background, integration trends and a way forward. J Netw Comput Appl 103050. https://doi.org/10.1016/j.jnca.2021.103050
13. Lin X, Li J, Wu J, Liang H, Yang W (2019) Making knowledge tradable in edge-AI enabled IoT: a consortium blockchain-based efficient and incentive approach. IEEE Trans Ind Inf 15(12):6367–6378. https://doi.org/10.1109/TII.2019.2917307
14. Qazi S, Khawaja BA, Farooq QU (2022) IoT-equipped and AI-enabled next generation smart agriculture: a critical review, current challenges and future trends. IEEE Access 10:21219–21235. https://doi.org/10.1109/ACCESS.2022.3152544
15. Liu RW, Nie J, Garg S, Xiong Z, Zhang Y, Hossain MS (2021) Data-driven trajectory quality improvement for promoting intelligent vessel traffic services in 6G-enabled maritime IoT systems. IEEE Internet Things J 8(7):5374–5385. https://doi.org/10.1109/JIOT.2020.3028743
16. Malik A, Bhushan B, Kumar A, Chaganti R (2022) Opportunistic internet of things (OIoT): elucidating the active opportunities of opportunistic networks on the way to IoT. In: Sharma R, Sharma D (eds) New trends and applications in internet of things (IoT) and big data analytics. Intelligent systems reference library, vol 221. Springer, Cham. https://doi.org/10.1007/978-3-030-99329-0_14
17. Atan B, Basaran M, Calik N, Basaran ST, Akkuzu G, Durak-Ata L (2023) AI-empowered fast task execution decision for delay-sensitive IoT applications in edge computing networks. IEEE Access 11:1324–1334. https://doi.org/10.1109/ACCESS.2022.3232073
18. Ye L et al (2021) The challenges and emerging technologies for low-power artificial intelligence IoT systems. IEEE Trans Circuits Syst I Regul Pap 68(12):4821–4834. https://doi.org/10.1109/TCSI.2021.3095622
19. Cui Z, Jing X, Zhao P, Zhang W, Chen J (2021) A new subspace clustering strategy for AI-based data analysis in IoT system. IEEE Internet Things J 8(16):12540–12549. https://doi.org/10.1109/JIOT.2021.3056578
20. Song L, Hu X, Zhang G, Spachos P, Plataniotis KN, Wu H (2022) Networking systems of AI: on the convergence of computing and communications. IEEE Internet Things J 9(20):20352–20381. https://doi.org/10.1109/JIOT.2022.3172270

21. Wu Y (2021) Cloud-edge orchestration for the internet of things: architecture and AI-powered data processing. IEEE Internet Things J 8(16):12792–12805. https://doi.org/10.1109/JIOT.2020.3014845
22. Pan Q, Wu J, Bashir AK, Li J, Wu J (2022) Side-channel fuzzy analysis-based AI model extraction attack with information-theoretic perspective in intelligent IoT. IEEE Trans Fuzzy Syst 30(11):4642–4656. https://doi.org/10.1109/TFUZZ.2022.3172991
23. Malik A, Gautam S, Khatoon N, Sharma N, Kaushik I, Kumar S (2020) Analysis of black-hole attack with its mitigation techniques in ad-hoc network. In: Sagayam K, Bhushan B, Andrushia A, Albuquerque V (eds) Deep learning strategies for security enhancement in wireless sensor networks. IGI Global, pp 211–232. https://doi.org/10.4018/978-1-7998-5068-7.ch011
24. Bahalul Haque AKM, Bhushan B, Nawar A, Talha KR, Ayesha SJ (2022) Attacks and countermeasures in IoT based smart healthcare applications. In: Balas VE, Solanki VK, Kumar R (eds) Recent advances in internet of things and machine learning. Intelligent systems reference library, vol 215. Springer, Cham. https://doi.org/10.1007/978-3-030-90119-6_6
25. García-Magariño I, Muttukrishnan R, Lloret J (2019) Human-centric AI for trustworthy IoT systems with explainable multilayer perceptrons. IEEE Access 7:125562–125574. https://doi.org/10.1109/ACCESS.2019.2937521
26. Sutjarittham T, Habibi Gharakheili H, Kanhere SS, Sivaraman V (2019) Experiences with IoT and AI in a smart campus for optimizing classroom usage. IEEE Internet Things J 6(5):7595–7607. https://doi.org/10.1109/JIOT.2019.2902410
27. Alrubei SM, Ball E, Rigelsford JM (2022) The use of blockchain to support distributed AI implementation in IoT systems. IEEE Internet Things J 9(16):14790–14802. https://doi.org/10.1109/JIOT.2021.3064176
28. Onyema EM, Dalal S, Romero CAT, Seth B, Young P, Wajid MA (2022) Design of intrusion detection system based on cyborg intelligence for security of cloud network traffic of smart cities. J Cloud Comput 11(1):1–20
29. Vinugayathri (n.d.) AI and IoT blended—what it is and why it matters? Build offshore technology team in India. In: No Time. https://www.clariontech.com/blog/ai-and-iot-blended-what-it-is-and-why-it-matters
30. Chen M et al (2022) Wireless AI-powered IoT sensors for laboratory mice behavior recognition. IEEE Internet Things J 9(3):1899–1912. https://doi.org/10.1109/JIOT.2021.3090583
31. Firouzi F, Farahani B, Barzegari M, Daneshmand M (2022) AI-driven data monetization: the other face of data in IoT-based smart and connected health. IEEE Internet Things J 9(8):5581–5599. https://doi.org/10.1109/JIOT.2020.3027971
32. Wazid M, Das AK, Shetty S (2022) TACAS-IoT: trust aggregation certificate-based authentication scheme for edge-enabled IoT systems. IEEE Internet Things J 9(22):22643–22656. https://doi.org/10.1109/JIOT.2022.3181610
33. Sodhro AH et al (2021) Toward convergence of AI and IoT for energy-efficient communication in smart homes. IEEE Internet Things J 8(12):9664–9671. https://doi.org/10.1109/JIOT.2020.3023667
34. Malik A, Gautam S, Abidin S, Bhushan B (2019) Blockchain technology-future of IoT: including structure, limitations and various possible attacks. In: 2019 2nd international conference on intelligent computing, instrumentation and control technologies (ICICICT), Kannur, India, 2019, pp 1100–1104. https://doi.org/10.1109/ICICICT46008.2019.8993144
35. Jagatheesaperumal SK, Pham Q-V, Ruby R, Yang Z, Xu C, Zhang Z (2022) Explainable AI over the internet of things (IoT): overview, state-of-the-art and future directions. IEEE Open J Commun Soc 3:2106–2136. https://doi.org/10.1109/OJCOMS.2022.3215676
36. Chaudhry SA, Yahya K, Al-Turjman F, Yang M-H (2020) A secure and reliable device access control scheme for IoT based sensor cloud systems. IEEE Access 8:139244–139254. https://doi.org/10.1109/ACCESS.2020.3012121
37. Baccour E et al (2022) Pervasive AI for IoT applications: a survey on resource-efficient distributed artificial intelligence. IEEE Commun Surv Tutor 24(4):2366–2418. https://doi.org/10.1109/COMST.2022.3200740

38. Li J, Zhao Z, Li R, Zhang H (2019) AI-based two-stage intrusion detection for software defined IoT networks. IEEE Internet Things J 6(2):2093–2102. https://doi.org/10.1109/JIOT.2018.2883344
39. Kornaros G (2022) Hardware-assisted machine learning in resource-constrained IoT environments for security: review and future prospective. IEEE Access 10:58603–58622. https://doi.org/10.1109/ACCESS.2022.3179047
40. Qiu H, Zheng Q, Zhang T, Qiu T, Memmi G, Lu J (2021) Toward secure and efficient deep learning inference in dependable IoT systems. IEEE Internet Things J 8(5):3180–3188. https://doi.org/10.1109/JIOT.2020.3004498
41. Jacob S et al (2021) AI and IoT-enabled smart exoskeleton system for rehabilitation of paralyzed people in connected communities. IEEE Access 9:80340–80350. https://doi.org/10.1109/ACCESS.2021.3083093
42. Taimoor N, Rehman S (2022) Reliable and resilient AI and IoT-based personalised healthcare services: a survey. IEEE Access 10:535–563. https://doi.org/10.1109/ACCESS.2021.3137364
43. Figueredo K, Seed D, Wang C (2022) A scalable, standards-based approach for IoT data sharing and ecosystem monetization. IEEE Internet Things J 9(8):5645–5652. https://doi.org/10.1109/JIOT.2020.3023035
44. Malik A (2020) Steganography: step towards security and privacy of confidential data in insecure medium by using LSB and cover media. In: Proceedings of the international conference on innovative computing and communication (ICICC) 2021. Available at SSRN: https://ssrn.com/abstract=3747579; https://doi.org/10.2139/ssrn.3747579
45. Shi Y, Yang K, Jiang T, Zhang J, Letaief KB (2020) Communication-efficient edge AI: algorithms and systems. IEEE Commun Surv Tutor 22(4):2167–2191. https://doi.org/10.1109/COMST.2020.3007787
46. Mukhopadhyay SC, Tyagi SKS, Suryadevara NK, Piuri V, Scotti F, Zeadally S (2021) Artificial intelligence-based sensors for next generation IoT applications: a review. IEEE Sensors J 21(22):24920–24932. https://doi.org/10.1109/JSEN.2021.3055618
47. Sodhro AH, Pirbhulal S, Luo Z, Muhammad K, Zahid NZ (2021) Toward 6G architecture for energy-efficient communication in IoT-enabled smart automation systems. IEEE Internet Things J 8(7):5141–5148. https://doi.org/10.1109/JIOT.2020.3024715
48. Kumar A, Bhushan B, Malik A, Kumar R (2022) Protocols, solutions, and testbeds for cyber-attack prevention in industrial SCADA systems. In: Pattnaik PK, Kumar R, Pal S (eds) Internet of things and analytics for agriculture, vol 3. Studies in big data, vol 99. Springer, Singapore. https://doi.org/10.1007/978-981-16-6210-2_17
49. Yu W, Liu Y, Dillon T, Rahayu W, Mostafa F (2022) An integrated framework for health state monitoring in a smart factory employing IoT and big data techniques. IEEE Internet Things J 9(3):2443–2454. https://doi.org/10.1109/JIOT.2021.3096637
50. Sethi R, Bhushan B, Sharma N, Kumar R, Kaushik I (2020) Applicability of industrial IoT in diversified sectors: evolution, applications and challenges. Stud Big Data Multimed Technol Internet Things Environ 45–67. https://doi.org/10.1007/978-981-15-7965-3_4
51. Soret B et al (2022) Learning, computing, and trustworthiness in intelligent IoT environments: performance-energy tradeoffs. IEEE Trans Green Commun Netw 6(1):629–644. https://doi.org/10.1109/TGCN.2021.3138792
52. Liang Y, Samtani S, Guo B, Yu Z (2020) Behavioral biometrics for continuous authentication in the internet-of-things era: an artificial intelligence perspective. IEEE Internet Things J 7(9):9128–9143. https://doi.org/10.1109/JIOT.2020.3004077
53. Ghahramani M, Zhou M, Molter A, Pilla F (2022) IoT-based route recommendation for an intelligent waste management system. IEEE Internet Things J 9(14):11883–11892. https://doi.org/10.1109/JIOT.2021.3132126
54. Wang T, Lu Y, Wang J, Dai H-N, Zheng X, Jia W (2021) EIHDP: edge-intelligent hierarchical dynamic pricing based on cloud-edge-client collaboration for IoT systems. IEEE Trans Comput 70(8):1285–1298. https://doi.org/10.1109/TC.2021.3060484
55. Uprety A, Rawat DB (2021) Reinforcement learning for IoT security: a comprehensive survey. IEEE Internet Things J 8(11):8693–8706. https://doi.org/10.1109/JIOT.2020.3040957

56. Alladi T, Chamola V, Naren (2021) HARCI: a two-way authentication protocol for three entity healthcare IoT networks. IEEE J Sel Areas Commun 39(2):361–369. https://doi.org/10.1109/JSAC.2020.3020605
57. Shao C, Yang Y, Juneja S, Seetharam GT (2022) IoT data visualization for business intelligence in corporate finance. Inf Process Manag 59(1):102736. https://doi.org/10.1016/j.ipm.2021.102736
58. Chang Z, Liu S, Xiong X, Cai Z, Tu G (2021) A survey of recent advances in edge-computing-powered artificial intelligence of things. IEEE Internet Things J 8(18):13849–13875. https://doi.org/10.1109/JIOT.2021.3088875
59. Ullah FUM et al (2022) AI-assisted edge vision for violence detection in IoT-based industrial surveillance networks. IEEE Trans Industr Inf 18(8):5359–5370. https://doi.org/10.1109/TII.2021.3116377
60. Puri V, Kataria A, Solanki VK, Rani S (2022) AI-based botnet attack classification and detection in IoT devices. In: 2022 IEEE international conference on machine learning and applied network technologies (ICMLANT), Soyapango, El Salvador, 2022, pp 1–5. https://doi.org/10.1109/ICMLANT56191.2022.9996464
61. Kumar GT, Shashank KV (2022) Smart farming based on AI, edge computing and IoT. In: 2022 4th international conference on inventive research in computing applications (ICIRCA), Coimbatore, India, pp 324–327. https://doi.org/10.1109/ICIRCA54612.2022.9985023
62. Berger C (2022) Digital sovereignty and software engineering for the IoT-laden, AI/ML-driven Era. In: 2022 IEEE international conference on services computing (SCC), Barcelona, Spain, pp 353–355. https://doi.org/10.1109/SCC55611.2022.00059
63. Bundas M, Nadeau C, Nguyen T, Shantz J, Balduccini M, Son TC (2021) Towards a framework for characterizing the behavior of AI-enabled cyber-physical and IoT systems. In: 2021 IEEE 7th world forum on internet of things (WF-IoT), New Orleans, LA, USA, 2021, pp 551–556. https://doi.org/10.1109/WF-IoT51360.2021.9595077
64. Gautam S, Malik A, Singh N, Kumar S (2019) Recent advances and countermeasures against various attacks in IoT environment. In: 2019 2nd international conference on signal processing and communication (ICSPC), Coimbatore, India, 2019, pp 315–319. https://doi.org/10.1109/ICSPC46172.2019.8976527
65. Vuppalapati C, Ilapakurti A, Kedari S, Vuppalapati R, Vuppalapati J, Kedari S (2020) Stratification of, albeit mathematical optimization and artificial intelligent (AI) driven, high-risk elderly outpatients for priority house call visits—a framework to transform healthcare services from reactive to preventive. In: 2020 IEEE international conference on big data (Big Data), Atlanta, GA, USA, 2020, pp 4955–4960. https://doi.org/10.1109/BigData50022.2020.9378431
66. Chakraborty S, Chakravorty T, Bhatt V (2021) IoT and AI driven sustainable practices in airlines as enabler of passenger confidence, satisfaction and positive WOM: AI and IoT driven sustainable practice in airline. In: 2021 international conference on artificial intelligence and smart systems (ICAIS), Coimbatore, India, 2021, pp 1421–1425. https://doi.org/10.1109/ICAIS50930.2021.9395850
67. Savvidis P, Papakostas GA (2021) Remote crop sensing with IoT and AI on the edge. In: 2021 IEEE world AI IoT congress (AIIoT), Seattle, WA, USA, 2021, pp 0048–0054. https://doi.org/10.1109/AIIoT52608.2021.9454237
68. Khatoon N, Dilshad N, Song J (2022) Analysis of use cases enabling AI/ML to IOT service platforms. In: 2022 13th international conference on information and communication technology convergence (ICTC), Jeju Island, Republic of Korea, 2022, pp 1431–1436. https://doi.org/10.1109/ICTC55196.2022.9952990
69. Wajid MA, Zafar A (2021) Pestel analysis to identify key barriers to smart cities development in India. Neutrosophic Sets Syst 42:39–48
70. Lin Y-W, Lin Y-B, Liu C-Y, Lin J-Y, Shih Y-L (2020) Implementing AI as cyber IoT devices: the house valuation example. IEEE Trans Indus Inf 16(4):2612–2620. https://doi.org/10.1109/TII.2019.2951847
71. Williams A, Suler P, Vrbka J (2020) Business process optimization, cognitive decision-making algorithms, and artificial intelligence data-driven internet of things systems in sustainable smart manufacturing. J Self-Gov Manag Econ 8(4):39–48. https://doi.org/10.22381/JSME8420204

72. Polyakov EV, Mazhanov MS, Rolich AY, Voskov LS, Kachalova MV, Polyakov SV (2018, March) Investigation and development of the intelligent voice assistant for the Internet of Things using machine learning. In: 2018 Moscow workshop on electronic and networking technologies (MWENT). IEEE, pp. 1–5. https://doi.org/10.1109/MWENT.2018.8337236

AI Enabled Human and Machine Activity Monitoring in Industrial IoT Systems

Anindita Saha, Jayita Saha, Manjarini Mallik, and Chandreyee Chowdhury

Abstract With rapid progress in ICT, novel concepts like IoT and Industrial IoT, (IIoT) have emerged in the last decade, where the latter focuses on industrial applications such as manufacturing, agriculture, logistics, and pharmaceuticals. A significant improvement in sensor based Human Activity Recognition (HAR) is also evident especially after the upsurge of Industry 4.0. Activity recognition in IIoT serves the dual purpose of the development of proactive instruction systems in the assembly line for the employees as well as work quality monitoring. With flourishing RFID technology, context aware handlings of materials provide excellent services and improve system performance with cognitive-IIoT. Literature reveals that Machine Learning (ML) can analyze massive data generated in industry and utilize the extracted knowledge to make better decisions in complex situations. However, IIoT faces several security/privacy challenges which often leads to financial damages and information leakage. A thoroughly researched overview of the Industrial IoT, its uses in human activity identification and robotics, as well as highlighting the security challenges encountered in this rapidly emerging field, forms the major contribution of this chapter. In order to implement the application of ML and DL models, a dataset is prepared with the help of an Android application, and three standard ML as well as three standard DL models have been chosen to exhibit the performance metrics of all six classifiers. It is observed that KNN delivers an accuracy of 99% that establishes the significance of ML models in IIoT and its applications.

Keywords Activity recognition · Machinery activity · Security · Machine Learning · Deep Learning · Protocol

A. Saha
Techno Main Salt Lake, Kolkata, India

J. Saha (✉)
Dayananda Sagar University, Bengaluru, Kolkata, India
e-mail: gjai.2000@gmail.com

A. Saha · M. Mallik · C. Chowdhury
Jadavpur University, Kolkata, India

B. Bhushan et al. (eds.), *AI Models for Blockchain-Based Intelligent Networks in IoT Systems*, Engineering Cyber-Physical Systems and Critical Infrastructures 6,
https://doi.org/10.1007/978-3-031-31952-5_2

1 Introduction

The development of automated systems and light-weight smart sensing devices has made both industry and human life styles easier and more comfortable. The Internet of Things (IoT) is a network of various sensing devices that may gather important data, send that data into storage through a network, evaluate that data, and make decisions on their own. The entire system can function without any input from humans or computers. The manufacturing, agriculture, logistics, and other processes are improved by IoT-based new technologies. Smart sensors and actuators are becoming a crucial component of these technologies. Smart technologies and machine automation are useful to several sectors.

This type of industry's key activities includes real-time data analysis, anomaly detection, alert generation, and alert transmission to various destinations. These tasks can aid in making beneficial decisions more quickly and accurately. The Industrial IoT (IIoT) system is made up of a number of parts, such as various smart devices or sensor sets to collect data and store it, data transfer infrastructure, and edge devices enabling network-to-network communication.

There are numerous areas where IIoT technologies are crucial. Applications include automated packaging systems, intelligent robotics, preventive maintenance, smart wearable sensors, intelligent delivery systems, and automated vehicles, among others. One of the most significant uses of smart wearable sensing is the ability to track people's everyday activities, interactions with machines, and health. In many industries today, applications for intelligent robots are replacing human labor.

Smart robots can do tasks that have been predefined, such as handling industrial tools, creating items, and creating intelligent packaging. With real-time analysis, the complex interaction between a machine and a human can be lessened and efficiency can be raised. There are some literary works [1, 2] where IoT is used to closely monitor people's behaviors and activities in various industries. Monitoring machine activity is equally crucial to human activity as it is to any industry. If manual monitoring is used in the industry, the complexity increases. To maintain the health parameters of various devices without human involvement, automatic monitoring is required.

Few literary works [3] have smart sensors attached to the machines to record the machine's functioning and evaluate data in real time. Even the detailed time series data analysis of the past and present can be useful in predicting the future state of various tools and equipment.

Industries are using these sensors for providing real-time and control support but they are mostly based on complex and expensive wired solutions. For many cases, wired systems are unsuitable since cable separation is required. While improving reliability and communication, there was a necessity of wireless sensors through Wireless Sensor Network [4] which are a superior choice for dependable management and energy-efficient services for high asset protection applications due to their self-organizing and self-configuring characteristics.

For IIoT data, data prediction is a vital step. The possibility of future output data is predicted using machine learning algorithms and various statistical analyses based on historical or prior data. Apart from data forecasting and prediction, categorization is essential for both monitoring machine and human activities. IIoT data can be categorized and divided into various class labels to organize the data. Machine learning and deep learning approaches have a number of traits, including autonomy, the ability to handle numerous types and dimensions of input, and the ability to classify the outcome more well when the data has a nonlinear relationship. In research there have been some contributions [5] that have openly discussed the open source framework of ML. Industries including consumer demand forecasting in the energy sector, supply chain modifications, predictive maintenance, quality control, and increased production throughput—all stand to gain significantly from the machine learning implementation on Industrial Internet of Things data. All these works form a solid background for Artificial Intelligence (AI) enabled machine activity monitoring in IIoT.

When the labels are adequate and exist for training, few works [6, 7], are regarded as separate supervised approaches. Collecting real-time data and retaining their correct labels has become a difficult task. As a result, when ambiguous labels are present in the dataset, semi-supervised and unsupervised algorithms are applied. Deep learning algorithms also are utilized to identify these kinds of behaviors because machine learning requires human intervention.

Data is saved either on a local server or in cloud storage after being collected from various sources. Data must be maintained safely and protected from many unreliable dangers. The data must be securely sent to the customer, after processing. Therefore, secure communication is a crucial component of the IIoT. Literature has a number of works that discuss various ways to protect data and ensure secure connection. Few studies have taken into account various machine learning techniques to address various security difficulties and problems. Few studies have looked at using Blockchain [8] technology to secure data.

As security is one of the most crucial aspects of Industrial IoT, several security metrics have been explored along with blockchain application in relevance as discussed in [9]. Blockchain is widely used in IIoT as it possesses several beneficial traits such as data storage in form of block of chains in a variety of repository in a sequential manner, no involvement of third party in transactions with complete transparency, no alteration in the data structure of the stored repository, and ensuring utmost security in procuring, storing as well as retrieving sensitive data [10].

Therefore, monitoring both human and machine activity is a crucial use of the IIoT. Numerous recent studies have demonstrated that routinely keeping track of employees' efforts, working capacities, and physical fitness may result from regular monitoring of human daily work patterns, energy expenditure, and interaction with machines in the workplace. The smart sensor-based system's monitoring could aid in understanding the ultimate result and intriguing information.

The key motivation for this chapter, that surfaces from the above discussion, is to explore and carry out a thorough research on the various ML and DL based technologies that can be utilized in IIoT to improve reliability, effectiveness and efficiency in industrial procedures along with dealing with the security challenges that generate during real time operations.

The primary contributions of the chapter may be summarized as follows. It provides.

- An overview of basic taxonomy of HAR in IIoT.
- A Review on Machine activity monitoring in Industry using IoT concepts.
- A synopsis of ML and DL concepts and their applications in IIoT.
- An implementation of a sample case study, with the help of a collected dataset, to exhibit the performance of the ML and DL models through standard performance metrics.
- A thorough idea of the advanced application and protective measures from intruders in IIoT.
- A brief outline of the open research areas that incorporated this concept.

As a result, Sect. 2 of this chapter provides an overview of human activity monitoring in the IIoT. In the subsequent sections, several methods for keeping an eye on autonomous devices with smart sensors have been discussed in thorough details. Section 4 contains a brief discussion of the application of machine learning. Section 5 goes into great length about the significance of deep learning and its applications. The case study is discussed in Sect. 6. Section 7 discusses the various layer wise security strategies. Various open research questions based on prior efforts are briefly covered in Sect. 8. Section 9 concludes it in the end.

2 Human Activity Monitoring in IIoT

IoT in general refers to the interconnection of any device or "Things" across the internet which are identifiable through standard internet protocols (IPs). When IoT is applied over various industries, it is loosely called IIoT as it enables a holistic interconnection of industrial components empowered with IP. Thus the Industrial Internet of Things is an integral part of the evolution of general IoT that implements several applications to monitor human activities, as well, in relevant areas [11]. With the advancement of sensor technologies, and evolution of embedded systems with communication facilities, it is now possible to procure and process rich information to bridge the gap between real world and digital world. IIoT enables systems to access the data captured by various sensors deployed, in real time, exchange them between machines directly using various communication techniques such as Bluetooth or ZigBee, perform data analytics using standard Machine Learning or Deep Learning techniques, and deliver information to facilitate crucial industrial decision making and predictive analysis.

Being a key pillar of the Industrial Revolution, IIoT aims to establish a strong connection between industrial machines for proactive detection of existing or upcoming problems as well as monitoring or assessing worker productivity, especially in large manufacturing plants. Improving productivity of the plant is directly proportional to the worker's contribution on a regular basis, and IIoT can collect production data by tracking the daily activities of the workers through sensors and provide related predictions [12]. Human Activity Recognition has been prevalent to assess performance of individual workers in a dynamic and distributed working environment, through wearable sensors such as Smart Watch and Smart Phones. The powerful sensors within these devices can detect various activities such as walking, sitting, running, standing or sleeping. Data collected from such sensors can be processed locally though edge computers like Raspberry Pi or even sent to the cloud for further analysis and predictive measures. Through this, it is possible to track various movements of the worker like hand movements during operational activities or location change while across the plant [13]. The primary motivation of developing such individualized as well as generalized distributed solutions of detecting the activities of a worker in industry is to reduce operational hazards, failures, schedule tasks, along with maintaining and repairing cost of production. This leads the production mode of the several industries to convert into a "Smart Factory" that is technically advanced by using IIoT and cloud platform, improved manufacturing resources, skilled workers and enhanced QoS [14].

Earlier, computer vision based techniques were prevalent in industrial setup to identify activities of workers as discussed in [15], which required a substantial number of cameras to capture images of the subjects from different angles, and converted the images to 3D views using further photographic techniques like pose estimation and visual hulls. Further data processing was expensive which drove the attention of the researchers towards wearable technologies that were capable of monitoring the workers working along production lines or performing specialized industrial tasks. These wearable devices are equipped with plug and play facilities, need no skills to install or maintain, and can be easily integrated in any industrial unit with substantial battery backups for longevity. Researches reveal that wearable devices provide task related information of a worker, his general movement, usage of safety equipment by the worker in the plant, as well as understand the high level or low level KPI (a measurable unit to identify and measure the performance of a production process or a worker related to it). Wearable sensors are thus popular choices to recognize human activities as depicted in [16] that focuses on a "pick and place" task, typically performed in a manufacturing unit, to be recognized by a sensor embedded suit, with a glove embedding force sensors and using Hidden Markov Model (HMM) for classification. IIoT is encompassed by Industry 4.0 [17], that refers to a paradigm shift in industrial revolution, where ubiquitous computing and pervasive sensing are prevalent to make the operational processes highly effective and efficient. Sensor based activity recognition is desirable in industries of hand made products and craft where specific movements of fingers and position of the worker is crucial for tasks like casting a clay sculpture of molding, which requires a high resolution camera otherwise, to focus on different areas during production, as proposed in [18]. As per the authors of [19], activity monitoring of factory workers is indispensable to ensure safety at the shop floors, as they are directly exposed to threats due to occupational accidents such as abnormal vibration, falling objects, or other emergency detections during working hours. Further it is observed that IIoT can provide effective solutions to monitor, and evaluate the productivity and performance of the workers in the industry using wearable sensors in a distributed platform which

Fig. 1 General steps of human activity recognition

helps to compute performance metrics to improve their working capabilities [20]. Figure 1 depicts the general flow of HAR that includes all basic steps for predictive analysis.

3 Machine Activity Monitoring in IIoT

With the upsurge of IIoT in the last decade, traditional industries have successfully turned into Smart Industries through the development of intelligent systems that enable discrete machines to be empowered with computational capabilities. Irrespective of the Industry, machine activity monitoring is a crucial task as it forms the backbone of the ecosystem as a whole that provides a substantial idea about the condition of the equipment in advance and predicts whether corrective measurements are required or not. This aids in eliminating catastrophic failures in industry and saves endangered lives of workers. As in [21], the authors propose an intelligent system that identifies intrinsic faults in large scale machines such as electric generators that run wind turbines through feature extractions of significant features with several classifiers to improve maintenance routines. Researches [22] reveal that sensors deployed in machines under the IIoT ecosystem can detect anomalous behavior in them through several characteristics such as engine load, fuel usage and oil pressure which helps the workers to comprehend maximum up time for machines, understand inefficient usage if any and optimize their ineffectiveness, before or on time. Novel techniques such as AutoRegressive Integrated Moving Average (ARIMA) applied on time series data collected from various sensors deployed in slitting machines used in industries can also detect as well as predict possible failures along with quality defects, which has profound effect on overall operational procedures as depicted in [23]. Industries that involve essential commodities such as water purification systems which are used as input products for other industries like production of pharmaceutical products, also put much emphasis on data driven and model driven fault detectors which are heavily relied upon machine learning and data mining, as claimed by authors of [24]. As the sensors deployed in IIoT communicate using Wireless Sensor Network are susceptible to faults, frameworks have been developed to differentiate between error prone and active ones, with proposed algorithms such as HGAMP in [25]. Researches have proved that productivity in specialized industry like manufacturing can be increased by monitoring the conditions of essential features such as thermal signals and vibration, which reduces unexpected machine downtimes as well associated costs. The IIoT ecosystem allows the staff to monitor the conditions of the machines on a large scale and also estimate the remaining lifetime of the same.

As per the authors of [26], monitoring and visualization of a fleet of machines for lifetime testing, is possible under a restricted environment like Smart Maintenance Living Lab, which also may be implemented in a large- scale industrial environment. Authors of [27] have investigated the mathematical relationship among several critical factors and their causes in an IIoT environment, that takes place due to improper planning and production management. They claim that such critical failures in the production process can be detected early with a real time process monitoring application, which can be easily integrated with existing failure control systems and thus create an intelligent IIoT ecosystem for industries all across. Observations reveal that industries are eager to invest lump sum amount in predictive maintenance under IIoT, that monitors the actual condition of the machines, to alert the system about its failure in advance, and also predict whether corrective measures or not, as discussed in [28], since fault detection as well as condition monitoring can address fatal issues in big data management under IIoT ecosystem.

In the last few decades, robotics has been considered as a booming research field that has significantly shown great prospective among researchers. Industrial robots are flexible, efficient machines equipped with heterogeneous sensors and tools that are specially designed for repetitive and dangerous operational tasks. In [29], an IIoT based smart robotic warehouse is proposed with autonomous robots that helps to implement the concept of Smart warehousing for improved utilization of labor and infrastructure and thus improves logistics operational capabilities. The autonomous robots are made to deliver goods automatically for drop and pick up operations. Recently, Human–Robot Interaction (HRI) has also been considered as a good area of research especially in the context of HAR where ML algorithms can be applied to recognize skeletal activities under various situations [30]. Collaborative robots or Cobots deployed in manufacturing units enable robots to work with humans side by side and perform several operations in the industry at different levels. This unique combination of human intelligence, cognitive skills and dexterity to handle unexpected situations and robotic capabilities of high precision, strength and repeatability can be advantageous for improved efficiency and effectiveness in the industry. As depicted in [31], a joint assembly task of building a tower by combining six cubes with different bolts has been shown to be done by a human and a robot in a collaborative fashion. The workflow analysis was conducted by Hidden Markov Model with an accuracy of 95%.

4 Application of ML in IIoT

In the last few decades, Machine Learning (ML) has become the most effective and efficient way to implement Artificial Intelligence (AI). ML allows machines or computers to learn and detect patterns in a set of data and can use this knowledge to perform various operations such as classification, clustering, and future predictions which can be beneficial for decision making in future. It is evident that IIoT deals with different kinds of data such as unstructured data, meta data as well as transformed or value added data which is well handled by ML algorithms for predictive analysis. Researches reveal that the emergence and wide acceptance of ML has given satisfactory solutions to several industrial automation problems and made the flow of work easier by reducing errors, improving maintenance, detecting anomalies and curbing production, operational as well as maintenance costs [32]. As discussed in [33], ML can significantly contribute in conducting predictive maintenance, providing longest equipment life and improve reliability and hence they collected different types of data from several mechanical and electronic materials, and run popular ML algorithms such as support vector machine (SVM), k-nearest neighbor (kNN), decision tree (DT), random forest (RF), linear discrimination analysis (LDA) to obtain satisfactory values of performance metrics. In [34] the authors have evidently shown how production machines are enabled by fitting low cost sensor devices into them and conducting predictive maintenance with three types of ML models in action. The semi-supervised models like One-Class Support Vector Machine, Isolation Forest and Local Outlier Factor can detect and store data with anomalies, and this faulty data can be used to train clustering methods such as K-Means, Spectral clustering, Mean shift clustering, DBSCAN and Agglomerative clustering which clusters faults into groups. This creates a dataset of detected anomalies which is given as an input to supervised models such as Support Vector Machine (SVM), Logistic Regression (LR), Decision Tree (DT, Random Forest (RF), Naïve Bayes (NB) and K-Nearest Neighbor (KNN). Another essential application of ML in IIoT is anomaly detection that includes several hardware failures as well as potential attacks under IIoT infrastructure. The authors of [35] have analyzed 15 anomaly situations such as water device sabotage or hardware breakdown, and applied standard ML algorithms to classify malicious activities. As evident, IIoT devices are connected to the Internet and wireless infrastructure and are always susceptible to trust issues. ML models like KNN and SVM can cluster and classify the trustworthiness of the devices based on temporal, spatial and behavioral patterns of data retrieved from those IIoT devices [36]. As IIoT deals with interconnected smart objects with several sensors in action, connectivity among them is a crucial factor for the data to be exchanged continuously. The authors of [37] have proposed a ML based connectivity restoration strategy utilizing the Neural Network that also reduces cost related to energy which is a crucial factor in the industrial domain [38]. The CRrbf strategy proposed by them utilizes Radial Basis Function Neural Network clubbed with Kalman Filter to reduce the cost of energy which improves efficiency in the operational paradigm.

Machine learning algorithms have been effective in playing a significant role handling the enormous data produced by heterogeneous sensors which are complicated as well as diverse in nature with anomalies that create hindrance in data analytics in IIoT ecosystem. Supervised algorithms like Random Forest (RF), Logistic Regression (LR), Decision Tree (DT) and K-Nearest Neighbor (KNN) have been considered in [39] to be applied in benchmark datasets such as E-Coating Ultrafiltration Maintenance Dataset, Semiconductor Manufacturing Process Dataset and Demand Vs Response Data For IoT Analytics with several parameters to detect anomalies and it is observed that RF outperforms others with an accuracy of 98%.

Observations reveal that Cybersecurity plays a vital role in the IIoT paradigm, by safeguarding their fiscal transactions, across the Internet, without which industries might face several negative impacts. The authors of [40] have proposed a few methodological solutions that can be incorporated in IIoT to overcome the challenges and make the system less vulnerable against security threats.

As proposed in [3], ML algorithms can detect normal motor faults, inner as well as outer bearing defects apart from traditional faults found in induction machines, which helps to improve their lifetime in the long run. The authors also claim to detect cyber-attacks which are crucial in the IIoT environment with several heterogeneous sensors susceptible to various attacks.

Especially in large scale distributed servers under the IIoT ecosystem, it is essential to control information flow for achieving semantic security. As discussed in [41], a secure, effective and efficient scheme using secure KNN or seed-KNN has been proposed using encrypted training data maintained over multiple servers for industrial systems, rendering higher accuracies. The system ensures confidentiality of data flow, and enables homomorphic operations over highly encrypted data.

A Little more work such as [42] also monitors and distributes sensors throughout an industrial plant, which is a common scenario in the IIoT ecosystem. Fault identification is done with highest detection quality, from heterogeneous sensors through Fisher Wrapper feature selection method, and proves to be ideal for a system that processes the collected data closer to the sensors, i.e. the edge of IoT, eventually reducing common issues such as bandwidth, latency and privacy issues faced by centralized system. Literary works such as [43] depict that edge devices employed to process data locally can reduce communication costs as well reduce the delay to substantial levels. ML approaches such as RF can handle the feature selection and select minimal edge devices to ensure a data driven fault detection system for better effectiveness.

Some recent trends in research exhibits that IIoT faces several challenges that need to be addressed that can be solved by ML based countermeasures as depicted in [44]. The authors have highlighted the crucial roles of enabling technologies like Blockchain, IoT and cloud computing along with ML, that can be adopted as retaliatory measures for such security challenges prevalent in Industrial IoT. In fact the entire concept of Blockchain can be remodeled to the specific and accurate requirements of IoT, to develop Blockchain IoT or BIoT that form a crucial aspect of IIoT [45]. In recent research, researchers have attempted to integrate the concept Blockchain with IoT in different application scenarios that can provide new insights to this paradigm. Literary contributions such as [46] depict how significant improvements can be seen in the IoT as well as IIoT systems, using blockchain and the methods to combat the security challenges that enumerate during this integration.

5 Application of Deep Learning in IIoT

Huge volumes of data are handled by large-scale industrial networks, which increase the need for IIoT frameworks to have reliable, intelligent methods for big data analysis [47]. Due to their extensive learning and processing capabilities, deep learning (DL) algorithms have the ability to be utilized in IIoT networks.. This section discusses some well-known DL algorithms that are now employed in IIoT networks along with their theoretical foundations.

The most basic variety of deep learning is Deep FeedForward Neural Network. In DFNNs, the weights are initialized randomly and then, the gradient descent algorithm is commonly preferred to minimize the errors. The process of learning takes place through multiple executions of consecutive forward and backward propagations [48]. The input is processed towards the output layer through multiple hidden layers. Forward propagation compares the estimated output to the desired output of the matching input. To update the weights in the backward process, the rate of change of error is determined in relation to network parameters. The process repeats until the desired output is produced. In [49], researchers used DFNN to maximize the security capacity of an IIoT system by finding an optimal transmit power under the maximum transmit power constraint. The maximum confidentiality rate among secure applications was achieved by the non-linear mapping of the power allocation tuple of DFNN. DFNN has been utilized in [50] for classification and detection purposes with several training methods, with different groups of extracted features. In this experiment, the need to store previous information was observed when RNN outperformed DFNN in terms of accuracy.

CNNs are a specific type of neural network motivated by the human visual system. It consists of multiple convolutional layers, each responsible for extracting crucial features from input data. Pooling layers are responsible for dimensionality reduction without losing any important feature information. Finally, a fully connected layer is attached to the output layer to obtain the desired outputs. CNN is widely used in various fields, including IIoT, to understand important and complex features. In [51], researchers proposed a lightweight CNN for bearing fault diagnosis in the IIoT context. This fault diagnosis takes place by learning and selecting crucial features from the dataset. To understand the progress of production in IIoT manufacturing, it is important to understand the balance between old and recent data features. CNN has been used in the work [52] for the said purpose along with LSTM. In the IIoT, multi-objective interference is a serious challenge since multiple tasks are performed simultaneously. To address this challenge, an optimal resource allocation method was proposed in [53] using CNN, by extracting the optimal channel state for distinct applications. Researchers in [54] proposed a skin lesion classification framework by including transfer learning with CNN, to enhance the field of skin cancer related research.

RNNs are a specific type of neural network where the output from one step is fed into the subsequent stage's input. Hence, in a traditional neural network, all inputs and outputs are independent of each other, and hence, it fails to satisfy the requirement where previous knowledge is needed to process the journey from input to output. For example, the knowledge of previous words in a sentence may ease the prediction of the next word. This kind of important information in a sequence is stored in the hidden layer of RNN. The memory of RNN remembers all the information found. However, it cannot process very long sequences due to the heaviness of information, and from this point, the idea of LSTM arises. In wireless IIoT networks using the information storage properties of RNN, remote state estimation is possible by comparing current and previous states of the dynamic systems [55]. Also, the workflow of the cloud environment can be controlled using RNN by predicting the future workload and accordingly providing the required resources, as proposed in [56].

The LSTM model was designed to overcome the long-term dependency problem faced by RNN. This model contains feedback connections, which enables the process of continuous data sequence without treating each data point independently. Important information about previous data in the sequence is retained and used to process upcoming data points. Hence, LSTMs are found to perform well at processing sequences of data such as text, speech, and time-series data. Enormous data collected from multiple sensors can be analyzed and the anomalies can be detected by LSTM, as proposed in [57]. Researchers in [58] proposed a model based on LSTM and an auto-encoder to identify invasive events from the Internet Industrial Control Systems (IICS) networks. The LSTM trains the concealed behavior of valid as well as malicious activities, and after that, the decision engine can correctly identify normal and attacked events. LSTM has been used to classify malware applications also [59], with an action-value function for training an active learner. From Fig. 2, a basic idea about the applications of DL in IIoT can be understood in general.

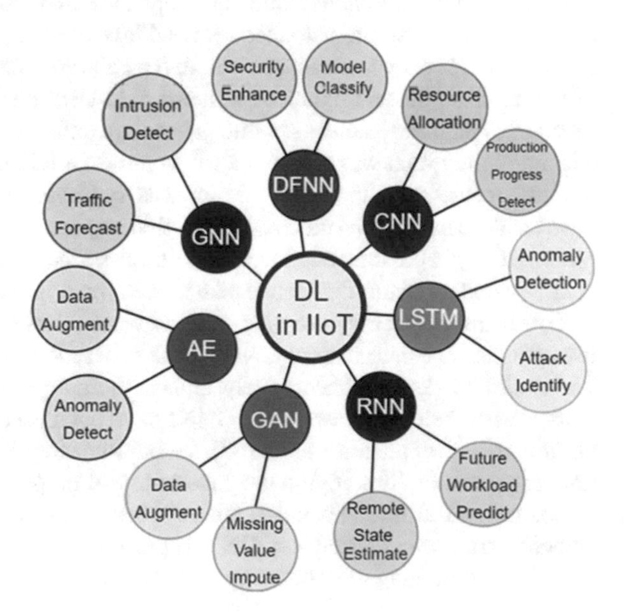

Fig. 2 Deep Learning in IIoT

GAN is a recent specialization of neural networks, proposed in 2014 by Goodfellow [60]. In this model, two neural networks, namely, generator and discriminator, participate in the zero-sum game (where one participant's gain causes another participant's loss). An input dataset is fed to the GAN model, and the generator randomly generates new data samples with a predefined shape. The discriminator investigates how close the generated data is to the original data and trains the generator by providing feedback to generate data in the next phases that is as close as possible to the original data by means of statistical representation. Three types of GAN were used in [61] in a federated learning-based IIoT architecture. On the participant side, data was triggered by two dynamic poisoning attacks by a simple GAN and a federated GAN. Then, on the server side, anomalies were avoided by aggregation using A3GAN. When IIoT deals with sensor data, among the huge amount of data, there can be some missing values due to hardware or network problems. In the work [62], GAN was used to interpret the missing data, which in turn helped with fault detection. Researchers have used conditional GAN as a part of determining the remaining useful life of complex systems [63]. CGAN was used to generate realistic artificial data by learning the data distribution of complex, multivariate fault data collected under different operating conditions and fault modes.

An autoencoder is a different kind of neural network used to treat unlabelled data. It finds the low-dimensional representations of the input data by utilizing the extreme non-linearity of neural networks. It consists of two parts: an encoder and a decoder. The encoder transforms the input from high-dimensional to a low-dimensional compact representation, and the decoder transforms that compact data into a high-dimensional output, again. Because a large amount of data is required to train a deep learning model, and sufficient data may not be collected for a variety of practical reasons, the autoencoder is widely used for data augmentation to meet the demand for deep learning models. In [64] a graph-autoencoder was proposed for group-anomaly detection for IIoT systems of large scale. In this method, multivariate time series data were converted into graph data and combined with previous knowledge, and the graph was reconstructed to obtain a reliable anomaly score value. In [65], the accuracy for anomaly detection was improved by 10% using an autoencoder-based feature extraction and feature selection approach. A recurrent autoencoder was proposed in [66] to enhance incomplete time-series datasets by generating missing data. A gated regulator distinguished between the original and fictitious information.

GNNs are a specific class of neural networks that process data by representing it as graphs. The core idea of GNNs is the utilization of pairwise message passing. The nodes of the graph iteratively update their representations by sharing information with their neighbor nodes. GNNs have been used for anomaly detection in a number of IIoT application areas [67]. Researchers in [68] used a variational GNN for traffic forecasting by studying the uncertainty of graph node representations, sensor attributes, and also the spatial–temporal correlations among sensors. An intrusion detection framework based on GNN was proposed in [69]. Many other deep learning models are emerging to improve cyber security [70].

Individual data samples were treated as nodes in the graph and grouped based on characteristics. This way, the chance of imbalanced categories was reduced even for a smaller number of training data, which in turn helps the intrusion detection method.

6 Implementation of HAR—A Sample Case Study

Activity recognition is implemented through Python, by comparing a few standard ML classifiers and analyzing their results for understanding predictive performances. Datasets have a profound role in such performance analysis because the choice of the same may render different results in different cases. Datasets may be classified into two broad categories (i) Benchmark datasets (ii) Dataset prepared by the researcher collected through various applications such as G-Sensor Logger and others. In this chapter, we have presented a case study of the implementation of HAR through a dataset prepared from data collected through the G-Sensor application, installed in the Smartphone.

6.1 Description of the Dataset

The dataset is prepared by collecting data from an application called G-SENSOR which is installed in a Smartphone. It captures input features of three axial accelerometers in the form of x axis, y axis and z axis or 3 dimensions, via the embedded accelerometer sensor in the phone. The raw data undergoes several steps of preprocessing that is essential to obtain desired performance metrics such as Accuracy of the model. For noise reduction, a low pass filter called Butterworth filter has been used. Further, the raw data is processed and statistical functions have been used onto them in order to obtain a featured dataset, containing 15 input features. The obtained dataset has been randomly partitioned into two sets of "Train" and "Test" where 80% of the subjects were selected for the former and the remaining 20% contributed to the latter. Python is used as a programming language along with Scikit learn toolkits in order to analyze the performance of the aforementioned classifiers on this dataset. The number of samples in each activity is shown in Fig. 3 graphically.

6.2 Machine Learning Models Applied on Collected Dataset

Three standard Machine Learning classifiers have been chosen such as KNN, and Logistic Regression (LR), and Support Vector Machine (SVM) for Activity Recognition. It is observed that KNN performs better than the other two with an accuracy of 99% Fig. 4.

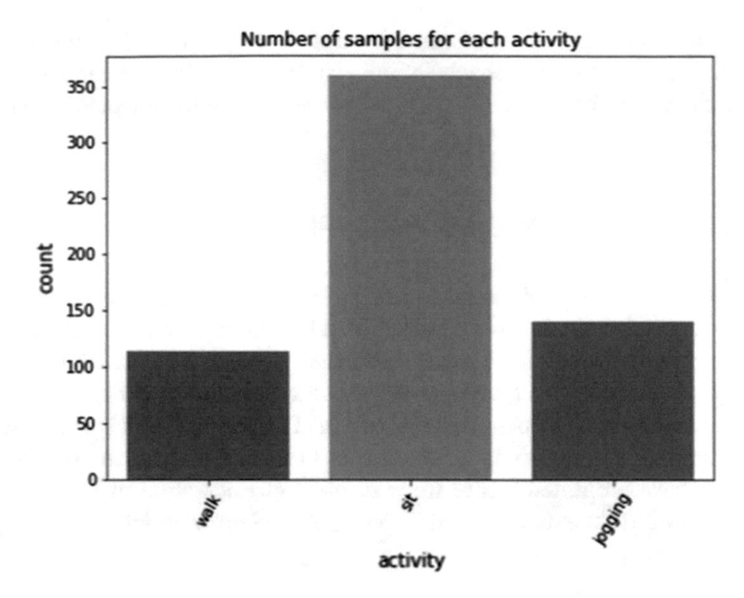

Fig. 3 Distribution of total number of samples per activity

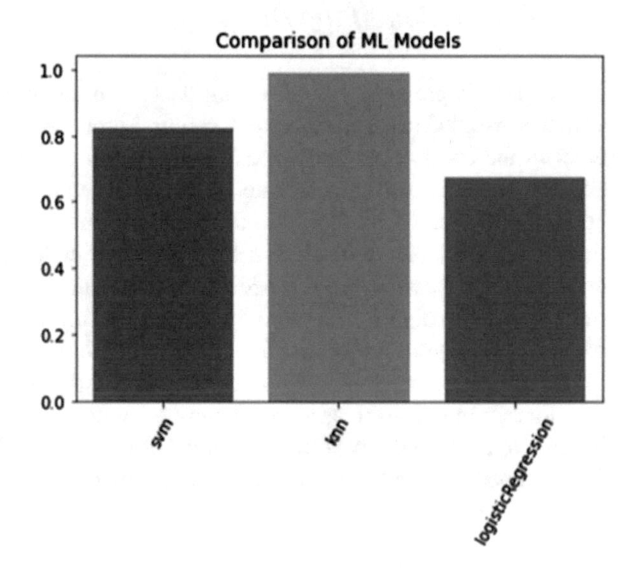

Fig. 4 Comparison of accuracies of different ML models

6.3 Deep Learning Models Applied on Collected Dataset

Experiments have been conducted on the same dataset using Deep Learning models for Activity Recognition such as Recurrent Neural Network (RNN), Long Short Term Memory (LSTM) and Convolutional Neural Network (CNN). The confusion matrix of each has been shown in Figs. 5, 6 and 7 respectively. Few results related

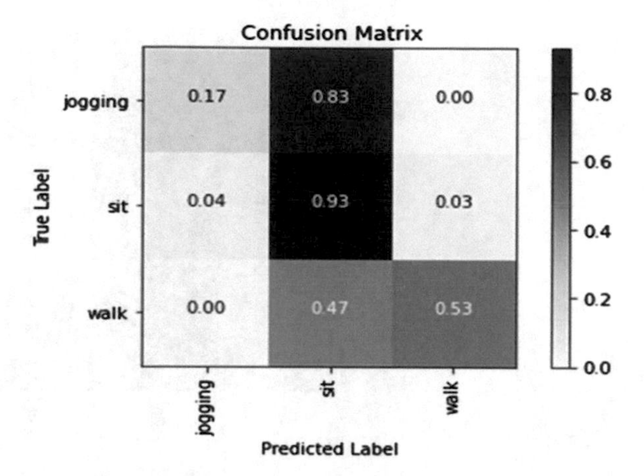

Fig. 5 Confusion matrix of RNN

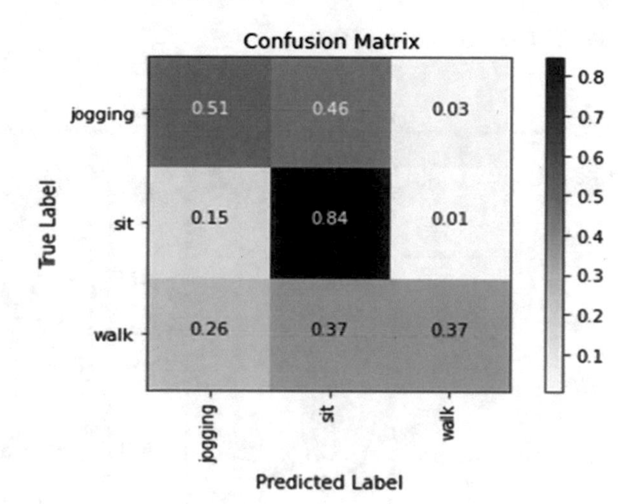

Fig. 6 Confusion matrix of LSTM

to model behaviors are depicted in Figs. 8, 9 and 10 respectively, with respective learning curves of loss and accuracy for each DL model.

A comparison of all three classifiers has been shown graphically in Fig. 11.

7 Security Approaches in IIoT

For the creation and upkeep of various IIoT-based applications, secure data storage and communication are essential components. Data is transferred between smart sensor embedded devices or any end devices like smartphones from other devices in

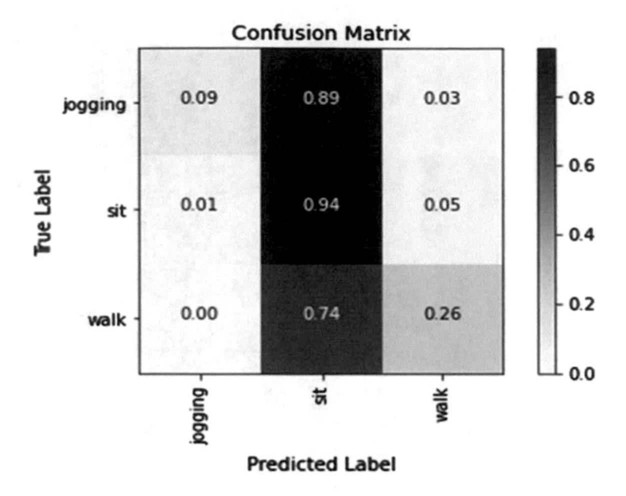

Fig. 7 Confusion matrix of CNN

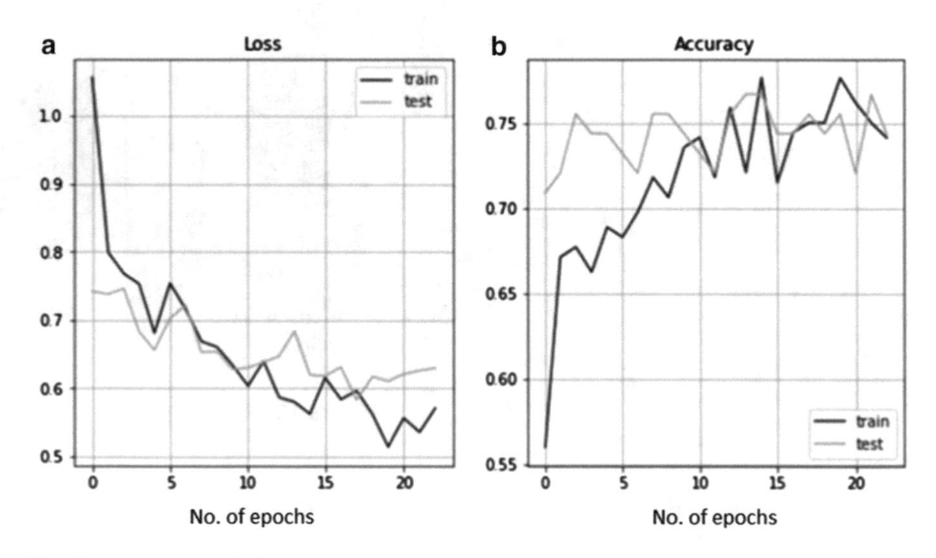

Fig. 8 **a** Data loss and Fig. 8, **b** Accuracy during training and testing using RNN

the same networks. In addition, the data transmits to entirely other networks. Real-time data exchange increases the likelihood that security adversaries would steal or alter sensitive data.

In IIoT [71], a number of security assaults are taking place. The current attacks can be divided into three categories: network attacks, software attacks, and physical attacks. Direct attacks on IoT hardware components are a type of physical attack. Attackers frequently gain direct access to all of these hardware components in an effort to compromise several important data and pieces of information. It comprises

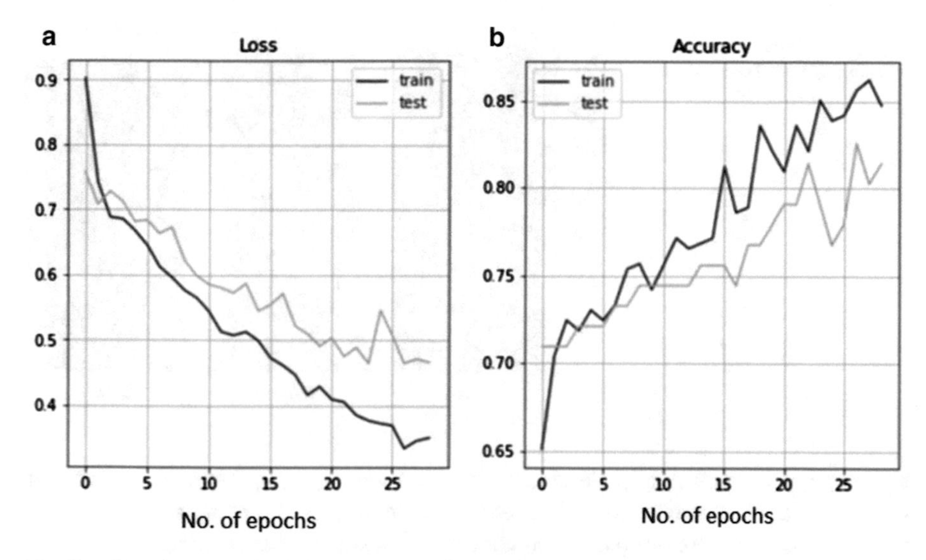

Fig. 9 **a** Data loss and Fig. 9, **b** Accuracy during training and testing using LSTM

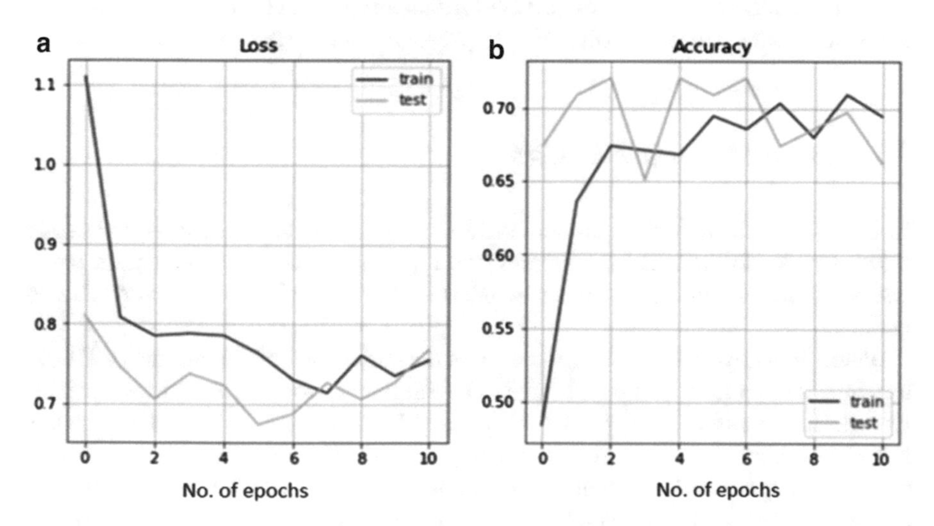

Fig. 10 **a** Data loss and Fig. 9, **b** Accuracy during training and testing using CNN

malicious behavior, denial-of-service attacks, RF interface jamming, code injection, physical tampering, fake node injection, and so on. Physical assault can cause the node to be closed, resources to be harmed, and communication issues. Network assault is a different type of attack.

Data is transmitted between devices across a network, which can be inside the same network or one that is not. Even if the attacker is far from the network, they can still readily carry out any form of security assault. There are numerous sorts of network assaults, including middleware, routing, and encryption attacks. Wormhole

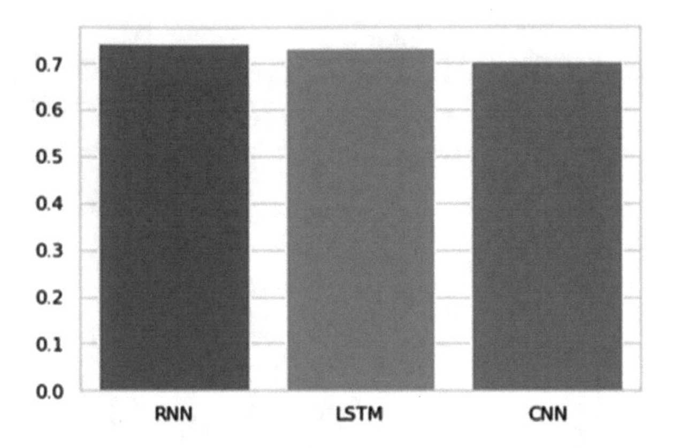

Fig. 11 Comparison model of accuracies of DL models

attacks, Sybil attacks, sinkhole attacks, etc. are among them. Due to this form of assault, data will be changed and messages could be erased, posing a threat to data security. The software or application under assault is specifically targeted.

This type of attack takes place as a result of security risks in the software. This form of assault includes a number of security vulnerabilities, including viruses, worms, spyware, adware, etc. It is challenging to preserve data privacy and consistency.

7.1 Layer wise Attack in IIoT

In the IOT architecture, there are four different layers. Because the functions of each layer vary, so do the security concerns and types of attacks. The perception layer, network layer, and application layer are the three key layers that are being considered here.

Perception Layer: The perception layer is the initial layer of the Internet of Things architecture and functions as a bridge between the network layer and the devices. Devices have multiple sensors, and this layer collects the sensing data. Through this layer, the actuators are also communicating with the surroundings. This layer also receives other network communication-related commands. This layer is subject to numerous threats, some of which include physical assault, denial-of-service attack, eavesdropping, capturing nodes, malicious and false nodes, etc. [72].

Several methods are used to safeguard the data against these types of attacks. A few are covered in [73], including raising an alarm when unwanted node access occurs and gaining access to a private identifying key when a message is being transmitted over a network. Another option is error correction code, which can handle damaged messages.

Network Layer: The following layer is the network layer, where devices primarily maintain network communication. Data is sent from the perception layer to the following layer and vice versa. For this layer, wireless communication is primarily utilized. Different wireless communication protocols, edge devices or gateways between local networks, and other external big networks are the main technical components [71]. Due to network connection, this layer experiences few threats. The Mim attack typically occurs during message transmission between two trustworthy parties. Replay attacks are when crucial messages are altered and then transmitted to the target. The other attack is a routing attack, in which security concerns might change route information.

The many security breaches and the IoT network's subsequent responses are outlined in [74]. In this work, two lightweight protocols specifically designed for compact sensor-based devices are taken into consideration for future discussion. A brief explanation of the three main attacks that can be made on the topology, resources, and traffic of the RPL and 6LoWPAN protocols. Finally, using the DODAG acyclic graph technique, they have devised a strategy to stop ranking attacks on the network layer.

Application Layer: Data enters the application layer from various devices via the network and perception layers. In this layer, the software applications are present. Security risks are primarily seeking to directly attack program code here.

References [75, 76] discuss several security threats and associated preventative measures. Several protocols, including MQTT and AMQP, are utilized at the application layer. Data security, data integrity, and data confidentiality must all be upheld. Different security risks are present at the application layer, although only a handful are significant, such as code injection, phishing, and DoS. Different authentication protective procedures are used to provide a dependable and secure solution to various security issues.

Different security concerns and their preventive remedies are described in literature across a variety of applications. Different anomalies attacks in IoT are identified by the authors [6]. The suggested model has two layers, where they will first determine whether certain events or behaviors are abnormal or not. If any activity is abnormal, the suggested model can determine the type of the abnormal activity at the following level. To find these, a number of machine learning algorithms are used. The decision tree outperforms the competition.

We have primarily concentrated on this chapter's discussion on monitoring Human Activity and Machine Activity among various IIoT applications.

7.2 Secure Human Activity and Machinery Activity Monitoring

Monitoring human activity and the status of various equipment tools is a crucial task in the present smart technology-based industry, as was addressed in earlier sections. There is a high probability of assault occurring while data is being sent between various devices both inside and outside of the same network. Several literary works exist where the authors examine various security issues, risks, and attacks that may occur while human activity is being monitored. In order to resolve those problems and challenges, other pertinent solutions were also explored.

In order to continue monitoring human activity in a secure manner, user access and authorization are crucial, especially in the IIoT. The authors of [7] thought that one possible option was to use soft biometrics. To securely examine the data, various machine learning and deep learning methods are applied. The system performs better when a neural network and an optimization method are combined.

The authors in [77] explore many attack techniques, in particular data location assault and how adversarial attack will impact performance. The CNN-based deep learning method aids in defending data from three major hostile attacks. Adversarial vulnerability in sensor-based time series data is addressed by the suggested study.

The suggested work [78] has identified various sitting postures and a robot that goes with them. Important data are at great risk of attack during data transit. Therefore, the encryption decryption based key method is used in the data transfer between the sensor device and the PC. When communicating between a desktop and a smartphone, another two factor authentication method based on a one-time password and the Diffie-Hellman algorithm is used. Here, the Hidden Markov Model is used to improve system performance.

Different security assaults in manufacturing and production, where tools are fitted with IoT sensors, were covered by the authors in [79]. To protect against attacks like sinkholes and black holes, among others, secure data transfer mechanisms based on thresholds and genetic algorithms are used.

The proposed approach in [80] this work tracks the state of gas-insulated switchgear in online mode. The most crucial role is to locate any problems or anomalies that may exist and to raise an alarm. In order to create a more secure and dependable communication for this form of defect analysis and cyber assault, the XGBoost algorithm is utilized here.

The adoption of IoT in business enhances several industrial industries. Reference [81] addresses secure machine monitoring and real-time data analysis. For communication, simple protocols are designed. The proposed technique is considered to be shown using dos and clone assaults. To address various security issues and enhance system performance, various layer wise authentication and encryption techniques are deployed.

8 Open Research Issues

Though the applications of ML and DL techniques have greatly improved the performance of IIoT systems, there are a few open problems that still remain. These are discussed as follows.

- *Data Annotation*—Machine learning techniques learn patterns from the data. Such approaches perform effectively when they can learn well, that is, when data distribution for all the activity classes are balanced, the data is free from outliers and for all the representative real-life conditions, the system data are available. This is not easy to achieve. Thus, most of the works reported on ML and DL based IIoT show their performance on either their own dataset or on a specific benchmark dataset collected for a specific purpose. So, data distribution should be given importance. Works should be designed to detect and correct labeling errors and unbiased data distribution.

- *Privacy of the ML based Systems*—Privacy is fast becoming a major research challenge as most of the systems are becoming available online through utilizing the IIoT paradigm. It has many aspects. For instance, for human activity monitoring applications, the users should be aware of the fact that anonymity is not enough to hide privacy sensitive behavior information. Thus, novel privacy models are needed to be designed that can hide the sensitive data without hampering the usability of the data for training the learning models. In fact, the learning models should be designed in a way so as to protect against model inversion attacks. Federated learning is one such approach that should be investigated for sensor based IIoT systems.
- *Reproducibility of the results*—Most ML and DL based applications suffer from this drawback. Even when appreciable accuracy values are reported in research publications, in real life, many of the systems fail to work beyond the research laboratories. Many times the data are collected for a constrained ambience ignoring many of the common industry problems or uncertainties. This limits the applicability of the work and the reliability of the systems for real life scenarios. Thus, benchmark datasets should be designed for realistic environments so as to promote research solutions to become available commercially. The ML and DL based works should compare with state-of-the art on the common baseline in order to prove their applicability.
- *Network Bandwidth*—As more and more systems are connected to the Internet, the Internet traffic increases manifold. So, providing a scalable architecture becomes a necessity. The challenge becomes even more crucial for real time systems. This calls for a unique research need of executing analytical tasks for the activity monitoring closer to the source rather than at a remote server over the Internet. In fact, strategic offloading of IIoT tasks to cloud servers and to an edge closer to the source could be explored. Edge servers are less powerful but have several benefits, such as, saving network bandwidth and hence, improved communication speed and inherent location sensitivity. However, due to their lower computational ability, the need for a cloud server for sophisticated learning tasks cannot be avoided completely. Thus, a runtime balance could be maintained by the IIoT based systems through pre-training of the learning models so as to execute common tasks on edges while a cloud server would be invoked to handle the scheduling/migration of the tasks among the edge.

9 Conclusion

This chapter presents a comprehensive overview of the activity monitoring IIoT systems from the machine learning and deep learning perspectives. The role of activity monitoring for humans and machines to provide context awareness to many applications adds an interesting dimension to this chapter. The literature reviewed here presents the immense potential of the learning mechanisms to be applied to a wide range of IIoT systems including the application domains such as, smart robotics and smart healthcare. The role of context awareness in improving performance of such applications is also discussed in this chapter. A case study has been presented on human activity recognition where the entire workflow has been shown from scratch. The importance of every phase of machine learning based analysis and the significance of data preprocessing mechanisms has been highlighted in this chapter which is crucial for getting proper output. The open research challenges are also highlighted that could be investigated further by the readers.

References

1. Roitberg A, Somani N, Perzylo A, Rickert M, Knoll A (2015) Multimodal human activity recognition for industrial manufacturing processes in robotic workcells. In: Proceedings of the 2015 ACM on international conference on multimodal interaction, pp 259–266
2. Reining C, Niemann F, Moya Rueda F, Fink GA, ten Hompel M (2019) Human activity recognition for production and logistics—a systematic literature review. Information 10(8):245
3. Tran MQ, Elsisi M, Mahmoud K, Liu MK, Lehtonen M, Darwish MM (2021) Experimental setup for online fault diagnosis of induction machines via promising IoT and machine learning: towards industry 4.0 empowerment. IEEE access 9:115429–115441
4. Bhushan B, Sahoo G (2020) Requirements, protocols, and security challenges in wireless sensor networks: an industrial perspective. In: Handbook of computer networks and cyber security. Springer, Cham, pp 683–713
5. Khan AI, Al-Badi A (2020) Open source machine learning frameworks for industrial internet of things. Procedia Comput Sci 170:571–577
6. Ullah I, Mahmoud QH (2020) A two-level flow-based anomalous activity detection system for IoT networks. Electronics 9(3):530
7. Priyadarshini I, Sharma R, Bhatt D, Al-Numay M (2022) Human activity recognition in cyber-physical systems using optimized machine learning techniques. Clust Comput 1–17
8. Wang Q, Zhu X, Ni Y, Gu L, Zhu H (2020) Blockchain for the IoT and industrial IoT: a review. Internet of Things 10:100081
9. Sethi R, Bhushan B, Sharma N, Kumar R, Kaushik I (2021) Applicability of industrial IoT in diversified sectors: evolution, applications and challenges. Multimed Technol Internet Things Environ 45–67
10. Sharma N, Bhushan B, Kaushik I, Debnath NC (2021) Applicability of blockchain technology in healthcare industry: applications, challenges, and solutions. In: Efficient data handling for massive internet of medical things. Springer, Cham, pp 339–370
11. Jaidka H, Sharma N, Singh R (2020) Evolution of IoT to IIoT: applications & challenges. In: Proceedings of the international conference on innovative computing & communications (ICICC)
12. Serpanos D, Wolf M (2018) Industrial internet of things. In: Internet-of-Things (IoT) systems. Springer, Cham, pp 37–54

13. Georgakopoulos D, Jayaraman PP (2016) Internet of things: from internet scale sensing to smart services. Computing 98(10):1041–1058
14. Chen B, Wan J, Shu L, Li P, Mukherjee M, Yin B (2017) Smart factory of industry 4.0: key technologies, application case, and challenges. IEEE Access 6:6505–6519
15. Mehrizi R, Peng X, Xu X, Zhang S, Metaxas D, Li K (2018) A computer vision based method for 3D posture estimation of symmetrical lifting. J Biomech 69:40–46
16. Malaisé A, Maurice P, Colas F, Charpillet F, Ivaldi S (2018) Activity recognition with multiple wearable sensors for industrial applications. In: ACHI 2018-eleventh international conference on advances in computer-human interactions
17. Lu Y (2017) Industry 4.0: a survey on technologies, applications and open research issues. J Ind Inf Integr 6:1–10
18. Bordel B, Alcarria R, Robles T (2022) Recognizing human activities in Industry 4.0 scenarios through an analysis-modeling-recognition algorithm and context labels. Integr Comput Aided Eng (Preprint) 1–21
19. Nwakanma CI, Islam FB, Maharani MP, Lee JM, Kim DS (2021) Detection and classification of human activity for emergency response in smart factory shop floor. Appl Sci 11(8):3662
20. Forkan ARM, Montori F, Georgakopoulos D, Jayaraman PP, Yavari A, Morshed A (2019) An industrial IoT solution for evaluating workers' performance via activity recognition. In: 2019 IEEE 39th international conference on distributed computing systems (ICDCS). IEEE, pp 1393–1403
21. de Sousa PHF, Navar de Medeiros M, Almeida JS, Rebouças Filho PP, de Albuquerque VHC (2019) Intelligent incipient fault detection in wind turbines based on industrial IoT environment. J Artif Intell Syst 1(1):1–19
22. Shah G, Tiwari A (2018) Anomaly detection in IIoT: a case study using machine learning. In: Proceedings of the ACM India joint international conference on data science and management of data, pp 295–300
23. Kanawaday A, Sane A (2017) Machine learning for predictive maintenance of industrial machines using IoT sensor data. In: 2017 8th IEEE international conference on software engineering and service science (ICSESS). IEEE, pp 87–90
24. Garmaroodi MSS, Farivar F, Haghighi MS, Shoorehdeli MA, Jolfaei A (2020) Detection of anomalies in industrial IoT systems by data mining: study of christ osmotron water purification system. IEEE Internet Things J 8(13):10280–10287
25. Chetot L, Egan M, Gorce JM (2021) Joint identification and channel estimation for fault detection in industrial IoT with correlated sensors. IEEE Access 9:116692–116701
26. Moens P, Bracke V, Soete C, Vanden Hautte S, Nieves Avendano D, Ooijevaar T, Devos S, Volckaert B, Van Hoecke S (2020) Scalable fleet monitoring and visualization for smart machine maintenance and industrial IoT applications. Sensors 20(15):4308
27. Ahmad S, Badwelan A, Ghaleb AM, Qamhan A, Sharaf M, Alatefi M, Moohialdin A (2018) Analyzing critical failures in a production process: is industrial IoT the solution? Wirel Commun Mob Comput
28. Yu W, Dillon T, Mostafa F, Rahayu W, Liu Y (2019) A global manufacturing big data ecosystem for fault detection in predictive maintenance. IEEE Trans Industr Inf 16(1):183–192
29. Lee CK (2018) Development of an industrial Internet of Things (IIoT) based smart robotic warehouse management system. In: International conference on information resources management (CONF-IRM). Association For Information Systems
30. Roitberg A, Perzylo A, Somani N, Giuliani M, Rickert M, Knoll A (2014) Human activity recognition in the context of industrial human-robot interaction. In: Signal and information processing association annual summit and conference (APSIPA), 2014 Asia-Pacific. IEEE, pp 1–10
31. Lenz C, Sotzek A, Röder T, Radrich H, Knoll A, Huber M, Glasauer S (2011) Human workflow analysis using 3d occupancy grid hand tracking in a human-robot collaboration scenario. In: 2011 IEEE/RSJ international conference on intelligent robots and systems. IEEE, pp 3375–3380

32. Ambika P (2020) Machine learning and deep learning algorithms on the Industrial Internet of Things (IIoT). Adv Comput 117(1):321–338
33. Cakir M, Guvenc MA, Mistikoglu S (2021) The experimental application of popular machine learning algorithms on predictive maintenance and the design of IIoT based condition monitoring system. Comput Ind Eng 151:106948
34. Strauß P, Schmitz M, Wöstmann R, Deuse J (2018) Enabling of predictive maintenance in the brownfield through low-cost sensors, an IIoT-architecture and machine learning. In: 2018 IEEE international conference on big data (big data). IEEE, pp 1474–1483
35. Selim GEI, Hemdan EZZ, Shehata AM, El-Fishawy NA (2021) Anomaly events classification and detection system in critical industrial internet of things infrastructure using machine learning algorithms. Multimedi Tools Appl 80(8):12619–12640
36. Khan MA, Alghamdi NS (2021) A neutrosophic WPM-based machine learning model for device trust in industrial internet of things. J Ambient Intell Humanized Comput 1–15
37. Wang J, Zhang H, Ruan Z, Wang T, Wang X (2020) A machine learning based connectivity restoration strategy for industrial IoTs. IEEE Access 8:71136–71145
38. Mallick A, Saha A, Chowdhury C, Chattopadhyay S (2019) Energy efficient routing protocol for ambient assisted living environment. Wirel Pers Commun 109(2):1333–1355
39. DS BN, Dondeti V, Balakrishna S (2022) Comparative analysis of machine learning-based algorithms for detection of anomalies in IIoT. Int J Inf Retr Res (IJIRR) 12(1):1–55
40. Babbar G, Bhushan B (2020) Framework and methodological solutions for cyber security in Industry 4.0. In: Proceedings of the international conference on innovative computing & communications (ICICC)
41. Yang H, Liang S, Ni J, Li H, Shen XS (2020) Secure and efficient k NN classification for industrial Internet of Things. IEEE Internet Things J 7(11):10945–10954
42. Marino R, Wisultschew C, Otero A, Lanza-Gutierrez JM, Portilla J, de la Torre E (2020) A machine-learning-based distributed system for fault diagnosis with scalable detection quality in industrial IoT. IEEE Internet Things J 8(6):4339–4352
43. Liu P, Zhang Y, Wu H, Fu T (2020) Optimization of edge-PLC-based fault diagnosis with random forest in industrial Internet of Things. IEEE Internet Things J 7(10):9664–9674
44. Attri T, Bhushan B (2021) Enabling technologies, attacks, and machine learning-based countermeasures for IoT and IIoT. In: Integration of WSNs into Internet of Things. CRC Press, pp 249–272
45. Bhushan B, Sahoo C, Sinha P, Khamparia A (2021) Unification of blockchain and Internet of Things (BIoT): requirements, working model, challenges and future directions. Wirel Netw 27(1):55–90
46. Saxena S, Bhushan B, Ahad MA (2021) Blockchain based solutions to secure IoT: background, integration trends and a way forward. J Netw Comput Appl 181:103050
47. Tiwari R, Sharma N, Kaushik I, Tiwari A, Bhushan B (2019) Evolution of IoT & data analytics using deep learning. In: 2019 international conference on computing, communication, and intelligent systems (ICCCIS). IEEE, pp 418–423
48. Orimoloye LO, Sung MC, Ma T, Johnson JE (2020) Comparing the effectiveness of deep feedforward neural networks and shallow architectures for predicting stock price indices. Expert Syst Appl 139:112828
49. Mukherjee A, Goswami P, Yang L, Sah Tyagi SK, Samal UC, Mohapatra SK (2020) Deep neural network-based clustering technique for secure IIoT. Neural Comput Appl 32(20):16109–16117
50. Jayalaxmi PLS, Saha R, Kumar G, Kim TH (2022) Machine and deep learning amalgamation for feature extraction in Industrial Internet-of-Things. Comput Electr Eng 97:107610
51. Goswami P, Mukherjee A, Chatterjee P, Yang L (2021) An optimal resource allocation method for IIoT network. In: Adjunct proceedings of the 2021 international conference on distributed computing and networking, pp 31–36
52. Liu C, Zhu H, Tang D, Nie Q, Li S, Zhang Y, Liu X (2022) A transfer learning CNN-LSTM network-based production progress prediction approach in IIoT-enabled manufacturing. Int J Prod Res 1–24

53. Wang Y, Yan J, Sun Q, Jiang Q, Zhou Y (2020) Bearing intelligent fault diagnosis in the industrial Internet of Things context: a lightweight convolutional neural network. IEEE Access 8:87329–87340
54. Khamparia A, Singh PK, Rani P, Samanta D, Khanna A, Bhushan B (2021) An internet of health things-driven deep learning framework for detection and classification of skin cancer using transfer learning. Trans Emerg Telecommun Technol 32(7):e3963
55. Cai S, Lau VK (2021) RNN-based learning of nonlinear dynamic system using wireless IIoT networks. IEEE Internet Things J 8(14):11177–11192
56. Senthilkumar P, Rajesh K (2021) Design of a model based engineering deep learning scheduler in cloud computing environment using Industrial Internet of Things (IIOT). J Ambient Intell Humanized Comput 1–9
57. Wu D, Jiang Z, Xie X, Wei X, Yu W, Li R (2019) LSTM learning with Bayesian and Gaussian processing for anomaly detection in industrial IoT. IEEE Trans Industr Inf 16(8):5244–5253
58. Khan IA, Keshk M, Pi D, Khan N, Hussain Y, Soliman H (2022) Enhancing IIoT networks protection: a robust security model for attack detection in internet industrial control systems. Ad Hoc Netw 134:102930
59. Khowaja SA, Khuwaja P (2021) Q-learning and LSTM based deep active learning strategy for malware defense in industrial IoT applications. Multimed Tools Appl 80(10):14637–14663
60. Goodfellow I, Pouget-Abadie J, Mirza M, Xu B, Warde-Farley D, Ozair S, Courville A, Bengio Y (2014) Generative adversarial nets. Adv Neural Inf Process Syst 27
61. Taheri R, Shojafar M, Alazab M, Tafazolli R (2020) FED-IIoT: a robust federated malware detection architecture in industrial IoT. IEEE Trans Industr Inf 17(12):8442–8452
62. Dzaferagic M, Marchetti N, Macaluso I (2021) Fault detection and classification in Industrial IoT in case of missing sensor data. IEEE Internet Things J
63. Behera S, Misra R (2021) Generative adversarial networks based remaining useful life estimation for IIoT. Comput Electr Eng 92:107195
64. Feng Y, Chen J, Liu Z, Lv H, Wang J (2022) Full graph autoencoder for one-class group anomaly detection of IIoT system. IEEE Internet Things J
65. Huang Z, Wu Y, Tempini N, Lin H, Yin H (2022) An energy-efficient and trustworthy unsupervised anomaly detection framework (EATU) for IIoT. ACM Trans Sens Netw (TOSN)
66. Zhao J, Qiu J, Sun D, Chen B (2021) RAEF: an imputation framework based on a gated regulator autoencoder for incomplete IIoT time-series data. Complexity
67. Wu Y, Dai HN, Tang H (2021) Graph neural networks for anomaly detection in industrial internet of things. IEEE Internet of Things J
68. Zhang Y, Yang C, Huang K, Li Y (2022) Intrusion detection of industrial internet-of-things based on reconstructed graph neural networks. IEEE Trans Netw Sci Eng
69. Zhou F, Yang Q, Zhong T, Chen D, Zhang N (2020) Variational graph neural networks for road traffic prediction in intelligent transportation systems. IEEE Trans Industr Inf 17(4):2802–2812
70. Malhotra L, Bhushan B, Singh RV (2021) Artificial intelligence and deep learning-based solutions to enhance cyber security. In: Proceedings of the international conference on innovative computing & communication (ICICC)
71. Atlam HF, Wills GB (2020) IoT security, privacy, safety and ethics. In: Digital twin technologies and smart cities. Springer, Cham, pp 123–149
72. Serror M, Hack S, Henze M, Schuba M, Wehrle K (2020) Challenges and opportunities in securing the industrial internet of things. IEEE Trans Industr Inf 17(5):2985–2996
73. Khattak HA, Shah MA, Khan S, Ali I, Imran M (2019) Perception layer security in Internet of Things. Futur Gener Comput Syst 100:144–164
74. Sharma R, Pandey N, Khatri SK (2017) Analysis of IoT security at network layer. In: 2017 6th international conference on reliability, infocom technologies and optimization (trends and future directions) (ICRITO). IEEE, pp 585–590
75. Swamy SN, Jadhav D, Kulkarni N (2017) Security threats in the application layer in IOT applications. In: 2017 international conference on i-SMAC (iot in social, mobile, analytics and cloud) (i-SMAC). IEEE, pp 477–480

76. Nebbione G, Calzarossa MC (2020) Security of IoT application layer protocols: challenges and findings. Future Internet 12(3):55
77. Yang Z, Zhao Y, Yan W (2020) Adversarial vulnerability in doppler-based human activity recognition. In: 2020 international joint conference on neural networks (IJCNN). IEEE, pp 1–7
78. Tariq M, Majeed H, Beg MO, Khan FA, Derhab A (2019) Accurate detection of sitting posture activities in a secure IoT based assisted living environment. Futur Gener Comput Syst 92:745–757
79. Qureshi KN, Rana SS, Ahmed A, Jeon G (2020) A novel and secure attacks detection framework for smart cities industrial internet of things. Sustain Cities Soc 61:102343
80. Elsisi M, Tran MQ, Mahmoud K, Mansour DEA, Lehtonen M, Darwish MM (2021) Towards secured online monitoring for digitalized GIS against cyber-attacks based on IoT and machine learning. IEEE Access 9:78415–78427
81. Tedeschi S, Emmanouilidis C, Mehnen J, Roy R (2019) A design approach to IoT endpoint security for production machinery monitoring. Sensors 19(10):2355

AI Model for Blockchain Based Industrial IoT and Big Data

Lipsa Das, Vimal Bibhu, Rajasvaran Logeswaran, Khushi Dadhich, and Bhuvi Sharma

Abstract In this recent era of technology, blockchain and the Internet of Things (IoT) are regarded as highly effective, popular technologies. Blockchain is a decentralised database that is used for various transactional purposes. In contrast to IoT, which relates to the spread of linked machines by offering information on the Web, it offers unique directions for storage and management of information. Even though blockchain requires real-time data application and IoT describes processes to store and manage information overloads safely and effectively, a hybrid of these two technologies is promising. The Industrial IoT (IIoT) is a result of the Industrial IoT's significant impact on the manufacturing industry, which is undergoing a digital revolution. Blockchain for Industrial Internet of Things (BIoT) is the name given to such a combination of IIoT and blockchain. Also, over the recent years, big data generated by IoT devices has attracted significant interest across a range of scientific and engineering fields. Big data has many benefits and uses, but there are also many challenges that must be overcome for a higher level of service, such as big data analytics, big data management, and big data privacy and security. Big data services and applications have the great potential to be improved by blockchain due to its decentralisation and security features. This chapter discusses the IIoT domain and the pertinent IIoT challenges. A brief overview of blockchain technology applications is also provided before the intersection of IIoT and blockchain is then explored.

L. Das (✉) · V. Bibhu · K. Dadhich · B. Sharma
Amity University, Greater Noida, Uttar Pradesh, India
e-mail: lipsaentc9@gmail.com

V. Bibhu
e-mail: vimalbibhu@gmail.com

K. Dadhich
e-mail: khushidadhich08@gmail.com

B. Sharma
e-mail: bhuvisharma86@gmail.com

R. Logeswaran
Asia Pacific University of Technology and Innovation, Kuala Lumpur, Malaysia
e-mail: loges@ieee.org

B. Bhushan et al. (eds.), *AI Models for Blockchain-Based Intelligent Networks in IoT Systems*, Engineering Cyber-Physical Systems and Critical Infrastructures 6,
https://doi.org/10.1007/978-3-031-31952-5_3

The chapter provides a thorough overview of IIoT as well as blockchain for big data, with a focus on contemporary methods, possibilities, and future perspectives of these cutting-edge ideas.

Keywords IoT · Blockchain · Bigdata

1 Introduction

The Internet of Things (IoT) comprises of machines and devices with networking capabilities that may collect data from their surroundings using a variety of sensors, share it with other gadgets through Internet, and autonomously analyse the information they receive without the need for human interaction [1, 2]. Intelligent IoT resident appliances, for instance, could utilise sensors so as to sense when home gadgets should be turned on or off, and can perform this automatically, saving energy and simplifying people's lives. IoT encompasses a broad range of technologies, including sensors, modern smart goods such as smart televisions, wearable electronics, implants (such as for identification or medical purposes in humans and animals), and tracking chips for animals [3]. Applications for IoT extends widely, from household goods, medical, farming, power, and transportation to other industries that stand to gain from its integration. However, IoT is normally more centred on smart products that benefit end-users, such as a smartwatch that tracks heart rate, making it more consumer-oriented. Although the number of IoT based gadgets is growing at an incredible speed, there are numerous difficulties associated with the same [4]. IoT devices typically only have a few processor cores and are susceptible to security breaches [5]. Security is therefore a key consideration when creating an IoT-based system.

Four key areas can be used to categorise attacks against IoT-based systems [6]. If a malicious person physically approaches the network and attempts harmful functionality, it is referred to as a physical attack. Most frequent types of physical attacks include tampering with IoT devices, radio frequency signal jamming, side-channel attacks, and malicious code injection. Researchers employ the physical unclonable function (PUF) [7] for the authentication of IoT devices to defend against physical attacks. In those systems, the challenge-response technique is primarily used for authentication. The benefit of employing PUF is that it becomes infeasible to replicate the device's precise micro-structure. The second type of assault is when the attacker attempts to modify the network of the IoT, where it is known as a network attack. This attack can be carried out by the attacker even if they are far from the network. The network attack is the foundation for assaults including traffic assails, RFID (radio-frequency identification) spoofing, Sybil attacks, and man-in-the-middle attacks. This assails can be stopped by using a secure hash algorithm and authentication methods [8]. Software assaults and data attacks, respectively, are the third and fourth types of attacks. When launching a software assault, the attacker considers the weaknesses of the software already there in IoT system. A data attack, on the other hand, involves illegal access to the data and inconsistent data. The privacy-preserving blockchain technology effectively thwarts data breaches [9].

Industrial IoT (IIoT) is the application of IoT technologies in the manufacturing and industrial operations to improve productivity and efficiency [10]. Industries can profit from the Internet of Things by improving data collection, device communications, hardware maintenance and monitoring quality control, supply chain transparency, power management, and total reduced costs. Everything can benefit from being connected to the Internet [11] because sensors can gather data, interpret it, and transfer it to servers with powerful computing power for further analysis and decision-making [12]. From there, other IoT devices can put those decisions into action. For instance, a crop irrigation system could employ moisture sensors to gather information about soil moisture, send it to a server to determine when and how much watering is required, and then water the crops automatically. It can also access meteorological data from the Internet and base decisions thereon.

This chapter explores the IIoT frameworks to extend its usability and other factors such as monitoring, operations and network communication. The overview of cyber physical systems, including the layers of IIoT such as edge level, platform level, and enterprise level, are discussed with respect to the monitoring, reliability, and collaboration under the application scenarios. The challenges such as resource limitations, service disruption, security and personal data privacy, fusion information technology with things, and requirement of smart operating system environment for operations, are discussed and the possible solutions are mentioned to reduce the issues. Further, the details of the smart contracts are given to explore the significance and applications in the field of IIoT. Finally, the big data analytics processes and its applications in the field of education, urban planning, agriculture, service delivery, healthcare and industry are discussed in detail, including the challenges and their available possible solutions, by integrating the artificial intelligence (AI) and blockchain technologies.

The organization of this chapter is as follows: Sect. 2, presents the work that focuses on the IIoT framework, while Sect. 3 discusses the challenges for IIoT. Section 4 introduces the concept of smart contracts and its significance in the application of IIoT, whereas Sect. 5 describes the application of smart contracts in blockchain technology. In Sect. 6, big data analytics is discussed along with its applications, and in Sect. 7 presents the research related to the challenges of big data. Finally, the conclusions are drawn in Sect. 8.

2 IIoT Framework

In a limited sense, IIoT is regarded as an IoT sub-part and is formalized as a communication-based network that offers ubiquitous interconnection for industrial intelligent items and have high reliability, low delay, and high controllability. Through the integration of IIoT and digital technology, the conventional production architecture will eventually give rise to cyber-physical systems (CPS) [13]. To monitor or handle units that are present physically in the world, such as actual people, operational environments, and equipment, the CPS integrates the computing abilities, interactions, and repositories [14]. Therefore, IIoT is considered as the commercial CPS embodiment in a wide sense from a networks point of view [15].

The IIoT study built on CPS is more comprehensive, offering a fresh perspective on safety and other concerns. Figure 1 depicts IIoT from a CPS perspective [16, 17]. Through data contact, the physical and cyber systems come together as one. The primary components of the physical systems are all types of networked manufacturing equipment (e.g., mechanical arms, detectors and many more) and the intelligent goods created, which together make up a full visible production line. The primary components of the electronic/cyber systems are abstract data paradigms that connect to the physical elements through mapping. The capacity to manage digital content is offered for every type of information analysis and application. Self-behaviour is the capacity to autonomously combine all the data and produce judgements for the physical system. The IIoT requires close monitoring across a field network, internal software programmes, and network functions that are concerned with punctuality, reliability, and the quality of service (QoS). Virtual management and self-behaviour uses a distributed, collaborative capability to feel, comprehend, affect the IIoT and achieve the goals of the particular application, such as self-configuration, adjustment, and optimization [18, 19].

As seen in Fig. 2, the IIoT can be conceptualised as having a three-tier design [20, 21]. Edge level, platform level, and enterprise level are the components of a fundamental IIoT architect. Below is a description of each layer's primary duties:

- **Layer 1—Edge level**: The edge layer resolves the conflict between information technology (IT) and operational technology (OT) protocols, connects to the production network at the field level, and then uses scattered edge resources to

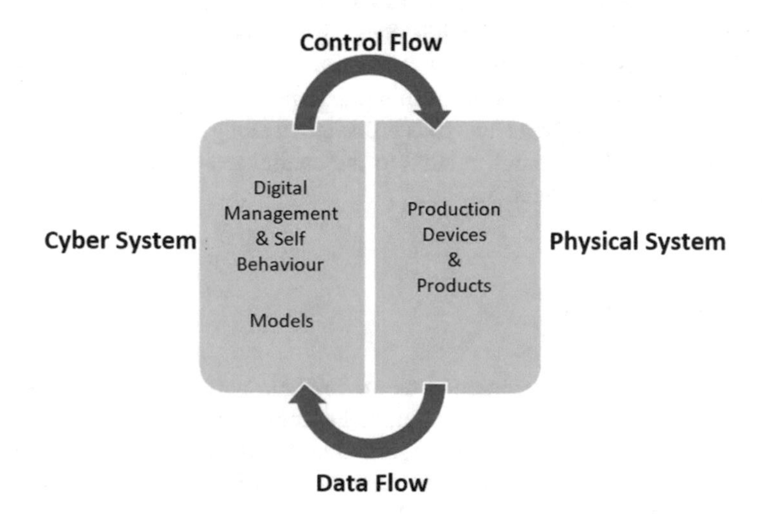

Fig. 1 Cyber-physical system (CPS) viewpoint of IIoT

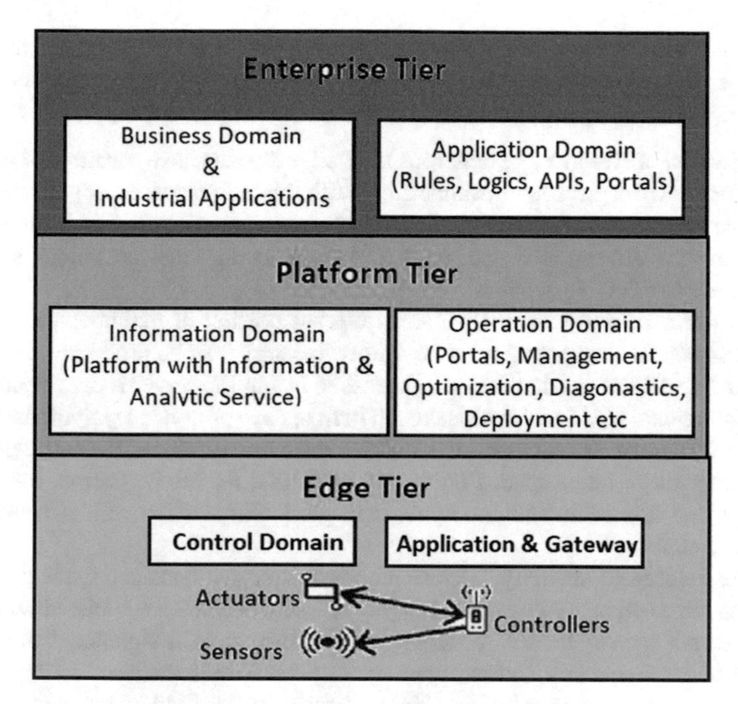

Fig. 2 IIoT three tier framework

support computing for the underlying intelligent devices. Particularly, by establishing controllers that handle command issuance and information awareness, several control domains can be built. Gateways can then be utilised to make capabilities available.

- **Layer 2—Platform level**: Depending on the type of information the edge layer supplies, the platform level is split into the data world and the operational area. To help boost the value of industrial information, the domain will offer information management and analysis facilities to the data sector. Specific manufacturing services including surveillance, diagnostics, prognostics, and optimization are included in the operations domains. These services facilitate automated production.
- **Layer 3—Enterprise level**: The small abilities offered by both domain and edge are combined and wrapped. Additionally, it offers open capabilities for business-related apps and third-party applications.

3 Challenges for IIoT

The IIoT, a typical industry 4.0 application scenario, is working to advance the advancement of all spheres of life while juggling many technological hurdles. The continuous development of IIoT is hampered by a lack of intelligent and efficient operating systems, limiting equipment assets, service interruptions, security concerns, the merging of IT and OT, and service interruptions [22].

- **Limitation of Resources**: Numerous IIoT edge devices have constrained storage, network and other resource availability. With limited resources, edge devices are unable to carry out many tasks that demand computational power and cannot store a lot of data. An essential issue for IIoT is how to store the vast amounts of data that the edge devices acquire.
- **Disruption of Service**: In IIoT, a significant portion of activities that require high computational power must be sent to the server to be processed due to the limited resources of edge devices. One issue is that the server is overloaded with computational work, which causes each firm to have prolonged wait times. On the other side, if the server comes under attack from hostile actors, the entire network service will be interrupted. Due to this difficulty, the IIoT system's robustness is low and it is challenging to apply it in particular settings with strict security requirements.
- **Issues related to Security**: Lack of authentication methods and inadequate data privacy protection are two of the biggest security problems associated with IIoT. To prevent the disclosure of their private information, a significant portion of edge nodes frequently conceal their own personal information. Identity cannot be verified when personally identifiable information is hidden, and nodes cannot rely on one another. Once a harmful attack has been launched, it is impossible to determine who the perpetrator is and to stop it in time. The value of data has received a lot of attention with the development of big data. Numerous IIoT devices gather a variety of information, including a sizable amount of personal data regarding user privacy. The foundation of using data is ensuring its security and the privacy of its owners.
- **Fusion of IT and OT**: As IIoT technology advances, more tasks can be undertaken by robots, and more manual tasks can be automated by software [23]. Continuous IT and OT convergence can enhance industrial processes and offer advantages like boosted productivity and security. However, the benefits of IIoT cannot be fully tapped into given the state of IT and OT integration. IT focuses more on modernising system architecture and attempting to increase system efficiency by utilising cutting-edge technology. OT is more concerned with preserving the system's stability. Blindly pursuing new technology upgrades will introduce unstable elements into the system. Productivity will be impacted if stability is overemphasised and new technology is not adopted. Therefore, one of the biggest challenges for the growth of all IIoT companies is how to encourage the deep integration of IT and OT.
- **Absence of a Smart and Effective Operating System**: A tailored IIoT operating system is created to work within the specific constraints of the IIoT devices (including memory, size, power and processing capacity). The system has strict demands for device security, memory utilization, actual efficiency, energy efficiency, confidentiality, and data storage. On the one hand, the growing variety of

devices and the growing number of devices in the IIoT allow intelligent systems to reduce the burden of administrators and increase management efficacy. On the other hand, feedback data can be acquired through a range of tools to help decision-makers make more sensible decisions. The IIoT application also has strict requirements for service quality. There will soon be a safe, sophisticated, and functional IIoT special operating system.

One of the potential technologies that can address numerous problems of trust and attack, etc. in IoT and IIoT-based systems is blockchain [24], which uses smart contracts [25]. Blockchain is a distributed, immutable ledger and a decentralised peer-to-peer network [26, 27]. It is made up of a number of interconnected blocks, each of which contains data made up of transactions, a timestamp, and a hash of the block before it that was used to build a hash for the most recent block [28]. A wide range of applications are ready for innovation and transformation thanks to blockchain technology. It is a transactional database or digital ledger. Data are transferred among a network of several computers or nodes, each of which runs specialised software to ensure that all the data remain consistent. This is how the blockchain process differs from traditional databases. Traditional databases keep the data on a central server, and if needed, a central authority (a reputable one) changes or erases the information [29]. Because the data is spread over numerous computers (or nodes) and that each bit of information (or block) is connected to the one before it, it is difficult for an enemy to change (or erase) anything from a blockchain-based system. In the meantime, the blockchain supports maintaining trust in the decentralised network by securely storing and validating sensitive data [30].

4 Smart Contracts

The concept of smart contacts was proposed by Nick Szabo in 1994 as "a computerised transaction mechanism that performs the term of contracts". In terms of the blockchain, smart contracts are pieces of code, scripts, or programmes that are stored there [31]. The scripts are composed using a sophisticated language. Smart contract scripts typically use "IF… this THEN… this" statements that are always operating in a secure environment. The integrity of code and data, as well as correct execution, are provided by the closed environment. These contracts also communicate with other contracts and systems that use smart contracts.

In general, deterministic state machine models serve as the foundation for smart contract processes in blockchain-based systems. Smart contracts cannot have a nondeterministic flow. Because the nondeterministic model generates diverse outputs (or states) for the same inputs, consensus cannot be reached in a nondeterministic system. In the permissionless or permissioned blockchain-based smart contract-based system, all the nodes in a network agree on the contracts. These agreements specify a set of regulations that are kept on each node in the blockchain network.

These scripts are then carried out by blockchain nodes to carry out particular operations or transactions in a network [32, 33]. In the state machine-based smart contract's systematic architecture, vertices in this directed graph stand for states, while edges show how states change over time. Transitions between states are made using the set of actions (authentication, no action, violation, permission, etc.). For instance, if the action is authentication and the machine is in state 0, it will automatically move to state 1; similarly, if the action is violation and the machine is in state 1, it will automatically move to state 2, and so on.

Because there are not enough applicable digital systems and theoretical methodologies at the beginning, the smart contracts are not initially produced well. There are virtually no digital assets accessible at that time that can be controlled directly by contracts. The second significant issue with contracts is the restriction on computational law. The computational law requires extremely efficient code execution, but the lack of readily available fast processors restricts this process. Finally, the development of smart contracts is slowed down by the absence of a trustworthy environment [34]. Decentralized blockchain systems have offered a safe and reliable environment for smart contract execution in recent years. Technology is evolving very quickly, and at the same time, attackers are becoming too intelligent. Attackers attempt to introduce flaws into smart contracts when it comes to them. Nikolic et al. [35] performed the security analysis in the smart contracts in 2018 using the MAIAN tool. They gathered 1 million smart contracts for this, and the results showed that 34,200 of them were vulnerable, taking 10 s to execute on average. There are numerous formal verification tools available now that can check smart contracts and identify any faults or errors that may be present. The Oyente tool is frequently used by researchers and programmers to examine the vulnerability in smart contract programming. It typically captures re-entrancy, transaction ordering reliance and time-stamp dependence. The symbolic variables are used in place of the programme variables each time the smart contracts are checked using the Oyente tool. The contract byte code's control flow graph is created by examining several paths [36].

Smart contracts offer the advantages of cost savings, trust, and quicker, more precise execution. Many applications employ the smart contracts in accordance with their needs and requirements because of their distinctive qualities. The following are a few examples of the smart contracts' varied range:

- The trade settlement system's ability to handle transactions efficiently is now hampered by the expensive and dangerous settlement process. The settlement process is time-consuming and ineffective since it involves a variety of successive practises. For instance, it usually takes longer than 20 days to settle a leveraged debt. When the transactions are delayed until the predefined conditions are satisfied, smart contracts can be quite useful. Additionally, they lessen risk, operational costs, and settlement times [37].
- The existing mortgage financing system is based on antiquated procedures and structures, much like the trade settlement. There is a lot of documentation involved and there are numerous middlemen in the loan approval. Each financial related document is manually distributed to various departments, where it is then tracked

and signed by the appropriate authorities [38]. As a result, this entire process involves a lot of human work and occasionally is weak due to human error. Smart contracts can be incorporated to digitise and automate the entire process.
- Real estate professionals are involved in the land registry process and go over records of buyer and seller agreements as well as development information. Due to the third party's involvement and manual process, land records may be altered, which ultimately results in fraud where a third party may receive a significant additional payment. The afore-mentioned issue can be resolved by blockchain systems with smart contracts, where the blockchain token contains the pertinent property information, such as buyer and seller IDs and legal descriptions [39].
- The insurance sector faces a number of difficulties, including high operational expenses and sluggish claim processing. The insurance industry may use blockchain and smart contracts to speed up transactions even more, requiring just a few seconds to settle claims and move money from one party to another [40]. In the insurance sector, where smart sensors are installed on vehicles or homes and transmit real-time data to authorities, IIoT devices are essential. The smart contracts are automatically triggered to settle the claim after the user's credentials have been verified.

5 Significance of Smart Contracts in IIoT

Without the assistance of a reliable trusted third party (TTP), the smart contract executes autonomously, independently and transparently (in accordance with the terms of the consensus), and transmits the possessions among the two parties. Smart (digital) contracts' dissociation of TTP reduces the additional costs and time involved with physical contracts [41]. The following is a very basic explanation of the primary benefits of smart contracts, which include computerization, visibility, authentication, increased safety, and trust-ness.

Because the information and logic of the smart contracts are transparent to all network members, the smart contract system fosters a culture of trust (which is according to the permissioned and permissionless network structure). The participants in the commercial IoT systems are dispersed throughout the world, thus the industrial sectors must ensure complete system transparency [42, 43]. The ideal method for doing this is an IIoT system based on smart contracts. In the commercial field, a third party that upholds security and a confidence level is typically used. The TTP resulted in increased expenses in the commercial fields. The smart contract eliminates the two parties while simultaneously establishing belief in the overall atmosphere of lack of trust. Contracts are becoming more automated as a result of the industrial sector's adoption of smart contracts. Additionally, once a contract is implemented on the blockchain network (assuming majority consensus), it cannot be changed. More speed and efficiency are offered by smart contracts. Smart contracts allow for more transactions to be processed per unit of time than conventional contracts do [44]. Because the IIoT domain (like the medical industry) obtains huge amounts of information, smart contracts are the best solution for processing it.

6 Application of Smart Contracts in Block Chain Technology

An essential component of blockchain technology is smart contracts. Real-world contracts and smart contracts are identical in the sense that they represent that a written agreement between two or more parties exists [45, 46]. The main distinction is that smart contracts are entirely digital. A blockchain contains a machine software which is deposited/stored there so that desktops/laptops/ computers (or nodes) can access and run to update the ledger. When specific terms and circumstances are satisfied and verified, smart contracts are self-executing codes. They gain several intriguing qualities by being kept on a blockchain, such as immutability and distribution. Figure 3 shows how blockchain is integrated with the IoT-based system. Authentic data owners transmit the IoT data they have collected to the blockchain for storage, and users retrieve the data from the blockchain using the necessary access control mechanisms.

The devices that collect data from various IoT devices retain a lot of personal information, which must be safe and accessible only to the owner and permitted and

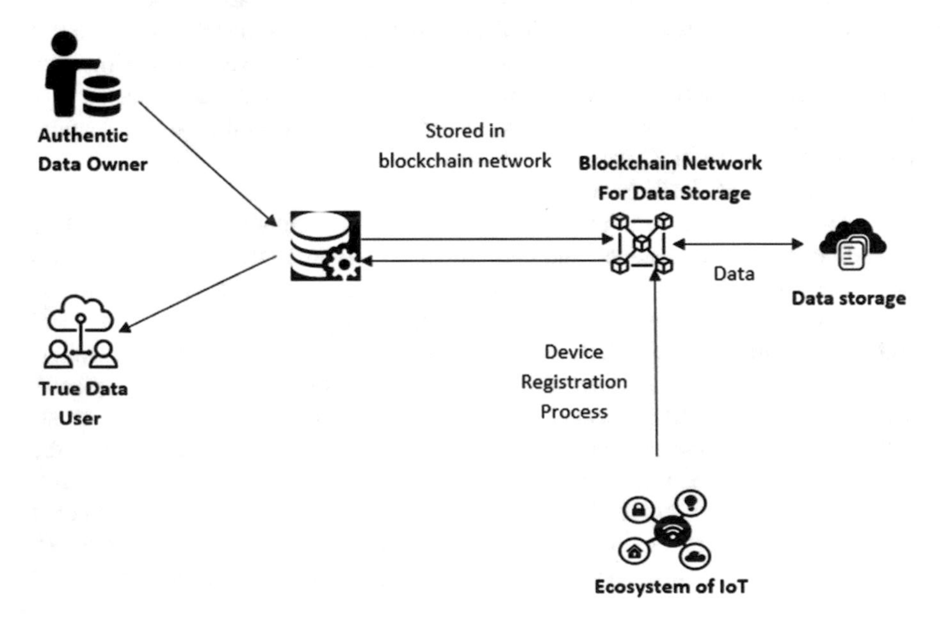

Fig. 3 Blockchain, smart contracts, and IoT-based system integration [47, 48]

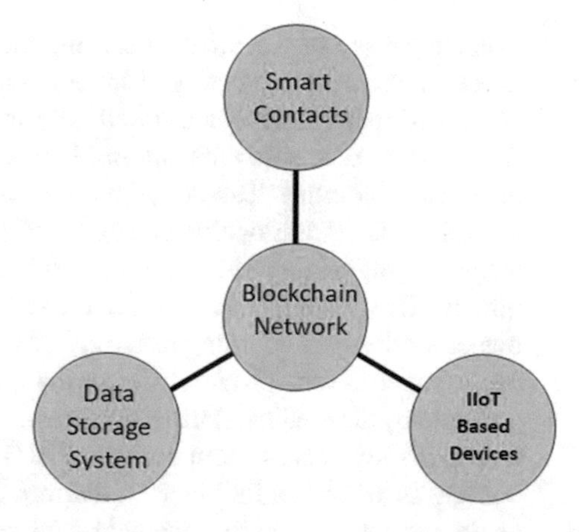

Fig. 4 Fusion of blockchain technology with IIoT devices and smart contracts [47, 48]

validated stakeholders. The owner can control the sharing of their information on their conditions with the use of smart contracts, and the blockchain is among the probable options for ensuring data security and integrity. Three key advantages of blockchain technology in IIoT include decentralisation, permanent archives, and nonrepudiation of recorded data [49]. Figure 4 illustrates how smart contracts working on an IIoT-based system often operate. The smart contract processes and stores the data from the IIoT devices on the blockchain.

7 Domain-Specific Use of Blockchain Technology

As discussed above, the blockchain technology can be used to address a number of problems with the conventional systems. Some examples of how blockchain can be used in different domains is given below:

- **Health Sector**: This has emerged as one of the most significant social and economic issues because of the continuously growing global population. There are no hospitals in some locations, and many have very few resources. Wearable medical technology is crucial to treating patients who are far away. These tools measure the patient continually and gather the necessary health information. The benefit of this method is that doctors and the medical staff can remotely view ad analyse the information at any time. But maintaining these wearable devices' security and privacy is a difficult undertaking. A few of the problems this industry faces could be resolved by implementing blockchain technology. As an extension of this, the researchers in [50] developed a new paradigm for supervising distant sufferers utilising a blockchain system based on patient agents. The wearable sensors gather the health-related data, which is then sent to the sufferer's agent who divides it in manageable units prior to storing it on the blockchain network. For the private health-data management system, Xia et al. [51] proposed a blockchain-based secure data-sharing method. The proposed protocol made use of key-agreement protocol and identity-based authentication. The authors have suggested using blockchain-based solutions to manage patient health information and promote safe data sharing between hospitals, insurance providers, and healthcare facilities. The confidentiality and privacy of health data are ensured by this model. Additionally, Sun et al. [52] presented an attribute-based signature technique to maintain the decentralised health data based on the blockchain technology. This system both validates the validity of the health data and protects the owner's privacy. According to Griggs et al. [53], the allowed blockchain domain might supervise the distant sufferers thanks to the smart contracts. Information is gathered by the medical sensors and sent to the smart based contracts for additional determination. This system uses the PBFT (Practical Byzantine Fault Tolerance) consensus technique for block validation. The authors in [54] proposed a patient-centric method of sharing medical data between various hospitals and research facilities. The security of the patient's information is assured by the system.

- **Supply Chain Management**: A product in the supply chain management system is made up of various units that are supplied directly through various producers and it is little costly to find minimal quality parts if a manufacturer or other organisation submits them during the entire process (or if they are forged). The afore-mentioned problem is resolved by integrating blockchain with IoT. Every component has a unique ID that is kept in the blockchain network along with a timestamp and may be tracked at any moment. The blockchain mechanism was selected by the authors for the validation of each component supplied by the manufacturer. To track the products, the researchers combined the blockchain with IoT and made use of the Ethereum platform. With the provided approach, the supply chain management system achieves data provenance. According to the researchers, integrating IoT with blockchain technology can speed up the supply chain process while lowering costs and risks. Additionally, Machine Learning (ML)-Blockchain method are coupled in [55] to enhance the quality and its services. The researchers in [56] coupled RFID with the blockchain mechanism to offer a traceability solution for an agrifood supply chain. Transparency and product traceability are provided by the system. The issues with the supply chain system include the twofold marginalisation and knowledge asymmetry. To address this issue, Nakasumi [57] provided a homomorphic encryption-based approach for the privacy and security of the client's information. The supply chain system's blockchain-based secured information-sharing mechanism was proposed by the researchers. To make the recommended system more secure, the authors developed a two-phase validation process (transaction and block validation). To illustrate visibility in the supply chain system, smart contracts were developed and put into use on the Ethereum blockchain technology.
- **Internet of Things (IoT)**: IoT devices often focus on data and receive and upload data from a large number of other devices. Attackers may target both the device and the data. The attacker transmits the network with the fabricated data [58]. The IoT-based system may contain sensitive and private data [59]. Data privacy and integrity are therefore of utmost importance to these systems [60]. This issue may

be resolved by blockchain technology. Some Research presented the virus detection system built on the blockchain network, which is additionally based on statistical analysis. To reduce the false-positive rate, they applied the numerous labelling function and fuzzy comparisons technique. In order to control IoT devices, it is suggested to adopt a blockchain-based system for access control. Their solution consists of a blockchain network, management hub, managers, agent node, smart contracts, and wireless sensor network. The system maintains transparency and is scalable. The authors of [61] suggested using the cloud blockchain for the IoT system to handle the resource allocation strategy. An issue with this strategy is that the contracts for allocation of resources are not examined. The IoT is equally relevant in the 5G period, when thousands of devices are linked. To address the 5G's privacy issue, a blockchain data sharing and preserving scheme was proposed by Fan et al. [62].

8 Big Data Analytics

The Term "Big Data" has recently been used to describe datasets that grow to sizes which makes it challenging to handle them with traditional database management systems. These data sets cannot be gathered, saved, managed or analysed in an acceptable length of time using conventional computing tools and datastores [22]. One data collection might contain anywhere between terabytes (TB) to several petabytes (PB) of data. The massive amounts of data might be difficult to obtain, store, search, share, analyse, and show as a result. Businesses are currently actively undertaking big data analytics, i.e. studying enormous volumes of incredibly detailed information with sophisticated tools, in order to uncover as much information as possible to benefit the business. However, it becomes more difficult to manage a larger amount of data [63].

9 Big Data Applications

Big data analytics benefits all sectors that have access to large amounts of data. Some examples are described below.

- **Education Sector**: Education will undoubtedly undergo a transformation because to big data. Eight years back, the discipline of studying analytics emerged, and is predominantly used in Learning Management Systems (LMS). This marked the beginning of the data gold rush in education. The goal of learning analytics is to enhance learning and learning environments by measuring and analysing student data. Education is becoming more engaging, accessible, and inexpensive thanks to big data. A thorough study was conducted, which includes a quantifiable equivalence of the various methods used by 35 charter educational institutes in New York City. One among from 5 topmost strategies associated with quantifiable educational potency involves the usage of information to make decisions [64]. The obtained information can be taken into the usage for creating the highly efficient educational plans, beginning with the delivery for going through, writing down, STEM (expanded as "Science, Technology, Engineering, and Mathematics"), and high-level university courses.

A significant trend toward the widespread deployment of educational activities on the Web is expected to produce a growing amount of precise data about student performance and enable the school management to manage students' wellbeing more pro-actively. Every mouse-click and key-press that every student makes is being recorded as well as analysed by e-learning platforms. Hence, universities now have access to massive goldmines of data on how to follow students with their higher studies. The educational system must, however, continue to concentrate on its main goal of really assisting students in learning and improving their presentation [65]. This technology is helping education and connected businesses today to grasp things that was not previously realised. Big data and predictive analytics can be used to create logical databases that could provide university administrators with quick, useful data about student enrolments. This knowledge aids them in making informed enrolment decisions, allocating personnel and financial resources to increase enrolment from ongoing markets, and building new ones. Big data and predictive analytics can help candidates shortlist universities that are the best fit for their profile, from the perspective of a student. Furthermore, with the development of online learning systems, there is an overwhelming amount of data on online learning that is both seductive and mind-boggling. However, it can be challenging to extract this information and determine the location to look in the big data environment in order to uncover insightful information. Most schools aim to address issues like student retention, to help students stay on track, give them feedback on their learning or try to spot students who are in danger of lagging behind or quitting. To understand the value of big data, it is generally necessary to have defined student KPIs (key performance indicators). More critically, many organisations aspire to match the outputs of big data analysis with their overarching organisational goals. The main obstacle is figuring out how to leverage o the large data to enhance learning.

- **Urban Planning**: The old town planning, design and its management approaches, are no longer functional due to the increasing intricacies of modular and smart cities. While old conventional planning methods largely relied on fixed and sector methods, involving only a fairly limited number of residents and stakeholders in important choices, life within major cities and urban regions has grown more dynamic. Social scientists have gathered extensive and detailed data about cities and their inhabitants thanks to the rise of big data. This technology is now a rapidly expanding source for making high-quality, evidence-based decisions by reviewing recent or prior events. Big data-informed modern town design goes beyond the backdated perspective by merging cutting-edge data analytics into the modern process of planning and design.

Big data has recently been used to make compelling arguments for urban planning. Intelligent transportation is achieved by the analysis and visualisation of real-time, detailed road network data gathering and the merging of high-quality spatial data; environmental modelling is made possible by ubiquitous sensor networks that collect data [66]; energy conservation is made possible by patterns; smart materials are made possible by the new materials genome initiative [67]; computational social science is made possible by a new method that is quickly gaining popularity due to the previously reduced amount of gaining information [68], native area security is made possible by analysis of logged information and other events, which can be used to identify threats.

Big Data Knowledge Urban Design is expected to mature into a structure to assist with urban management, planning, and design. Urban governance, intelligent design computation, urban intricacy, public design science, and evidence-informed urban design are the five work streams. Big data knowledge is also anticipated to make urban designers and planners more educated and vigilant [68]. Urban data processing and utilisation will be substantially improved by big data-informed urban design, which will also help formalise expertise for the design stage. The team's methodologies will be included into an interactive planning support system at the applied research level, giving designers, decision-makers, constituents, and other stakeholders a tool by providing a different way to see planning consequences. Communities may strengthen laws, better allocate limited resources, foresee future demands, and potentially control the threat of flooding that low-lying cities face during rainy seasons by making their operations more data-driven. In fact, such technology is the most recent force behind city economies and urban planning in the future.

- **Agricultural Sector**: The primary focus in the agricultural industry is on identifying the difficulties that farmers are now encountering. To better their operations in terms of food output, plant health, etc., they must comprehend what farm data is and how to use it. Additionally, there is a learning curve for the tools and technology that farmers use to gather, transfer, and interpret their data. Many farmers are already using their data to their advantage, but many others are only beginning to walk the journey. It is time to recognize the big data revolution in agricultural and start understanding how information may be used to manage an individual's or business' operations. Even though it may be a complex subject, a farmer who is aware of it can adapt and apply it to increase productivity, yields, and profitability. The use of big data in the agri-food sector, where there are several technologies for data collecting and analytics that can be used by anyone involved in the food systems, such as farmers and huge corporations, has significantly transformed farming operations in recent years.

  Researchers must closely monitor the digital transformation in agriculture in order to map the opportunities and limitations of big data applied to food and agriculture, which is a vast study topic for big data scholarship. It should be anticipated that this will bring food studies and data scholarship into dialogue over the significance of big data in the society. The benefit of using big data for agriculture is that it forces experts in agriculture, hydrology, dairy, aquaculture, and big data analytics to interact across domains in order to identify the key issues that big data analysts are trying to address for agricultural communities, discuss potential solutions, and spot opportunities. According to researchers, the following are some of the major problems and issues [69]:

- Big data, which affects the entire supply chain, is anticipated to have a significant impact on smart farming.
- Large volumes of data are generated by smart sensors and gadgets, giving decision-makers newfound power.
- Both traditional and non-traditional players' roles and power structures are predicted to undergo significant changes as a result of big data.
- Governance (including data ownership, privacy, and security) and business strategies should be the main topics of future research.

A digital revolution is taking place in agriculture. Big data are being used to drive real-time operational choices, rethink corporate processes, and provide predictive insights into farming operations. Big data is anticipated to lead to significant changes in the power dynamics and roles of many participants in the present and upcoming supply chain network of eatables. Current stakeholder landscape shows an exciting interplay between large tech corporations, venture capitalists (VCs), and frequently small established businesses and new freshmen in corporate world. Nothing is more important as compared to the availability of eatables. USA is known for its wheat fields and citrus groves. However, nearly a third of the eatables that is produced is misspent annually, according to McKinsey & Co.

Agriculture is being heavily impacted by big data. Sensors on fields and crops are beginning to offer incredibly precise information on state of the soil, also comprehensive data on wind, the need for fertiliser, the availability of water, and pest infestations. More specifically, drones with IoT and RFID sensors are assisting on-farm pest management that have already greatly increased crop yields all over the world. A newly established company in Bulgaria is now working on creating a web-based tool that will assist farmers in processing photographs taken by drones and identifying potential crop health issues in real-time. The platform runs on a cloud infrastructure and doesn't need any particular hardware to be installed on the local workstation of the farmer (or a fast internet connection) [70].

- **Public Services Delivery**: Big data solutions provide governments the ability to enhance the effectiveness, effectiveness, and invention of delivery of services and policymaking. For instance, the important query at hand is: What significance does big data have for government? Connecting public sector data sources is expected to enhance the delivery of government services trying to save money, exposing scam, and help public organisations serve their constituents more effectively. The government can use the data to carry out present tasks more efficiently, more effectively, and to carry out new tasks that it doesn't already carry out. Modern information technologies have also greatly enhanced our capacity to gather, store, process it, and integrate huge amounts of information on the basis of actual behaviour, precise locations, and frequently on the ground of modern sensor gadgets supervision the events we engage in on a daily basis, including both human and environmental activity. This contrasts with the conventional sources, such as the user surveys that governments routinely carry out to gather information about citizens' needs, perspectives, attitudes, and behaviours. For instance, the Dutch tax authorities once found that people who were divorcing were far more prone to make errors in the tax self-assessments when they cross-matched their data sets. Similarly, a classic example of how Google was capable enough to anticipate flu epidemics and migration based on the search terms that users input a few years ago, but accuracy was questioned at the time.

Some believe that this trajectory involves not just a revolution in the way the public is served, but also the creation of a actual knowledge-based resource for the government that can be shared or even sold to private interests, thereby fostering innovation and national economic development. As a result, it is anticipated that in the future, the position of a policy advisor will essentially be reduced to that of a gatherer of available or known data, after which some algorithm will be used to determine the policy solution, of which there can only be one. In fact, if taken to its logical conclusion, information systems management will completely replace policy analysis. It is anticipated that the analogue way of doing things may be abandoned and usher in a new digitally powered era that offers real opportunities for new digital realities and new forms of governance. The administrative profession as a whole will vanish into history. This alternative viewpoint transforms data analytics into a tool that the public service adviser can use from a knowledge-based standpoint. The remaining difficulty will be in persuading public sector officials to comprehend both the potential and the constraints of data analytics. By doing this, it is anticipated that governance will grow more complex, public officials' roles will become more significant in a variety of authorising situations, and citizens will become active coproducers of policy rather than passive consumers of it [71]. Finally, it is important to remember that a balance between the benefits of data sharing and the rights of people to privacy and secrecy must be established in order to realize the full the promise of big data. Big Data has the potential to strengthen both government and private services, but only if these issues are resolved to allay public scepticism.

- **Healthcare**: The world of healthcare is undergoing a big-data revolution. Big data has the potential to revolutionise the healthcare industry by accelerating innovation and value creation, but before stakeholders can fully realise its potential, the sector must go through fundamental transformations. The greatly expanded amount of data being collected by Web-enabled medical data systems, which are currently standard in the majority of contemporary hospitals around the world. Medical organizations have been compiling years worth of research and development data into medical databases, and payors and providers have been digitising patient records to make managing and transferring patients' healthcare easier. As a result, it is now generally accepted that in the years to come, the use of IT will help to lower healthcare costs while enhancing the quality and delivery of care. This will be achieved by personalisation and prevention healthcare in order to easily adapt it to more comprehensive (home-based) continual supervision of sufferer. A McKinsey study [72] found the adopting big data for healthcare might prevent up to $300 billion in economic losses for the US alone. By today's standards, this number may be considerably greater. For examples, the US government and other public collaborators continue to share their extensive knowledge bases in healthcare, including data from scientific trials and patient statistics from public insurance programmes [73]. Simultaneously, technological developments have simplified the grouping, repository, and analysis of information from numerous sources. The healthcare sector will benefit greatly from this because of the data.

Although the rise of big data may seem to be mostly driven by healthcare expenditures, clinical trends have also been important. A few years ago, doctors would typically use their judgement when deciding on a course of therapy, but recently there has been a shift towards evidence-based medicine, which entails thoroughly determining clinical information and choosing a course of action according to the finest accessible data. Specifically, the ability to combine different sets of big data information frequently offers the required high reliability proof, as subtle differences in sub-populations, like the people with gluten allergies, can be so uncommon that it becomes difficult to detect in minimal/small samples. A new method of determining the risk of cardiovascular disease using ML has been developed by Google analysts and its health-tech subsidiary Verily. AI-powered ML is steadily gaining ground in the healthcare sector. The software can precisely infer information, such as person's age, blood pressure (BP) and detection of smoking habits, from scans of a patient's eye. Then, it is used for forecasting their likelihood of experiencing a main heart event, such as a cardiac arrest, with approximately the same precision as the top methodologies employed. Because they are using data that has been gathered for one clinical purpose and pulling more from it than presently done, it shows how AI can assist improve current diagnostic tools. Due to the algorithm's lack of a blood test, it may be rapid as well as simpler for medical experts to assess a person's heart risk. Although it may seem strange, the concept of evaluating the condition of your eyes to determine the health rate of your heart is supported by a body of proven research. Google admits, the research offers more than just a fresh way to assess risk of heart; it highlights the path to new AI-powered model for scientific innovation.

With such capacity to predict medical issues before they occur, some significant and inventive work is being harnessed through partnerships between medical and data specialists. One of these collaborations, the Pittsburgh Health Data Alliance in the US, for instance, explores the potential for using information from a several places, including medicinal and insurance tracks, wearable sensors, genetic information, and also media platforms usage, to create a thorough portrait of the person and provide personalised medical care [74].

However, a poorly managed patient data management process enhances the probability that a hospital or healthcare provider would experience a negative impact on patient care. Improperly handled patient data too can raise the possibility of inaccurate patient diagnosis or treatment, as well as the loss or corruption of test results. Another issue is the possibility that the same patient may have distinct datasets at two separate stages in their medical journey, such as their general practitioner and a specialist. So, by putting dataset of medicine on blockchain, practitioners would have a single, immutable tool to utilise when caring for the patients. Security is the main advantage that the blockchain can bring to the medical industry. For instance, more than 100 million patient records were compromised by hackers in the US in 2015, putting patients at danger of identity theft. Insurers including Anthem, UCLA Health, and numerous others lost the data. Doctors also will need numerous authorised "signatures" or permissions from other network members under blockchain in order to gain entry into medical reports. A Blockchain-Medical system will give permission to providers so as to distribute/share reports with governmental organisations, insurance firms, business owners, and any additional sector with an involvement in peoples' health without the risk factors caused by the weak link in dealing with multiple departments. Regrettably, the adoption of the Medical-Blockchain system is slow due to the need to update the current database architecture, train and hire new workers, and persuade executives that the blockchain technology is worth the financial investment [75].

The integration of blockchain with AI, IoT and ML would open up new prospects for the economies of e-health. Blockchain technology might provide a shareable domain which decentralises medical conversations and ensures authenticity, integrity, privacy, and entry management. Additionally, it might introduce fresh value-based healthcare and payment schemes. More specifically, the rise of linked medical equipment and the necessity to prevent information violations make this decentralized technology, with pervasive privacy architecture, the ideal substructure for advanced healthcare interoperability and new digital health processes. A crucial point to remember is that blockchain technology may not be the solution to all of the problems facing the healthcare sector, it has the ability to protect many dollars through streamlining current processes and eradicating some expensive gatekeepers who prevent potential sufferers from gaining entry into essential healthcare. As a result, this sector should develop blockchain consortiums to promote collaborations and define standards for eventual widespread application across all healthcare use cases. Important factors to take into account for the healthcare industry include information interoperability, strong authentication systems, insured detection and prevention, the drug supply chain provenance, and control of such crimes [76]. A blockchain ecosystem with use cases related to healthcare, such as health data exchange, smart medicine and investment management, insurance settlement, and online payments, is expected to be developed within the next five to 10 years.

- **Industry**: Any firm that can successfully maximise its usage of data will have a competitive advantage over its rivals. Data is the lifeblood of any organisation. Some industries have been struck more severely than others in the last few of years by disruptive new technology and transformative discoveries. More quickly than others, some industries adopt new technologies. Financial services, retail, and communications are some of the industry leaders in big data. And as already seen, these leaders from other industries have a lot to learn from these precedent-setters.

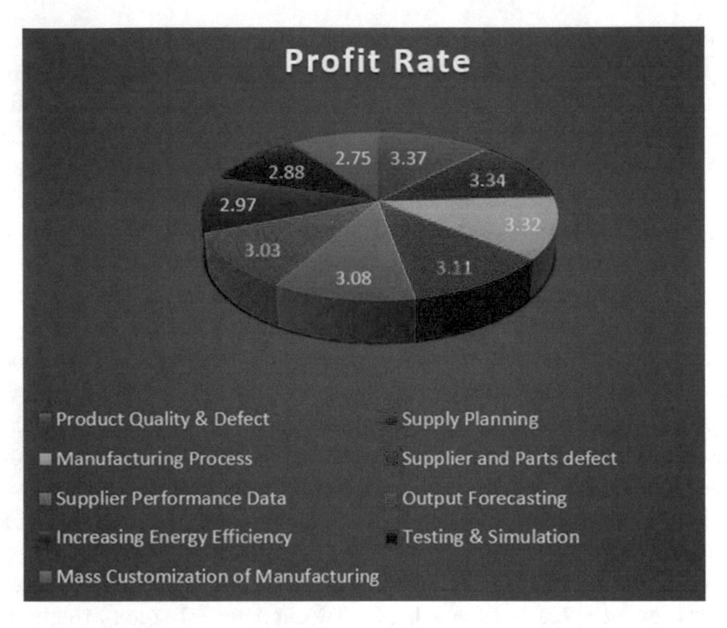

Fig. 5 Rating of big data benefits within the manufacturing industry

For instance, the manufacturing sector can benefit from the effective analysis of big data use cases by reducing processing errors, improving customer service, raising production quality, boosting productivity, enhancing value chain activities, and saving time and money. As an example, Tata Consultancy Services asked manufacturers to rank the following big data advantages on a scale of one to five in one of its studies, and the findings are displayed in Fig. 5. Figure 5 shows that the manufacturing industry's use of big data is only constrained by its imagination and the data that is already accessible.

It is crucial to remember that a big data use case can give analytics a focus by defining the parameters for the different forms of data that can be valuable and selecting how to model that data using, for instance, Hadoop analytics. This can help with complex issues that are harder to answer than "who is likely to buy more merchandise in the United States," such as "where is the next large market for my product." Furthermore, by studying weather data such as tornado, earthquake, hurricane, etc., a corporation may be able to foresee potential delays in the supply chain by plotting them on a map. Additionally, the use of predictive analytics enables a business to estimate the likelihood of delays. As a result, the organisation can use analytics data to identify backup supply and create backup plans to ensure that production is not interrupted by natural disasters.

10 Big Data Challenges

It is a technology that is still in development, but like other new technology, big data still faces number of important problems and intricacies which must be overcome before it can be widely used by end users. Control, data authenticity, and monetization are a few of these challenges [77].

- **Big Data Controls**: Deciding on who handles the architecture when several people are involved, is the biggest problem with big data. How does an international company exchange data globally, for instance? How to identify which copy is the most recent if there are numerous copies? How are system administrator roles at each regional office to be balanced? How to distribute control of the ecosystem infrastructure among the enterprises in an industry consortium? If those businesses are rivals, this is very challenging. Why is there just no data "out there" that everyone can access and parse as a single, shared source of truth?
- **Big Data Trust**: How much can the data be trusted? This is the second most crucial issue. How to establish ownership of the data if it is self-produced? How to confirm the source identity if the information is from someone else? What about malicious conduct, crashes, bits flip, errors and machines crashes? How to sell or purchase other people's rights to data? A universal data marketplace has long been a goal; how might this be achieved?
- **Data Security Requirements for Big Data**: Enterprises' data security and compliance settings are experiencing huge challenges due to the big data's ever-growing data volumes and the lengthy compliance data retention requirements. Technologies that offer the strength and potential of a big data platform must therefore be used to address the demand for data security. As a remedy, IBM is presently providing IBM Security Guardian Big Data Intelligence. Organizations can become more flexible if they can swiftly create a safe, optimised data lake that stores a huge volume of historical data for a long time in order to give new, enhanced analytical insights and virtually real-time reporting while reducing operating and performance costs.

11 Integrating AI, Big Data and Blockchain Technologies

The guaranteed technique to address and manage most of the primary problems and challenges related to big data is to integrate it along with blockchain technology. With the growth of bitcoin and other cryptocurrencies, blockchain technology gained attention. Every single Bitcoin transaction must go via the block chain, which is effectively a ledger, so as to be confirmed by the large number of other peer to peer users. This implies that each transactional activity is open to the public and many other users on the block chain may see everything that every Bitcoin user does. With the technology, digital transactions and record-keeping are incredibly safe. Although cryptocurrencies were where blockchain initially saw use, the idea may be used to various kinds of transactions, including those in the agricultural sector. Every blockchain is really just a database in terms of technicality, but one that has "blue ocean" benefits like immutability, audit trails and native assets [78]. It is anticipated that blockchain technology would boost confidence in the accuracy of the data. As blockchain technology spreads, you will see improvements in immutable entries, consensus-driven timestamping, audit trails, and transparency identifying the source of data (such a sensor or a kiosk). The shared data layer of blockchains will create a whole new set of options for AI skills and insights, in addition to data integrity, which is a crucial component. The Chief Technology Officer (CTO) of Big-ChainDB, Trent McConaghy, did a nice job of articulating the advantages of decentralized/shared control, particularly as a foundation for AI. Epstein, in his paper, claims that under these circumstances, the following is possible [79]:

- Extra data, leading to enhanced modelling abilitie.
- The creation of completely new models using qualitatively new data. Due to the inherent immutability, training and testing data as well as the models they generate can be more confidently used. In this regard, the future appears promising as the concept of big data grows through the transition from private data silos to shared data layers facilitated by blockchain. Those who possessed the data held sway during the early days of big data. Power in the big data and blockchain era will be held by those with the most access to data (public blockchains will ultimately triumph over private blockchains) and the ability to quickly learn the most valuable insights. Two important ramifications result from this would be [79]:

  - Organizations will not own or have access to customer data stored in corporate databases. On an identity blockchain, it will be represented as tokens or coins that will belong to each individual. Future customers will permit access to others as required.
  - Anyone will be able to access transaction data. The information regarding the transaction on a specific blockchain is available to anybody. Transitioning from private systems to open blockchains, just having the data actually is not significant benefit. The advantage is in how the data is interpretted. In a world powered by blockchain, all rivals consult the same ledger.

12 Conclusion

For improving system in terms of productivity and efficiency, Industry 4.0 will connect cutting-edge technologies like blockchain, AI, ML, cloud computing and IoT, where the data or the information is the lifeline of any organisations. Nearly all segments of the economy, which includes retail, healthcare, the financial industry, and government, are already using big data. Any business that can utilize big data to fix unanswered queries about its processes may benefit from it. Big data is frequently desired across all business areas. Big data and data analytics are progressively offering high corporate insights and knowledge in addition to quicker and much more focused reactions, which opens up fascinating new ways to operate. It is very important to keep in mind that in order to fully grasp the potential of big data, a balance must be struck between the advantages of information sharing and the rights of individuals to privacy and confidentiality. The enormous volume of information that is generated by IoT based devices needs to be handled carefully to avoid misuse, to determine its meaning and use in decision-making to support business expansion. IoT and big data analytics must work together. Due to the complicated and varied nature of the gadgets being used, data collected from IoT gadgets is highly vulnerable to security risks. Since antivirus software cannot be installed on IIoT devices to guard them against malicious purposes because of the limited memory and lower power consumption of the devices, IoT devices are prone to attacks and, once compromised, can cause damage at higher levels. Blockchain technology is a distributed ledger that enables safe data transfer between parties and can be used to effectively increase device security. When used in the fields of healthcare, retail, industrial production, and banking, AI refers to a set of techniques that can imitate human intelligence. By creating a ML approach with the aid of AI algorithms, machines can be made to learn from the data produced by IoT devices. This makes it possible to categorise patterns and find anomalies in the data produced by the gadgets and related sensors.

This chapter will help various researchers to gain a thorough understanding of the current state of the Blockchain technology and its difficulties, as well as to identify possible solutions. There are still a lot of obstacles in the way of using blockchain technology in IIoT. Addressing these issues and using them in conjunction with the developing IIoT technologies are some of the future research directions.

References

1. Tan L, Wang N (2010) Future internet: the internet of things. In: Proceedings of the 2010 3rd international conference on advanced computer theory and engineering (ICACTE), vol 5, Chengdu, China, September, pp V5–V376
2. Khan MA, Salah K (2018) IoT security: review, blockchain solutions, and open challenges. Futur Gener Comput Syst 82:395–411
3. Atzori L, Iera A, Morabito G (2010) The internet of things: a survey. Comput Netw 54(15):2787–2805
4. Reyna A, Martín C, Chen J, Soler E, Díaz M (2018) On blockchain and its integration with IoT, challenges and opportunities. Future Generat Comput Syst 88:173–190
5. Grover S, Feamster N (2016) The internet of unpatched things. In: Proceedings of FTC PrivacyCon
6. Sengupta J, Ruj S, Das Bit S (2020) A comprehensive survey on attacks, security issues and blockchain solutions for IoT and IIoT. J Netw Comput Appl 149, Article ID 102481
7. Mahmood K, Shamshad S, Rana M et al (2021) PUF enable lightweight key-exchange and mutual authentication protocol for multi-server based D2D communication. J Inf Secur Appl 61, Article ID 102900
8. Mishra D, Vijayakumar P, Sureshkumar V, Amin R, Islam SH, Gope P (2018) Efficient authentication protocol for secure multimedia communications in IoT-enabled wireless sensor networks. Multimedia Tools Appl 77(14):18295–18325
9. Dwivedi SK, Amin R, Vollala S (2021) Blockchain-based secured IPFS-enable event storage technique with authentication protocol in VANET. IEEE/CAA J Automatica Sinica 8(12)

10. Xu LD, He W, Li S (2014) Internet of things in industries: a survey. IEEE Trans Industr Inf 10(4):2233–2243
11. Huh S, Cho S, Kim S (2017) Managing IoT devices using blockchain platform. In: Proceedings of the 2017 19th international conference on advanced communication technology (ICACT), IEEE, PyeongChang, Korea, March, pp 464–467
12. Mistry I, Tanwar S, Tyagi S, Kumar N (2020) Blockchain for 5G-enabled IoT for industrial automation: a systematic review, solutions, and challenges. Mech Syst Signal Process 135, Article ID 106382
13. Lee J, Bagheri B, Kao H-A (2015) A cyber-physical systems architecture for industry 4.0-based manufacturing systems. Manuf Lett 3:18–23
14. Yoon CY (2019) Measurement model of smart factory technology in manufacturing fields based on IIoT and CPS. In: Proceedings of international conference on artificial intelligence, robotics and control, pp 80–84
15. Sisinni E, Saifullah A, Han S, Jennehag U, Gidlund M (2018) Industrial internet of things: challenges, opportunities, and directions. IEEE Trans Ind Inf 14(11):4724–4734
16. Xu H, Yu W, Griffith D, Golmie N (2018) A survey on industrial internet of things: a cyber-physical systems perspective. IEEE Access 6:78238–78259
17. Jeschke S, Brecher C, Meisen T, Özdemir D, Eschert T (2017) Industrial internet of things and cyber manufacturing systems. Springer Int., Cham, Switzerland, pp 3–19
18. Li J-Q, Yu FR, Deng G, Luo C, Ming Z, Yan Q (2017) Industrial internet: a survey on the enabling technologies, applications, and challenges. IEEE Commun Surveys Tuts 19(3):1504–1526, 3rd Quart
19. Das L et al (2022) Application of blockchain technology in an IoT-integrated framework. In: Information security practices for the internet of things, 5G, and next-generation wireless networks, IGI Global, pp 131–151. https://doi.org/10.4018/978-1-6684-3921-0.ch007
20. Khan WZ, Rehman MH, Zangoti HM, Afzal MK, Armi N, Salah K (2020) Industrial internet of things: recent advances, enabling technologies and open challenges. Comput Elect Eng 81, Art no 106522 [Online]. https://www.sciencedirect.com/science/article/pii/S0045790618329550
21. Bhattarai S, Bleakley G, Buchheit M, Byers C, Chigani A (Ind. Internet Consortium, Boston, MA, USA) (2019) The industrial internet of things volume G1: reference architecture [Online]. https://www.iiconsortium.org/IIRA.htm
22. Sethi R, Bhushan B, Sharma N, Kumar R, Kaushik I (2020) Applicability of industrial IoT in diversified sectors: evolution, applications and challenges. Stud Big Data Multimedia Technol Internet Things Environ 45–67. https://doi.org/10.1007/978-981-15-7965-3_4
23. Madaan G, Bhushan B, Kumar R (2020) Blockchain-based cyberthreat mitigation systems for smart vehicles and industrial automation. Stud Big Data Multimedia Technol Internet Things Environ 13–32. https://doi.org/10.1007/978-981-15-7965-3_2
24. Ray PP, Dash D, Salah K, Kumar N (2020) Blockchain for IoT-based healthcare: background, consensus, platforms, and use cases. IEEE Syst J 15
25. Saha S, Sutrala AK, Das AK, Kumar N, Rodrigues JJ (2020) On the design of blockchain-based access control protocol for IoT-enabled healthcare applications. In: Proceedings of the ICC 2020–2020 IEEE international conference on communications (ICC), IEEE, Dublin, Ireland, June, pp 1–6
26. Anusha R, Yousuff M, Bhushan B, Deepa J, Vijayashree J, Jayashree J (2022) Connecting blockchain with IoT—a review. In: Sharma DK, Peng SL, Sharma R, Zaitsev DA (eds) Microelectronics and telecommunication engineering. Lecture notes in networks and systems, vol 373. Springer, Singapore. https://doi.org/10.1007/978-981-16-8721-1_14
27. Zheng Z, Xie S, Dai HN, Chen X, Wang H (2018) Blockchain challenges and opportunities: a survey. Int J Web Grid Serv 14(4):352–375
28. Srinivas J, Das AK (2020) 9 lightweight security protocols for blockchain technology. In: Cyber defense mechanisms: security, privacy, and challenges, vol 131
29. Li X, Jiang P, Chen T, Luo X, Wen Q (2020) A survey on the security of blockchain systems. Futur Gener Comput Syst 107:841–853

30. Prieto J, Das AK, Ferretti S, Pinto A, Corchado JM (2020) Blockchain and applications. Springer, Berlin, Germany
31. Zheng Z, Xie S, Dai H-N et al (2020) An overview on smart contracts: challenges, advances and platforms. Futur Gener Comput Syst 105:475–491
32. Saxena S, Bhushan B, Ahad MA (2021) Blockchain based solutions to Secure Iot: background, integration trends and a way forward. J Netw Comput Appl 103050. https://doi.org/10.1016/j.jnca.2021.103050
33. Luu L, Chu DH, Olickel H, Saxena P, Hobor A (2016) Making smart contracts smarter. In: Proceedings of the 2016 ACM SIGSAC conference on computer and communications security, Vienna, Austria, October, pp 254–269
34. Liu J, Liu Z (2019) A survey on security verification of blockchain smart contracts. IEEE Access 7:77894–77904
35. Nikolić I, Kolluri A, Sergey I, Saxena P, Hobor A (2018) Finding the greedy, prodigal, and suicidal contracts at scale. In: Proceedings of the 2018 annual computer security applications conference, San Juan, Puerto Rico, December, pp 653–663
36. Murray Y, Anisi DA (2019) Survey of formal verification methods for smart contracts on blockchain. In: Proceedings of the 2019 10th IFIP international conference on new technologies, mobility and security (NTMS), IEEE, Canary Islands, Spain, June, pp 1–6
37. Chiu J, Koeppl TV (2019) Blockchain-based settlement for asset trading. Rev Financ Stud 32(5):1716–1753
38. Zhang Q, Zhu J, Wang Y (2020) Trustworthy dynamic target detection and automatic monitor scheme for mortgage loan with blockchain-based smart contract. In: Communications in computer and information science, Springer, Berlin, Germany, pp 415–427
39. Alam KM, Rahman JA, Tasnim A, Akther A (2020) A blockchain-based land title management system for Bangladesh. J King Saud Univ-Comput Inf Sci
40. Kar AK, Navin L (2020) Diffusion of blockchain in insurance industry: an analysis through the review of academic and trade literature. Telemat Inf 58(5), Article ID 101532
41. Wang S, Ouyang L, Yuan Y, Ni X, Han X, Wang F-Y (2019) Blockchain-enabled smart contracts: architecture, applications, and future trends. IEEE Trans Syst Man Cybernet Syst 49(11):2266–2277
42. Bhushan B, Sinha P, Sagayam KM, Andrew J (2021) Untangling blockchain technology: a survey on state of the art, security threats, privacy services, applications and future research directions. Comput Electr Eng 90:106897. https://doi.org/10.1016/j.compeleceng.2020.106897
43. Singh A, Parizi RM, Zhang Q, Choo K-KR, Dehghantanha A (2020) Blockchain smart contracts formalization: approaches and challenges to address vulnerabilities. Comput Secur 88. Article ID 101654
44. Kemmoe VY, Stone W, Kim J, Kim D, Son J (2020) Recent advances in smart contracts: a technical overview and state of the art. IEEE Access 8:117782–117801
45. Bhushan B, Sahoo G (2020) Requirements, protocols, and security challenges in wireless sensor networks: an industrial perspective. Handb Comput Netw Cyber Secur 683–713. https://doi.org/10.1007/978-3-030-22277-2_27
46. Aggarwal S, Kumar N (2020) Blockchain 2.0: smart contracts. Adv Comput 121:301–322
47. Huo R et al (2022) A comprehensive survey on blockchain in industrial internet of things: motivations, research progresses, and future challenges. IEEE Commun Surv Tutorials 24(1):88–122, Firstquarter. https://doi.org/10.1109/COMST.2022.3141490
48. Dwivedi SK, Roy P, Karda C, Agrawal S, Amin R (2021) Blockchain-based internet of things and industrial IoT: a comprehensive survey. Secur Commun Netw 2021:21. Article ID 7142048. https://doi.org/10.1155/2021/7142048
49. Novo O (2018) Blockchain meets IoT: an architecture for scalable access management in IoT. IEEE Internet Things J 5(2):1184–1195
50. Uddin MA, Stranieri A, Gondal I, Balasubramanian V (2018) A patient agent to manage blockchains for remote patient monitoring. Stud Health Technol Inf 254:105–115

51. Xia Q, Sifah E, Smahi A, Amofa S, Zhang X (2017) BBDS: blockchain-based data sharing for electronic medical records in cloud environments. Information 8(2):44
52. Sun Y, Zhang R, Wang X, Gao K, Liu L (2018) A decentralizing attribute-based signature for healthcare blockchain. In: Proceedings of the 2018 27th international conference on computer communication and networks (ICCCN), IEEE, Hangzhou, China, July, pp 1–9
53. Chen J, Ma X, Du M, Wang Z (2018) A blockchain application for medical information sharing. In: Proceedings of the 2018 IEEE international symposium on innovation and entrepreneurship (TEMS-ISIE), IEEE, Beijing, China, April, pp 1–7
54. Griggs KN, Ossipova O, Kohlios CP, Baccarini AN, Howson EA, Hayajneh T (2018) Healthcare blockchain system using smart contracts for secure automated remote patient monitoring. J Med Syst 42(7):130
55. Li Z, Guo H, Wang WM et al (2019) A blockchain and automl approach for open and automated customer service. IEEE Trans Ind Inf 15(6):3642–3651
56. Tian F (2016) An agri-food supply chain traceability system for China based on RFID & blockchain technology. In: Proceedings of the 2016 13th international conference on service systems and service management (ICSSSM), IEEE, Kunming, China, June, pp 1–6
57. Nakasumi M (2017) Information sharing for supply chain management based on block chain technology. In: Proceedings of the 2017 IEEE 19th conference on business informatics. IEEE, Thessaloniki, Greece, July, pp 140–149
58. Malik A, Gautam S, Abidin S, Bhushan B (2019) Blockchain technology-future of IoT: including structure, limitations and various possible attacks. In: 2019 2nd international conference on intelligent computing, instrumentation and control technologies (ICICICT). https://doi.org/10.1109/icicict46008.2019.8993144
59. Chaganti R, Bhushan B, Ravi V (2023) A survey on blockchain solutions in DDoS attacks mitigation: techniques, open challenges and future directions. Comput Commun 197:96–112. https://doi.org/10.1016/j.comcom.2022.10.026
60. Haque AKMB, Bhushan B, Hasan M, Zihad MM (2022) Revolutionizing the industrial internet of things using blockchain: a unified approach. In: Balas VE, Solanki VK, Kumar R (eds) Recent advances in internet of things and machine learning. Intelligent systems reference library, vol 215. Springer, Cham. https://doi.org/10.1007/978-3-030-90119-6_5
61. Li Z, Yang Z, Xie S (2019) Computing resource trading for edge-cloud-assisted internet of things. IEEE Trans Industr Inf 15(6):3661–3669
62. Saha A, Amin R, Kunal S, Vollala S, Dwivedi SK (2019) Blockchain technology based medical healthcare system with privacy issues. Secur Priv 2(5):e83
63. Rabah K, Nairobi K (2018) Convergence of AI, IoT, big data and blockchain: a review. Lake Inst J 1(1)
64. Dobbie W, Fryer R Jr (2013) Getting beneath the veil of effective schools: evidence from New York City. Am Econ J Appl Econ 5(4):28–60
65. Corrin L, Kennedy G, and de Barba P (2017) Asking the right questions of big data in education. https://pursuit.unimelb.edu.au/articles/asking-the-right-questions-of-big-data-in-education
66. Onyema EM, Dalal S, Romero CAT, Seth B, Young P, Wajid MA (2022) Design of intrusion detection system based on cyborg intelligence for security of cloud network traffic of smart cities. J Cloud Comput 11(1):1–20
67. Wajid MA, Zafar A (2021) Pestel analysis to identify key barriers to smart cities development in India. Neutrosophic Sets Syst 42:39–48
68. Lazer D, Pentland A, Adamic L, Aral S, Barabasi AL, Brewer D, Christakis N, Contractor N, Fowler J, Gutmann M, Jebara T, King G, Macy M, Roy D, van Alstyne M (2009) Computational social science. Soc Sci 323:721–723. https://doi.org/10.1126/science.1167742
69. Wolfert S, Ge L, Verdouw C, Bogaardt MJ (2017) Big data in smart farming–a review. Agric Syst 153:69–80
70. Mihail_agrohelper (2018) Drones in agriculture: a tool for early pest detection. http://www.eagriculture.org/blog/drones-agriculture-tool-early-pest-detection
71. Lofgren K (2016) Big data and public service delivery—big hype? http://www.victoria.ac.nz/news/2016/09/big-data-and-public-service-delivery-big-hype

72. Manyika J, Chui M, Brown B, Bughin J, Dobbs R, Roxburgh C, Byers AH (2011) Big data: the next frontier for innovation, competition, and productivity. McKinsey Global Institute
73. Kayyali B, Knott D, van Kuiken S (2013) The big-data revolution in US health care: accelerating value and innovation. McKinsey & Company. http://www.mckinsey.com/industries/healthcare-systems-and-services/our-insights/the-big-datarevolution-in-us-health-care
74. Marr B (2015) How big data is changing healthcare. Forbes/Tech. https://www.forbes.com/sites/bernardmarr/2015/04/21/how-big-data-is-changinghealthcare/#6643a6972873
75. Smyth D (2016) Why blockchain? What can it do for big data? http://bigdata-madesimple.com/why-blockchain-what-can-it-do-for-big-data-2/
76. Rabah K (2017) Challenges & opportunities for blockchain powered healthcare systems: a review. Mara Res J Med Health Sci 1(1):45–52
77. McConaghy T (2016) Blockchains for big data from data audit trails to a universal data exchange. https://blog.bigchaindb.com/blockchains-for-big-data-from-data-audit-trails-to-a-universal-dataexchange-cf9956ec58ea
78. Rabah K (2016) Overview of blockchain as the engine of the 4th industrial revolution. Mara Res J Bus Manage 1(1):125–135
79. Epstein J (2017) When blockchain meets big data, the payoff will be huge. https://venturebeat.com/2017/07/30/when-blockchain-meets-big-data-the-payoff-will-be-huge/

SIMDPS: Smart Industrial Monitoring and Disaster Prevention System

Arushi Jain, David Velho, K. S. Sendhil Kumar, and U. Sai Sakthi

Abstract Safety regulations have been unable to keep up with the needs of rapid industrialization over the last few decades. We propose SIMDPS, a novel Arduino-based smart industry monitoring system with various gas, ambient, and disaster detection sensors, GPS (Global Positioning System) and Wi-Fi modules, an alerting mechanism, and a prediction model. This system can be installed in industries to maintain workplace safety and anticipate disasters beforehand. SIMDPS was deployed in our university's automotive testing workshops, and the readings gathered were used to evaluate its performance. The mean absolute error and root mean square error values were found to be 0.79 and 1.02 for the temperature sensor, 0.94 and 1.12 for the humidity sensor, and 1.07 and 1.23 for the carbon monoxide sensor, respectively, compared to field-proven devices. These low error values indicate the high accuracy of our proposed system. We also trained a multiple linear regression model on our dataset to achieve an accuracy of 87%. This system will help prevent tragedies and monitor the working conditions of industries to maintain safety and the peak efficiency of machinery.

Keywords Industry monitoring · Arduino · Safety alert · Disaster prediction · Linear regression

1 Introduction

India is a developing country, and despite the COVID-19 pandemic, has ranked second in the world on the Global Manufacturing Index for the year 2021, according to Cushman & Wakefield's Global Manufacturing Risk Index [1] and with the Federation of Indian Chambers of Commerce & Industry (FICCI) [2] expecting a 10%

A. Jain · D. Velho · K. S. Sendhil Kumar (✉)
Vellore Institute of Technology, Vellore, India
e-mail: sendhilkumar.ks@vit.ac.in

U. Sai Sakthi
Meenakshi Sundararajan Engineering College, Chennai, India

B. Bhushan et al. (eds.), *AI Models for Blockchain-Based Intelligent Networks in IoT Systems*, Engineering Cyber-Physical Systems and Critical Infrastructures 6,
https://doi.org/10.1007/978-3-031-31952-5_4

increase in production. The Indian government seeks to maintain this global standing with the help of specific schemes that help industries grow and maintain steady production. Regular safety checks and inspections must be carried out to maintain workplace safety, especially those with dangerous chemicals or heavy machinery. Due to the magnitude of the task, these safety checks are often not carried out diligently, or concerning safety violations might be overlooked in favor of profits.

According to the Union Labour Ministry of India, at least 8000 incidents killed over 6300 employees between 2014 and 2017. With the projected growth of the manufacturing sector, this number is only sighted to increase. Human error or negligence leads to minor issues being overlooked, that over time have devastating consequences. Safety regulations may be forgotten, or advanced monitoring systems omitted, due to the cost involved in deploying them on a large scale. Existing monitoring systems are bulky and very specialized, requiring trained operators to monitor the readings at all times, in addition to the high cost. The larger the factory or workplace, the more difficult it is to monitor the systems and safety of employees. Thus, there is a need for a cost-effective, automated system that can be easily deployed in multiple locations, requiring minimal maintenance. The system must be accurate and dynamic, and should provide some warning to workers in the event of danger. Additionally, a disaster prediction mechanism can help save lives by notifying authorities and workers about the impending disaster, allowing the issue to be rectified in time. Govindarajan et al. [3] proposed a system that monitors and regulates factors like temperature, humidity, and toxic gas levels to acceptable thresholds. On the detection of abnormal readings, the system will trigger an alert action that executes a pre-programmed task and informs the concerned authorities of the incident via email.

We propose SIMDPS, a real-time IoT-enabled monitoring and alerting system that scans various factors like temperature, humidity, light values, ambient pressure, smoke, flammable and toxic gases, vibration, and flame. The system observes the factors critical to the safety of workers and the efficient operation of machinery, given that high temperatures and humidity, among others, can damage the machines. Furthermore, it can detect the presence/motion of workers, which can prove helpful in locating trapped workers. The gas sensors are used to detect an increase in gaseous concentrations. SIMDPS is implemented using IoT, thus allowing us to send real-time data to a control center and alert authorities with minimal delay. It is comparable in accuracy to industry standard sensors, given the low error values. Through machine learning, we can detect the probability of a disaster occurring with 87% accuracy and trigger an alert accordingly. Our system is entirely automated and requires little to no human intervention.

The paper is organized as follows. Section 2 discusses the recent work done by other researchers in the field of Industrial Monitoring and Prediction. Section 3 contains our proposed methodology for data collection and testing. Section 4 details the various components of our system and their description. Section 5 describes the system flow, the circuit connection details and a list of modules and their description. Section 6 presents the implementation and results obtained and finally, Sect. 7 discusses our paper's conclusion and future scope.

2 Literature Survey

Multiple projects have described an IoT-enabled industrial monitoring system for detecting disasters. Venkata Subbaiah et al. [4] demonstrated using a Raspberry Pi for real-time weather monitoring, specifically temperature, pressure, humidity level, and smoke, using an MQ2 sensor. The authors theorized and implemented a system where real-time data is sent to a server for processing. Ganeshan and Kumar Singh [5] proposed a system for real-time monitoring and data transfer for controlling devices, that included actions and triggers. The authors also explore the feasibility of access control via facial recognition using CNNs (Convolution Neural Networks), which can help prevent break-ins.

In 2020, Kumar et al. [6] and Gupta et al. [7] built a system to monitor temperature and flammable gases using gas sensors and, with their integration with GSM and GPS modules, alert the concerned authorities of an anomaly via SMS. Peng and Wu [8] proposed a cloud-based event driven performance monitoring system built using Java and the J2EE platform. It included an active diagnosis and maintenance module which helped alert the users of abnormal behavior, resulting in a cost-effective and scalable system. Deekshath et al. [9] built a system for environmental data collection through certain specialized sensors like soil, humidity, temperature, and gas sensors, along with an accelerometer to measure ambient vibration. This data was sent to the cloud using an Arduino Wi-Fi module, from where an android app fetched the data to display to the user in the form of a visualization.

Kishore Kumar et al. [10] proposed a novel system that can detect and deal with dangerous situations by performing certain pre-programmed tasks. The system consisted of gas, temperature, humidity, flame, and PIR (Passive Infrared) motion sensors, along with a Wi-Fi module. The temperature and humidity sensors, gas sensors, flame sensors, and PIR motion sensors are mapped to trigger the buzzer, exhaust fan, water sprinkler, and LCD display, respectively. Information was also displayed via the LCD display. Solanki et al. [11] also proposed a similar system built around an Arduino MEGA that could be controlled via a web application. Aravind et al. [12] developed a real-time industrial surveillance system to control and monitor using the Internet of Things (IoT) wherein the system detects intruders and triggers an alarm, and reports on the number of intruders via a PIR motion sensor. The device could connect to the user's Wi-Fi and send data to a server via the HTTP (Hyper Text Transfer Protocol) protocol.

Rajalakshmi and Vidhya [13] and Lohith et al. [14] proposed an IoT system for the monitoring of toxic environments using sensors for the monitoring of gases like carbon monoxide, methane, etc. The data collected was stored on a server and was accessible using the internet, along with visualizations of the toxicity levels. Rupali and Mahajan [15] built an IoT system for the monitoring of temperature, humidity, and gas present in an industrial area via appropriate sensors. The data was collected to a server and displayed locally using an LCD screen. Prasanti and Venkataramana [16] proposed a similar system for monitoring that makes use of a GSM module for connectivity and alerting, the MQ6 and MQ9 gas sensors, and an LM35 temperature sensor that would trigger an alarm when abnormal readings were detected. Merchant and Ahire [17] built a system for industrial automation, data collection, and monitoring where the data was collected and processed on the IoT device. The authors used a Raspberry Pi, which ran the data collection software and the web server required to display monitoring information accessible through a webpage.

Georgewill and Ezeofor [18] developed a real-time interactive IoT system that allowed the users to be alerted via GSM SMS about certain readings and trigger devices on or off by sending a special code to the devices via SMS. Kavitha and Alagappan [19] built a real-time industrial monitoring system that used Google Cloud for data storage, processing, and visualization. The given system used the Google Cloud Sheets API for data storage and used said data to generate the visualizations. Liu et al. [20] proposed a system of nodes to monitor various parameters like temperature, relative humidity, pressure, etc., and transmit data to a gateway using the Zigbee protocol. This gateway transmitted the data to a control center via the Short Message Service (SMS) provided by GSM. Banhazi [21] built a user-friendly version of the monitoring kit used in Australia to measure air quality. It monitored factors like air temperature, relative humidity, and ammonia and carbon dioxide concentrations and could store the data for up to 30 days.

Okigbo et al. [22] proposed a low-power solution to transfer monitoring data using Bluetooth Low Energy (BLE) to an app that stores it on Google Fusion tables in real time. The authors also compared the performance of two popular IoT devices in this low-power setting. Dhingra et al. [23] developed a mobile air pollution monitoring system that can measure the concentrations of various gases along various travel routes and transmit the data to the cloud via a Wi-Fi module, along with an android app for the prediction of pollution levels along the selected navigation routes. Deshpande et al. [24] proposed an IoT-based monitoring system capable of alerting users when the values recorded by various sensors exceed the pre-set threshold limits. Emails and SMSs (Short Message Service) were generated and sent to notify the system administrator.

Okokpujie et al. [25] defined a system built around the Arduino UNO, capable of monitoring the ambient air quality using sensors like the MQ135 and displaying data using an LCD screen. The device could also connect to Wi-Fi using appropriate credentials and transmit data to the cloud. Das et al. [26] built a real-time IoT system for monitoring the environment using legacy sensors. It was constructed around the Arduino UNO and ESP8266 Wi-Fi shield and transmitted data to Thingspeak, an open-source data storage and visualization platform using the HTTP protocol. Simbeye [27] designed a similar system comprising of sensors for the six most common pollutants—CO, NO_2, O_3, SO_2, particulate matter, and lead, which operated as a part of a Wireless Sensor Network (WSN) using the LoRaWAN technology. LabWindows/CVI was used as the monitoring software which received the data and allowed for real-time measurement, control, and data processing with visual output.

Ziętek et al. [28] proposed a portable system for environmental monitoring in mines, consisting of an Arduino M0 Pro, gas sensors for CO and H_2S, and temperature and humidity sensors. Data was sent to a paired smartphone via a Bluetooth controller and stored on the attached SD card. The smartphone could transmit data to the cloud for storage and analysis. A custom case was 3D printed to house the components. Kanan et al. [29] built an automated monitoring system for civil construction workers embedded in their helmets. It consisted of sensors to track the location and status of the workers. Data was transmitted to a server via GSM, and the supervisor could view information via a mobile app or web browser. Kaur et al. [30] proposed an air quality monitoring system built using an Arduino microcontroller with various gas, temperature, and humidity sensors. The sensor readings were transmitted via the Zigbee protocol to a base station, where the readings were then monitored. The temperature and humidity readings are broadcasted via Bluetooth to allow public access via smartphones. An alert is sent via SMS whenever the CO concentration exceeds the safe limit. Charaim et al. [31] proposed a monitoring network of 20 devices spread over 200 square meters to detect propane leakage. The detection accuracy was 91%, with only seven false alarms in 3 days. Kanappan and Hariprasad [32] developed a similar system consisting of toxic gas detection, radiation, temperature and humidity sensors. The observed data is displayed via a local LCD screen, sent to a server for storage and processing via a Wi-Fi module, and viewed via a website. Elsisi et al. [33] proposed a system to monitor the readings from smart meters and account for discrepancies in the readings caused by the industrial environment and sought to validate the data by making use of a decision tree to perform classification and regression tasks. Ramamurthy et al. [34] built a plug-and-play system for the wireless transmission of sensor readings using an RF link, configurable based on the sensor data being monitored. Ray et al. [35] developed specific IoT features and protocols designed to combat disaster situations, along with relevant issues involving early warning, notification, and predictions, and discuss current challenges and recent trends. Kök et al. [36] designed a system to use big data analytics with machine learning to predict the air quality of cities using a model based on LTSM networks. Mehta et al. [37] proposed a system in use in certain parts of Karnataka where real-time sensor data from a live feed of CCTV cameras near the road and vehicular-based sensors is sent to a cloud-based platform via a 3G/4G interface, where it is analyzed and processed.

A smart city is defined as an urban area that uses a variety of Internet of Things (IoT) devices and sensors to collect data and improve operational efficiency. Haque et al. [38] and [39] described the physical and network architecture of smart cities, and their strengths and weaknesses, along with the emergence of blockchain in industries and its features to boost the security and efficiency of existing processes. Malik et al. [40] proposed a novel IoT network architecture and discussed the best approach for message transmission. Bhushan [41] highlighted the security concerns in modern IoT protocols and discussed the use case of various middleware implementations. Mehta et al. [42] detailed popular machine learning algorithms and their use in a smart city.

3 Proposed Methodology

SIMDPS was built around the popular Arduino platform, utilizing two Arduino UNOs and a plethora of sensors to monitor every aspect that might threaten workers' safety. To monitor toxic and volatile gases, we use the MQ family of sensors, namely the MQ2, MQ3, MQ5, and MQ7, which detect smoke, alcohol, combustible gases, and carbon monoxide, respectively. To monitor additional ambient conditions, we use an LDR to read light levels, a BMP-1809 for relative pressure, a DHT-11 for temperature and relative humidity, an SEN-16 to detect the presence of a fire, and an SW-420 to detect vibrations. To transmit the sensor data to the server for collection, we use an ESP8266 Wi-Fi card for the Arduino UNO and a NEO-6M GPS to get the unit's precise location. We also have a PIR motion sensor to track human movement, a buzzer, and a few colored LEDs to alert us of danger. The Arduino UNO consumes 5 V of power through USB, and all attached sensors draw their power from it. The combination of the Arduino and sensors makes a monitoring unit for SIMDPS.

SIMDPS is intended to have multiple units mounted in various parts of a building to cover more area. Each unit is configured to transmit recorded sensor and location data to a centralized server where a team can monitor all the units simultaneously. This data is stored for future analysis and prediction. Should the machine learning model output a prediction of imminent danger, the server sends a notification to the unit(s) reporting abnormal sensor readings to trigger their alarms and simultaneously sends an alert to the concerned authorities.

To verify our system's accuracy, we deploy a unit in our university's automotive workshop where it is exposed to varying levels of smoke, temperature, humidity, and even the occasional flame due to welding torches. We compare our system's temperature, relative humidity, and carbon monoxide sensors to industry standard devices like a TMP35 temperature sensor, a PCE-555 relative humidity sensor, and the HTC CO-01 portable carbon monoxide sensor, respectively. We record the sensor readings through the server and the corresponding industry standard device readings manually at regular intervals of 1 h. The data is then compiled into a CSV file to calculate the Mean Absolute Error (MAE) and the Root Mean Square Error (RMSE) for our sensors.

4 SIMDPS: Detailed Architecture

Our integrated system consists of the following major components (Fig. 1).

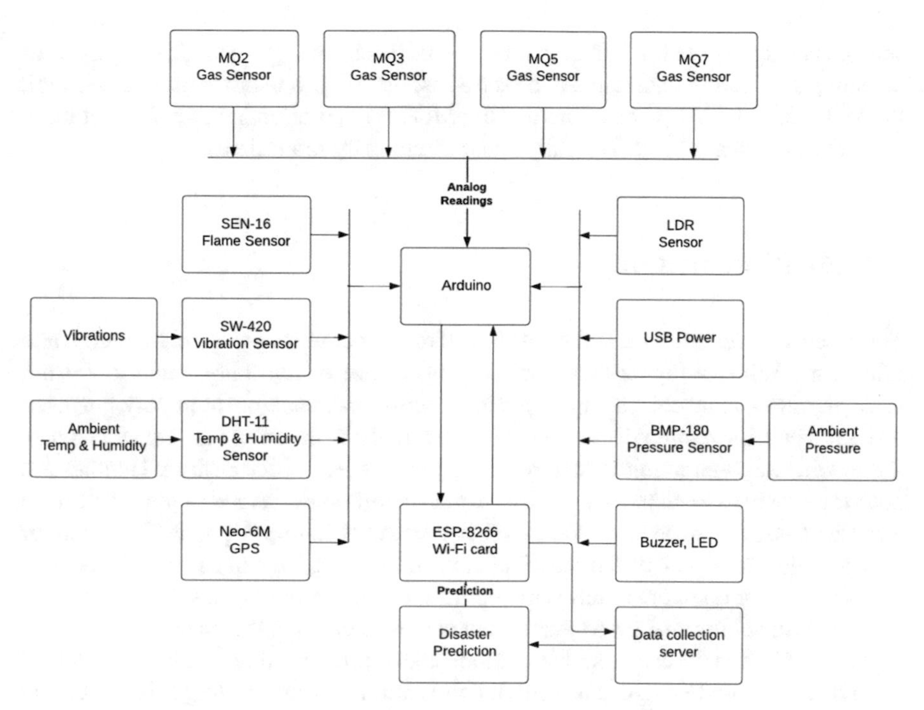

Fig. 1 Block diagram of SIMDPS

4.1 Gas Sensors

The system of gas sensors helps detect various gases that can potentially be dangerous to human health. Gaseous elements like carbon monoxide, flammable hydrocarbons, smoke, and gaseous alcohol can cause serious harm to life.

Gas sensors are electronic devices that identify the presence and quantity of certain gases or particulate matter in the air. We use gas sensors to detect the concentration of toxic gases past the legal limit. These sensors are generally installed in factories and other manufacturing plants to detect gas leaks or in homes to monitor the levels of carbon monoxide. Some of them are even available as compact units capable of detection and alerting through built-in alarms.

These gas sensors are commonly used as a part of a more extensive system of sensors like hazmat and security systems. They require regular calibration as they constantly interact with toxic gases. The frequency of these calibrations might be higher than they are for the average sensor.

The gas sensors are from the MQ family of sensors and are known as Metal Oxide Semiconductor type gas sensors or Chemiresistors. Gases are detected when the sensitive resistant metal encounters the gas, leading to a resistance change. We can obtain the concentration by measuring the voltage output from the sensor. The sensors cannot be used directly and must be used with breakout boards. Each of the sensors is responsible for detecting one type of hazardous gas—the MQ2 sensor for detecting smoke and flammable hydrocarbons, the MQ3 sensor for gaseous alcohol, the MQ5 for volatile hydrocarbons and the MQ7 for carbon monoxide. Each of these sensors has its readings individually and continuously monitored.

4.2 Ambient Sensors

These sensors monitor ambient conditions like temperature, pressure, humidity, and vibrations. While a higher-than-normal temperature or humidity reading will not affect us outright, machines require precise conditions to operate optimally, and even a slight deviation can be disastrous. An abnormally high temperature may result in a fire, whereas high humidity can result in parts of a machine rusting. Humans are not very sensitive to changes in pressure, but the pressure change might result from a gas leak, which can be hazardous. Excess vibrations signify the abnormal behavior of a machine or a part of the machine, and these vibrations need to be inspected. Every device or component has some optimal operating conditions that, if not met, will result in suboptimal performance and can even damage the machine.

The BMP180 can detect ambient atmospheric pressure that ranges from 300 to 1100 hPa. This wide range is helpful in a pressure build-up due to gas leakage. The DHT-11 is an integrated sensor that can measure temperature and relative humidity. The SW-420 vibration sensor is a high sensitivity non-directional vibration sensor that can be calibrated to detect a wide range of vibrations.

Some of these sensors must be warmed up and calibrated before reading data. This data cannot be read continuously from them, thus requiring us to read it in intervals of one or two seconds for best results.

4.3 Disaster Detection

A disaster is defined as a problem that causes widespread human, economic or material loss. Every manufacturing plant or facility sees its fair share of incidents that, if not resolved on time, have the potential to turn into a disaster. The concerned authorities should be notified immediately about the affected areas to facilitate rescue efforts. If workers are trapped inside the premises, their immediate recovery must be organized. The target area is often in complete darkness or may even contain a fire, depending on the severity of the situation.

To help save lives by detecting disasters as accurately and efficiently as possible, we use an LDR, a PIR motion sensor, and a flame sensor capable of detecting fires. An LDR or a Light Dependent Resistor is an electrical component whose resistance depends on the ambient light level. We employ LDRs to detect ambient light levels in a factory in case of a disaster. This helps us monitor if the area containing the device has its lighting system intact or is in complete darkness. We can also detect motion with the help of a PIR or 'Passive InfraRed' sensor. These sensors detect if a person has moved within or out of the sensor's range by measuring the infrared light from objects in the current field of view. Counting the number of people trapped in a room will be immensely helpful during rescue operations. PIRs (Passive infrared) can calibrate their sensitivity via two potentiometers on the sensor. The first potentiometer adjusts the time delay between detection and notification, and the second changes the sensitivity.

We want to keep the time delay to a minimum and sensitivity to medium. We also use the SEN16 flame sensor to detect the presence of a fire and similarly adjust its sensitivity to a suitable level.

4.4 Data Transmission and Alerting

The data read from the sensors needs to be sent to the server for further processing. We decide on a certain payload structure recognized as our device's data, which helps in easy data transfer and communication between the host IoT device and the server. The IoT device actively listens for the alert status and will display it and alert the employees accordingly.

This module deals with transmitting data from the local device to the server for further analysis and for alerting of a disaster. The device has an ESP8266 Wi-Fi interface shield through which data is sent to the server. We decided to use Wi-Fi as the wireless communication interface due to its popularity and convenience. We require a Wi-Fi hotspot to be present to which the device can connect. The device's exact location is obtained through the GPS module and is sent to the cloud along with other sensor information.

If the cloud servers trigger an alert due to abnormal sensor readings, the local device will alert the factory workers with buzzers and LEDs and have them evacuate the premises.

4.5 Disaster Prediction

Rescue teams are directed to the affected area to begin the rescue operation. Using the data obtained from the sensors, we can quickly ascertain the state of the trapped workers. However, this is a reaction, and we can only deal with the effects of the problem. If there were a way to be alerted before the disaster takes place, we would be able to rectify the issue before it escalates to a catastrophe. Therefore, we use the data collected from our sensors to train a Machine Learning model to predict the probability of a disaster occurring.

The data received from the local IoT device is stored and analyzed for any anomalies. Predictions are made for a fixed unit of time in the future, given the available sensor readings. We make use of Machine Learning to aid us in the prediction. We try to predict the outcome of a fire, given the temperature and humidity readings. A fire is more likely to occur when the temperature is high, and humidity is low.

Conversely, a fire is less likely to happen when the temperature is low, or the humidity is high. Given this relationship, we opt for a regression model—Multiple Linear Regression. The more data that is present, the more accurate the prediction will be.

5 System Description

5.1 Pin Configuration

- VCC, GND and A0 pins of the MQ2 are connected to pins 5 V, GND and A5 of the Arduino, respectively.
- VCC, GND and A0 pins of the MQ3 are connected to pins 5 V, GND and D8 of the Arduino respectively.
- VCC, GND and A0 pins of the MQ5 sensor are connected to pins 5 V, GND and A3 of the Arduino respectively.
- VCC, GND and A0 pins of the MQ7 are connected to pins 5 V, GND and A2 of the Arduino, respectively.
- The SW-420 vibration sensor has VCC, GND and Digital Output pins connected to the Arduino's 5 V, GND and D12 pins, respectively. The sensitivity adjustment must be made by adjusting the built-in potentiometer while the sensor is powered off.
- The Neo-6M GPS module has its VCC and GND pins connected to the Arduino's 5 V and GND pins and the Tx and Rx pins are connected to pins D5 and D4, respectively.
- The DHT-11 temperature and humidity sensor has VCC, GND and Data pins connected to the Arduino's 5 V, GND and D7 pins, respectively.
- The buzzer has its GND connected to the GND of the Arduino and its +ve terminal to pin D10 of the Arduino.
- The LDR is connected to the Arduino via the A1 analog pin.
- The LED is connected to the Arduino via the D11 PWM pin.
- The PIR sensor has its VCC, GND and output pins connected to the Arduino's 5 V, GND and D6 pins.
- The ESP8266 Wi-Fi chip has VCC and GND connected to the Arduino's 5 V and GND pins. The Tx pin is connected to the Arduino's D3 pin and the Rx pin of the ESP8266 is connected to the Arduino's D2 pin via a voltage divider with two 1-kilo-ohm resistors
- The BMP-180 has VCC, GND, SCL and SDA pins connected to the Arduino's 5 V, GND, A0 and A4 pins, respectively.
- The SEN16 flame sensor has VCC, GND and A0 connected to the Arduino's 5 V, GND and D12 pins, respectively.

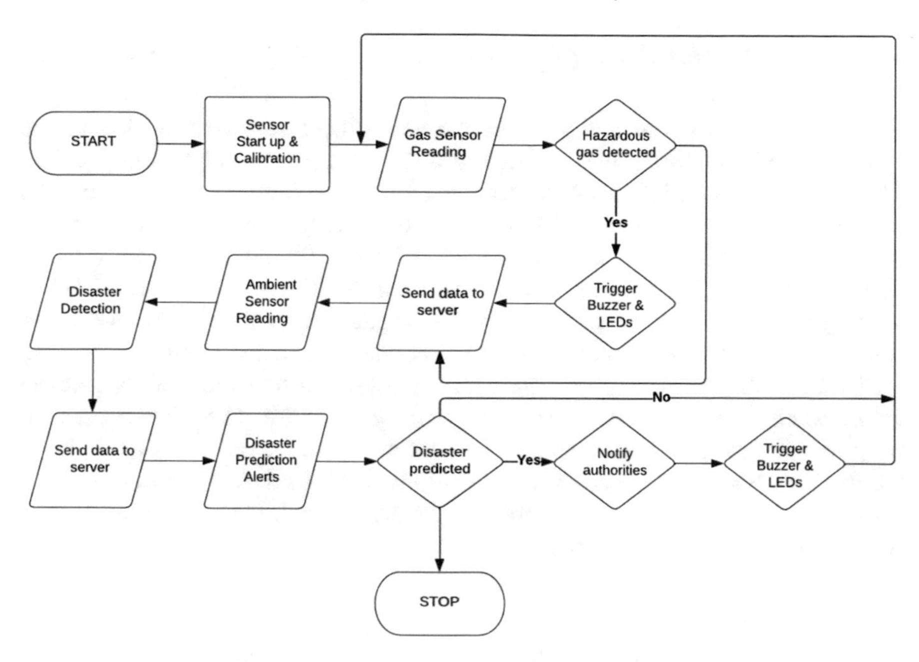

Fig. 2 Flowchart of SIMDPS

The block diagram in Fig. 2 illustrates these connections.

5.2 Hazardous Gas Detection Module

The gas sensors are connected to the Arduino UNO via their analog output pins. We prefer the analog output to the digital output as we need real-world values as input for our machine learning model. The digital output pins will give us a '1' or a '0' depending on the calibrated threshold set by the potentiometer. This is not desirable as we then need to calibrate the sensors per location/client, which is not feasible on a large scale. The gas sensors are used with the breakout boards, and the analog output pins are connected directly to the Arduino's analog pins. The +ve terminals of the gas sensors are connected to the +5 V of the Arduino via the breadboard, and the –ve terminals of the gas sensors to the GND of the Arduino.

The output analog value is directly proportional to the concentration of the gas in question—a high concentration will yield a higher output voltage, and a lower concentration will produce a lower output voltage. These values are read using the Arduino's 'analogRead' function. For best results, we read the sensor values at regular intervals of 5 s.

5.3 Ambient Monitoring Module

The BMP180 is a sensor that gives us the ambient pressure reading. It must be calibrated with the current sea level pressure and used as a constant during startup. The sensor takes a while to calibrate itself and will be ready after a rough calibration period of 30 s. The 5 V and GND of the Arduino are used for power and ground for the BMP, respectively. The two data pins of the BMP 180 are connected to the Arduino's analog pins. We read new values from the BMP every 1 s. The DHT11 sensor gives us temperature and humidity via a digital output pin. The Arduino's 5 V is used to power the chip, and the data pin is connected to a digital pin.

We query the DHT11 for readings every two seconds to avoid reading garbage values when the sensor is not ready. Similarly, with the SW-420 vibration sensor, we power it through the Arduino and connect its digital output to a digital pin on the Arduino. It must, however, have its sensitivity calibrated via the onboard potentiometer as it only has a digital output pin. We also query it for new readings every two seconds using the 'digitalRead' function.

5.4 Emergency Event Module

An LDR or a Light Dependent Resistor has its resistance change with changes in the ambient light levels. The LDR's signal pin is connected to an analog pin of the Arduino, and the ground pin is connected to the ground of the Arduino. The voltage drop across it will determine the ambient light level. The PIR motion sensor has three pins—+5 V, GND and output, and is powered by the Arduino's 5 V and GND. The output pin is connected to one of the Arduino's digital pins. We query the sensor for new values every three seconds. Similarly, for the SEN16, we power it through the Arduino and connect the analog output pin to the Arduino.

5.5 Data Transmission Module

We use the Neo-6M GPS module with our Arduino UNO to give us our latitude and longitude. It has an external antenna that must be connected for a stronger signal indoors, as the GPS module has a long initial startup period that depends on the signal's strength. The data pins of the GPS are connected to the PWM (Pulse Width Modulation) pins of the Arduino and 5 V and GND are connected as expected. The buzzer is also connected to a PWM pin and the LED to a digital pin of the Arduino. The ESP8266 data pins are connected to the Arduino's PWM pins and requires correct credentials to connect to a Wi-Fi network. The recorded sensor data is transmitted to our server via the HTTP protocol at a regular interval of 10 s. To avoid unauthorized access to our server, we can assign each device a username and a password for authorization with the web server.

5.6 Disaster Prediction Module

The data sent to the server is stored for analysis and prediction in our database. Our system was deployed in our university's automotive testing workshop, and the data collected during that period was used to evaluate the model's accuracy. We trained a multiple linear regression model on our dataset using this data. Uyanık and Güler [43] define regression analysis as a statistical technique to estimate the relationship between variables that have a reason and result in a relation. Regression models with one dependent variable and more than one independent variable are called multilinear or multiple linear regression models and are represented by the equation given below.

Multiple Linear Regression Equation

$$\widehat{Y} = \beta_0 + \beta_1 X_1 + \cdots + \beta_n X_n + \epsilon \quad (1)$$

From Eq. 1, $\widehat{Y}$ is the predicted value of our dependent variable, 0 is the y-intercept (when the other parameters are 0), X_1 is the regression coefficient of the first independent variable, X_n is the regression coefficient of the last independent variable and ϵ is the model error.

We considered the values of temperature and humidity and the resultant fire status, where temperature and humidity were the dependent variables and fire status was the independent variable. Our system gathered data for a week and was saved in a CSV file, aggregated by day. The data was split from 80 to 20, with 80% of the data used for training and 20% for testing.

The resultant accuracy of our multiple regression model is 87%, and the regression plot can be seen in the next section. Using this model, we can predict whether the temperature and humidity readings correspond to an anomaly for which the alert should be triggered. The prediction can be scheduled at regular intervals and was initiated on our server every 10 min by a cron job.

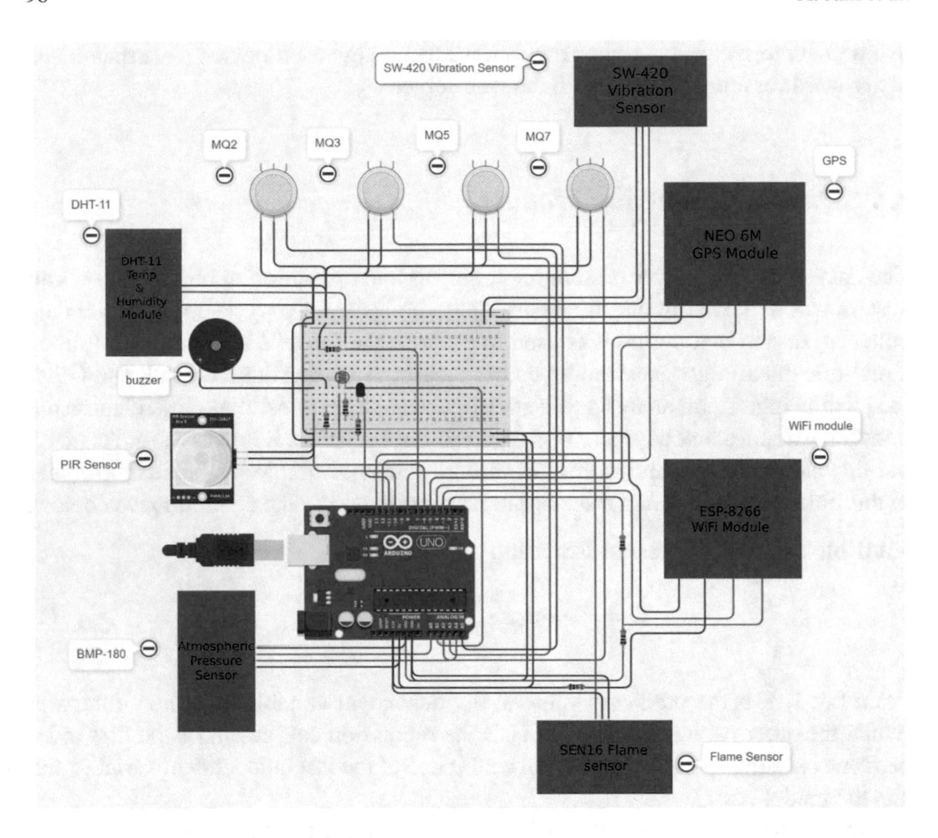

Fig. 3 Circuit diagram of SIMDPS (Fig. 3)

6 Results and Discussions

6.1 Implementation

The Arduino UNO is the core of SIMDPS and has all of the sensors connected to it. However, due to the UNO's limited flash memory, we were unable to fit the entire program on one board and had to distribute the sensors across two Arduino UNOs. Future work can include the usage of an Arduino MEGA, which has significantly more flash memory and GPIO pins and can accommodate the additional sensors.

Figure 4 and 5 show us the first Arduino UNO with the Neo-6 M GPS location sensor, BMP-180 pressure sensor, DHT-11 temperature and humidity sensor, PIR motion sensor, Buzzer and LEDs, along with the MQ2, MQ3, MQ5 gas sensors and the next with the MQ7 gas sensor.

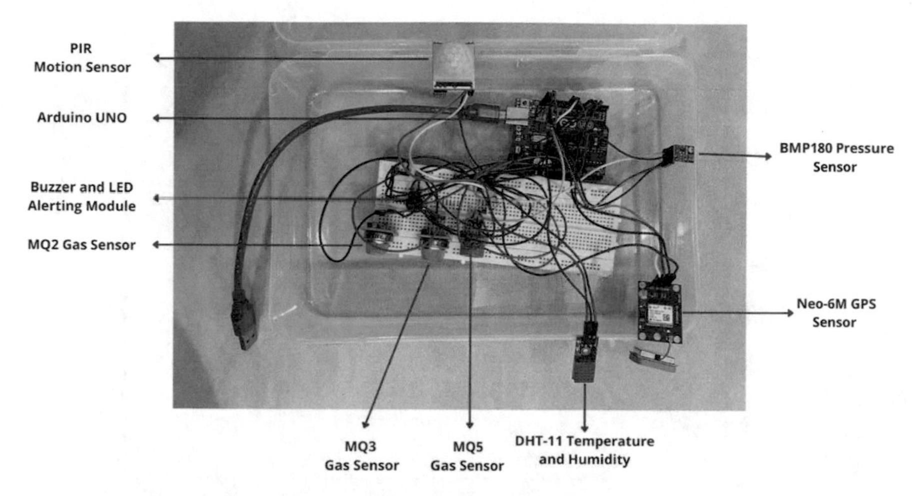

Fig. 4 SIMDPS implemented using an Arduino UNO

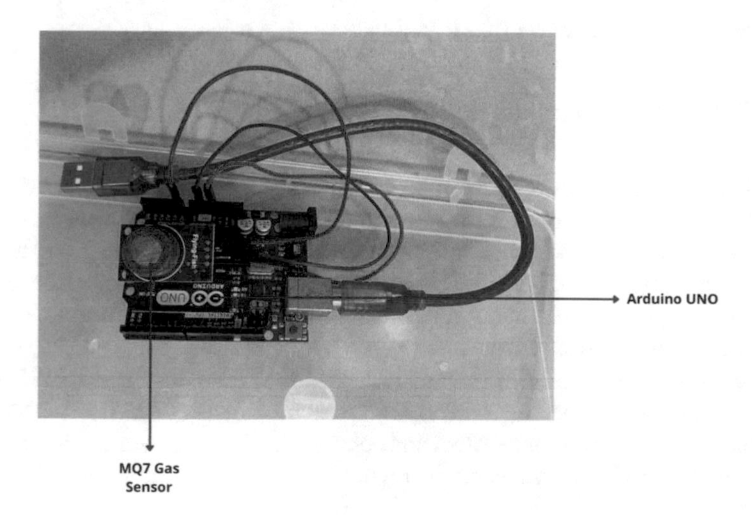

Fig. 5 SIMDPS carbon monoxide sensor implemented using an Arduino UNO

6.2 Testing the Accuracy of the Machine Learning Model

The data collected from our university's automotive workshop was exported to a CSV file, from which we observed around 100 data points for the readings of temperature, humidity, and carbon monoxide from the respective sensors of SIMDPS. We constructed, trained, and tested our machine learning model on the data collected using the 'scikit_learn' python package. The resultant model accuracy was found to be 87%, with the regression plot seen in Fig. 6. The closer the data points are to the straight line, the more accurate the prediction is.

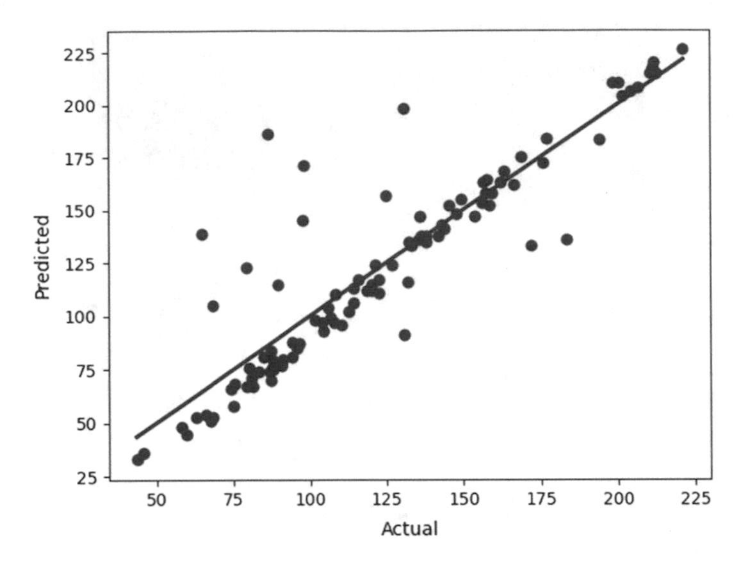

Fig. 6 Resultant regression plot for the trained multiple linear regression model

6.3 Testing the Accuracy of the Temperature Sensor

To find the accuracy of the DHT-11 temperature sensor that is used for the detection of the ambient temperature, we have compared it with the TMP35 sensor, an industry-standard chip that is used for measuring ambient temperature. We have conducted as many as 100 tests in our university's automotive workshops and recorded the readings to evaluate the performance of our sensor. To evaluate said performance, we calculate the Mean Absolute Error (MAE) and the Root Mean Square Error (RMSE) values. While comparing the ambient temperature readings, the MAE and RMSE values were 0.79 and 1.02, respectively. From the low error values of MAE and RMSE, we can infer the efficiency of our system and conclude that our system can be compared to devices that are industry standards. Figure 7 gives a visual representation of our comparison.

6.4 Testing the Accuracy of the Humidity Sensor

The performance comparison of our humidity sensor to the PCE-555, an industrial standard device, was carried out in the same conditions as the measurement of the temperature. At regularly spaced intervals, the humidity readings were taken from our DHT-11 sensor and from the PCE-555 and the MAE and RMSE were found to be 0.94 and 1.12, respectively. Figure 8 is a visual representation of the comparison for humidity.

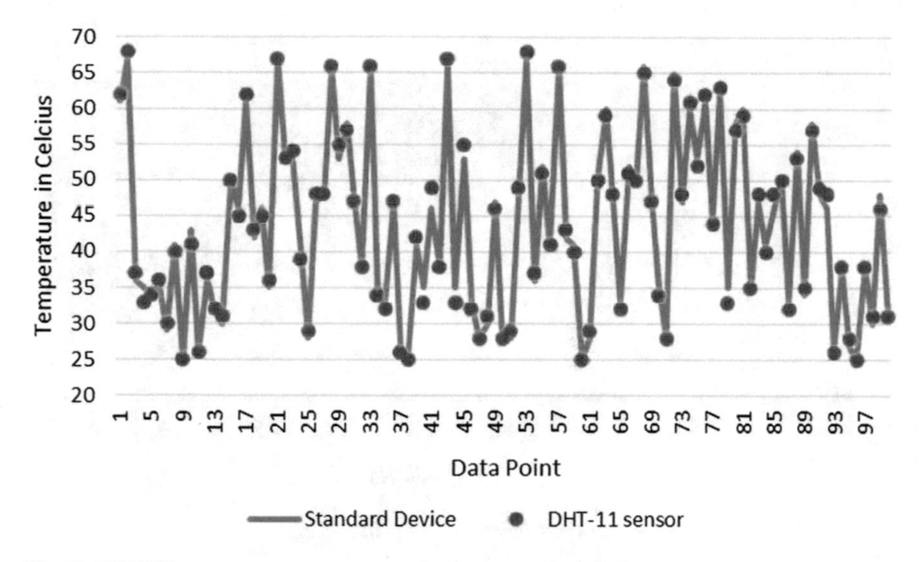

Fig. 7 SIMDPS temperature sensor versus industry standard device

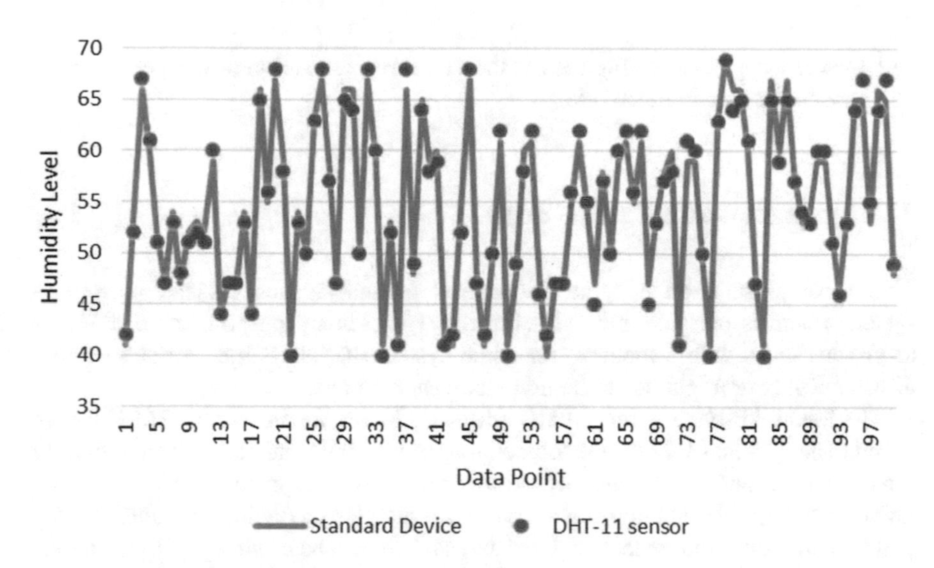

Fig. 8 SIMDPS humidity sensor versus industry standard device

6.5 Testing the Carbon Monoxide Sensor

The performance of the carbon monoxide sensor was also compared to the HTC CO-01 portable carbon monoxide sensor with the data collected from our university's automotive workshop. The Mean Absolute Error (MAE) and the Root Mean Square Error (RMSE) were 1.07 and 1.23, respectively. These scores are negatively oriented,

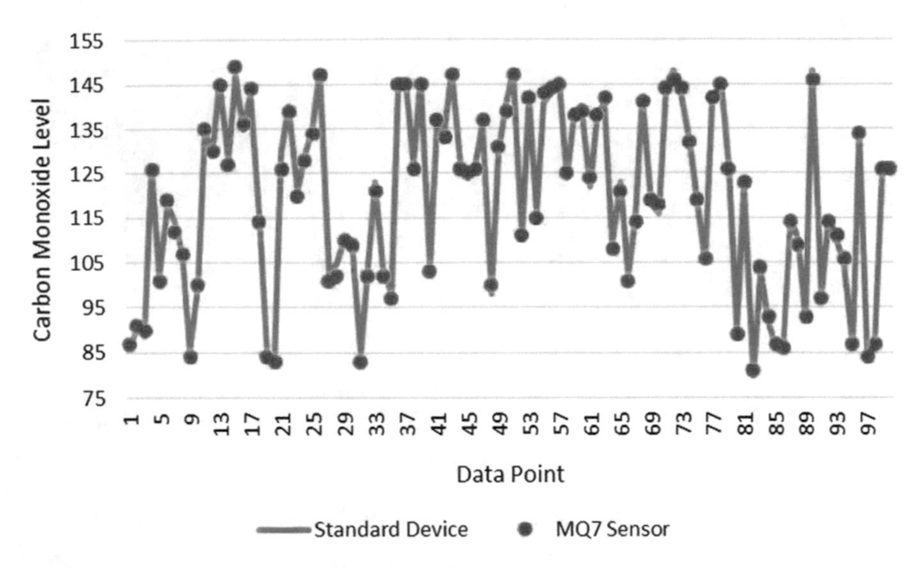

Fig. 9 SIMDPS carbon monoxide sensor versus industry standard device

and a lower value is better. Figure 9 is the visual representation of our performance comparison for carbon monoxide.

7 Conclusion and Future Scope

This paper presents an integrated industrial monitoring and disaster prevention system which is portable and cost-efficient. The primary purpose of SIMDPS is to provide users with a modern, real-time system that alerts the workers when a disaster occurs or might occur, based on current readings.

The Mean Absolute Error (MAE) and Root Mean Square Error (RMSE) were found to be 0.79 and 1.02 for the temperature sensor, 0.94 and 1.12 for the humidity sensor and 1.07 and 1.23 for the carbon monoxide sensor, respectively. This is the first low-cost IoT-enabled monitoring system with disaster prediction and alerting that will help to reduce industrial accidents and save lives. The accuracy of the machine learning model can be increased by collecting more data in different conditions.

Future research directions can extend to using different machine learning algorithms to train the model on more sensor readings including data from the vibration sensor and the various gas sensors, and comparing their results for more accurate predictions. The prediction model can also be shifted to execute on a Raspberry Pi or a similar device, using on-device AI for lower latency. The system, while designed to be an integrated disaster monitoring and prevention system, can also be integrated with a smart city to enhance existing services.

References

1. Cushman & Wakefield (2021) Global manufacturing risk index 2022. https://www.cushmanwakefield.com/en/insights/global-manufacturing-risk-index
2. FICCI quarterly survey on Indian manufacturing sector. https://ficci.in/SEDocument/20612/manufacturing-survey.pdf
3. Ga E, Gb N, Rc R (2021) Smart industry monitoring and controlling system using IoT. Smart Intell Comput Commun Technol 38:449
4. Venkata Subbaiah B, Venkata Sreekanth Reddy E, Abhishek K, Pavan Kumar Reddy, An IoT based smart industry monitoring system by using raspberry PI 3
5. Ganeshan C, Kumar Singh S (2018) Smart industrial system for monitoring, control and security using internet of things. In: 2018 2nd international conference on trends in electronics and informatics (ICOEI). IEEE
6. Kumar NS, Chandrasekaran G, Rajamanickam KP (2021) An integrated system for smart industrial monitoring system in the context of hazards based on the internet of things. Int J Safety Secur Eng 11(1):123–127
7. Gupta PK, Sai Koushik B, Deeban Chakravarthy V (2020) IoT based smart industry monitoring system by using Arduino with GSM. Int J Adv Sci Technol 29(05):9082–9088. http://sersc.org/journals/index.php/IJAST/article/view/18978
8. Peng Y, Wu IC (2021) A cloud-based monitoring system for performance analysis in IoT industry. J Supercomput 77:9266–9289. https://doi.org/10.1007/s11227-021-03640-8
9. Deekshath R et al (2018) IoT based environmental monitoring system using arduino UNO and thingspeak. Int J Sci Technol Eng 4(9):68–75
10. Kishore Kumar R, Nishanth N, Suriya Prakash SK, Dhanush Anand SB, IoT based industrial monitoring system using Arduino
11. Solanki S, Gaur D, Rasiq G (2021) IoT based industrial monitoring system. https://doi.org/10.1007/978-981-15-9873-9_28
12. Aravind R, Yadikiumarani, Meghna K, Divyashree R, Naregowda H (2021) IOT based real time data monitoring for industry. Int J Eng Res Technol (IJERT) NCCDS—2021 9(12)
13. Rajalakshmi R, Vidhya J (2019) Toxic environment monitoring using sensors based on IoT. Int J Recent Technol Eng
14. Lohith D, Kumar KV, Reddy IBK, Jeyaramya V, IoT Industry Protection using Arduino
15. Rupali SG, Mahajan P (2018) Home and industrial safety system for fire and gas leakage detection. Int Res J Eng Technol
16. Prasanti V, Venkataramana T, IoT Based Industrial Automation Control System Using Arduino
17. Merchant HK, Ahire DD, Industrial automation using IoT with Raspberry Pi
18. Georgewill O, Ezeofor C (2016) Design and implementation of SMS-based industrial/homes gas leakage monitoring & detection alarm system. Int J Eng Trends Technol 35:410–416. https://doi.org/10.14445/22315381/IJETT-V35P283
19. Kavitha BC, Alagappan V (2019) IoT based intelligent industry monitoring system 63–65. https://doi.org/10.1109/SPIN.2019.8711597
20. Liu JH, Chen YF, Lin TS, Chen CP, Chen PT, Wen TH, … Jiang JA (2012) An air quality monitoring system for urban areas based on the technology of wireless sensor networks. Int J Smart Sens Int Syst 5(1)
21. Banhazi TM (2009) User friendly air quality monitoring system. Appl Eng Agric 25(2):281–290
22. Okigbo CA, Seeam A, Guness SP, Bellekens X, Bekaroo G, Ramsurrun V (2020) Low cost air quality monitoring: comparing the energy consumption of an arduino against a raspberry Pi based system. In: Madhavje K, Soyjaudah S (eds) Proceedings of the 2nd international conference on intelligent and innovative computing applications. ICONIC'20, 24–25 Sept 2020, Plaine Magnien, Mauritius. ISBN 9781450375580 [Conference or Workshop Item]. https://doi.org/10.1145/3415088.3415124
23. Dhingra S, Madda RB, Gandomi AH, Patan R, Daneshmand M (2019) Internet of Things mobile–air pollution monitoring system (IoT-Mobair). IEEE Internet Things J 6(3):5577–5584

24. Deshpande A, Pitale P, Sanap S (2016) Industrial automation using Internet of Things (IOT). Int J Adv Res Comput Eng Technol (IJARCET) 5(2):266–269
25. Okokpujie KO, Noma-Osaghae E, Odusami M, John SN, Oluga O (2018) A smart air pollution monitoring system. Int J Civ Eng Technol (IJCIET) 9(9):799–809
26. Das A, Sarma MP, Sarma KK, Mastorakis N (2018) Design of an IoT based real time environment monitoring system using legacy sensors. In: 22nd international conference on circuits, systems, communications and computers (CSCC 2018), vol 210
27. Simbeye DS (2017) Industrial air pollution monitoring system based on wireless sensor networks. J Inf Sci Comput Technol, November 21
28. Ziętek B, Banasiewicz A, Zimroz R, Szrek J, Gola S, A portable environmental data-monitoring system for air hazard evaluation in deep underground mines
29. Kanan R, Elhassan O, Bensalem R, An IoT-based autonomous system for workers' safety in construction sites with real-time alarming, monitoring, and positioning strategies
30. Kaur N, Mahajan R, Bagai D (2016) Air quality monitoring system based on Arduino microcontroller
31. Charaim F, Erol YB, Pister K (2016) Wireless gas leakage detection and localization. Institute of Electrical and Electronics Engineers
32. Kanappan A, Hariprasad K (2017) Toxic gas and radiation detection monitoring using IoT. Int J Eng Res Technol
33. Elsisi M, Mahmoud K, Lehtonen M, Darwish MM (2021) Reliable industry 4.0 based on machine learning and IOT for analyzing, monitoring, and securing smart meters. Sensors 21(2):487
34. Ramamurthy H, Prabhu BS, Gadh R, Madni AM (2007) Wireless industrial monitoring and control using a smart sensor platform. IEEE Sens J 7(5):611–618
35. Ray PP, Mukherjee M, Shu L (2017) Internet of things for disaster management: state of-the-art and prospects. IEEE Access 5:18818–18835. https://doi.org/10.1109/ACCESS.2017.2752174
36. Kök İ, Şimşek MU, Özdemir S (2017) A deep learning model for air quality prediction in smart cities. In: 2017 IEEE international conference on big data (big data), pp 1983–1990. https://doi.org/10.1109/BigData.2017.8258144
37. Mehta Y, Manohara Pai MM, Mallissery S, Singh S (2016) Cloud enabled air quality detection, analysis and prediction—a smart city application for smart health. In: 2016 3rd MEC international conference on big data and smart city (ICBDSC), pp 1–7. https://doi.org/10.1109/ICBDSC.2016.7460380
38. Haque AKMB, Bhushan B, Dhiman G (2022) Conceptualizing smart city applications: requirements, architecture, security issues, and emerging trends. Expert Syst 39(5):e12753. https://doi.org/10.1111/exsy.12753
39. Haque AKMB, Bhushan B, Hasan M, Zihad MM (2022) Revolutionizing the industrial internet of things using blockchain: an unified approach. In: Balas VE, Solanki VK, Kumar R (eds) Recent advances in internet of things and machine learning. Intelligent systems reference library, vol 215. Springer, Cham. https://doi.org/10.1007/978-3-030-90119-6_5
40. Malik A, Bhushan B, Kumar A, Chaganti R (2022) Opportunistic internet of things (OIoT): elucidating the active opportunities of opportunistic networks on the way to IoT. In: Sharma R, Sharma D (eds) New trends and applications in internet of things (IoT) and big data analytics. Intelligent systems reference library, vol 221. Springer, Cham. https://doi.org/10.1007/978-3-030-99329-0_14
41. Bhushan B (2022) Middleware and security requirements for internet of things. In: Sharma DK, Peng SL, Sharma R, Zaitsev DA (eds) Micro-electronics and telecommunication engineering. ICMETE 2021. Lecture notes in networks and systems, vol 373. Springer, Singapore. https://doi.org/10.1007/978-981-16-8721-1_30

42. Mehta S, Bhushan B, Kumar R (2022) Machine learning approaches for smart city applications: emergence, challenges and opportunities. In: Balas VE, Solanki VK, Kumar R (eds) Recent advances in internet of things and machine learning. Intelligent systems reference library, vol 215. Springer, Cham. https://doi.org/10.1007/978-3-030-90119-6_12
43. Uyanık GK, Güler N (2013) A study on multiple linear regression analysis. Procedia Soc Behav Sci 106:234–240

An Artificial Intelligence Based Sustainable Approaches—IoT Systems for Smart Cities

N. Yuvaraj, K. Praghash, J. Logeshwaran, Geno Peter, and Albert Alexander Stonier

Abstract The Internet of Things (IoT) allows city officials to monitor the city in real time and communicate smoothly with the community. Smart cities need to provide the best possible service to their citizens on the most basic infrastructure. A successful smart city should make any service available to all in general and equally. In this paper, an artificial intelligence (AI) based smart sustainable IoT model was proposed to enhance the different services in the smart city environment. Where the list of services has begun to play a vital role in the construction of smart cities. The service tracks route priority and non-priority services to various locations. The proposed model provides a prominent place for both services. At a cut-off level, the proposed model achieved 94.97% of service recognition, 3.3% of service rejection, 94.06% of service accuracy, 95.86% of service precision, 94.34% of service recall, and 95.68% of F1-Score while compared with the existing models.

Keywords Smart cities · Electronic methods · Sensors · Internet of things · Artificial intelligence · Priority routing

N. Yuvaraj
Department of Computer Science and Engineering, Sri Shakthi Institute of Technology, Coimbatore, India

K. Praghash
Department of Electronics and Communication Engineering, Christ University, Bengaluru, India

J. Logeshwaran
Department of Electronics and Communication Engineering, Sri Eshwar College of Engineering, Coimbatore, India

G. Peter (✉)
CRISD, School of Engineering and Technology, University of Technology Sarawak, Sibu, Malaysia
e-mail: drgeno.peter@uts.edu.my

A. A. Stonier
School of Electrical Engineering, Vellore Institute of Technology, Vellore, India

B. Bhushan et al. (eds.), *AI Models for Blockchain-Based Intelligent Networks in IoT Systems*, Engineering Cyber-Physical Systems and Critical Infrastructures 6,
https://doi.org/10.1007/978-3-031-31952-5_5

1 Introduction

The Internet of Things is a rapidly evolving technology that makes everything smart. Its machine learning can now be clever by combining factory equipment, baby toys or clothing all with sensor technology [1]. But the existence of such data-intensive technology around us makes us vulnerable to security and privacy threats. There is potential for exploitation wherever the biggest amount of data is [2]. Despite the obvious benefits of this significant risk, many are reluctant to use such smart technology. During all this, the IoT is still growing and evolving. Still, the future of technology looks bleak [3].

Current IoT trends seem to focus on the use of processes and technologies that make it easier to respond quickly to changing security environments and move conversations towards management systems, devices, and time-tested positions [4–6]. Conversations with controllers move towards understanding what developers of IoT products are going to do when they are best set up. Cyber programs fail [8].

Regulators are beginning to understand that compromise, disease, etc. are inevitable. "The encouragement to implement consumer protection IoT devices is not very strong" [10]. The cost of doing so will not be reimbursed at the lower price points of these devices. The only way such protection can be enforced may be through regulation [11]. Except for the obvious, IoT continues to grow and use cases (yes, pet cameras, collars, window sensors, etc.) as costs shrink (Wi-Fi components, etc.), IoT providers will have to do a much better job of managing these issues, especially as the customization features get deeper (health data, financial information, etc.) can do better [13–15].

A few years ago, most of us only had to worry about protecting our computers. Next, we had to worry about protecting our smart phones [17]. Now we must take care of the safety of our cars, our home appliances, and our wearable devices. This is a constant security challenge as there are many devices that can be hacked. The IoT platform market is set to grow rapidly in the coming years, with the current leading platforms expanding and others entering the market. In the short term, companies are unlikely to dismiss the cloud for edge computing [18]. The day may come when there will be more computer power and more computers on the edge of the network than we have in datacenters [19]. Internet connectivity makes a very efficient contribution to smart cities. In the years to come, it can be expected to bring about a major shift, as we can already see the signs of change in various industry sectors. Its effectiveness and nature make it extremely helpful in daily activities for individuals, industry, and society [20]. Like other architectural expert plans, there are many challenges in implementing a smart city plan. These problems often turn out to be profoundly serious, and the chances of success depend on how a government handles them through various agencies [23–35].

The main contribution of this paper is,

- The authors attempt to transform the regular city into a smart city by gathering data about the city and its primary development areas. Specify its desired attributes, along with a timeline for release and cost range. Track and record every urban habit and learn from this case study and get to work on other initiatives to turn any city into a smart one.

- The spread of security, autonomous, and semi-autonomous driving solutions is made easier by the development of smart IoT systems that can communicate with each other and improve travel capabilities and levels.
- Enhance the service speed and service quality of the various priority-based services in the smart city.

The outline of the paper is presented below: Sect. 2 discusses the literature review. Section 3 proposed method. Section 4 evaluates the entire work and Sect. 5 concludes the work with possible directions of future scope.

2 Literature Review

In this section, various state-of-art security vulnerable detection methods are discussed here [7]. Discussed the basic security vulnerabilities of the Internet. This increases the number of devices behind your network firewall. Protecting IoT devices requires more than protecting real devices. Software that connects to those devices must provide security to applications and network connections [9]. Discussed one of the biggest challenges IoT lacks is industry collaboration to build an integrated IoT framework. Without a centrally shared site, companies have spent a lot of time producing solutions that are the same [12]. Discussed how the lack of cooperation reduced the acceptance of the IoT solution and created significant integration challenges. From the developer point of view, this has created a confusing and confusing mismatch of technology [16]. Discussed the smart cities' need to preserve their true compact nature even as they travel with modernity. The structure, economy, and community structure of a smart city must be prepared to handle any natural environment or emergency [21]. Discussed a smart city that will have integrated information and communication technologies to improve efficiency, tackle problems, and improve quality of life. This will make it possible to sustain the growth of the city's operations. Smartphone deployment, GPS technology, telemetric and rapid frameworks are the technologies required for smart city planning [22]. Discussed how transportation plays a key role in the design of any city. The main goal of smart city transport should be to enable its people to use their own vehicles and use well-designed public transport.

3 Proposed Model

The proposed strategy attempts to transform the regular city into a smart city. it Gather data about the city and its primary development areas. Specify its desired attributes, along with a timeline for release and cost range. When the project is finished, it is possible that the city current sustainable operating infrastructure and IT practices will be altered or improved. Track and record every urban habit. Learn from this case study and get to work on other initiatives to turn any city into a smart one. Authorities can make living in regular towns better by gradually upgrading them to smart cities.

3.1 Blockchain Based Security Management

There may be unneeded complications and performance bottlenecks while establishing a blockchain network. The goals of the proposed multi-layer network model are to lessen the burden of administration, increase the speed with which responses may be made, collect more data through connected channels, strengthen the safety of communications, and leave room for future growth. The basic layer of a network often consists of nodes and devices with widely varying processing speeds and energy reserves. The IoT network cluster leaders are in charge of administering a local authorization scheme. Devices that have been registered locally can use the program's authentication and authorization features.

Control nodes, authority nodes, edge computing nodes, and gateways are all examples of nodes that can be found in the logical second layer of a network. As soon as a lightweight consensus is implemented within the blockchain system, CH nodes will be able to safely communicate with one another. In this protocol layer, the permissioned HLF blockchain is built. When it comes to cellular networks, the base stations (BSs) are the pinnacle. This additional layer, made up of high-performance computers, may be organized as a set of distinct structures beneath the blockchain. This is the layer where asymmetric cryptography techniques can be used with the most security. By combining the global blockchain with other state-of-the-art security technologies, the anonymity and privacy of the top layer can be preserved.

Smart cities have many features, including smart healthcare systems, management, transportation systems, better surveillance for security, smart infrastructure, better jobs, and all the amenities and facilities to live comfortably. A city can be considered a smart city if it has the following features:

- Provide improved core services to the community in a reliable and cost-effective manner.
- Promoting the economic growth of the community and Efficient management of resources.

The Internet of Things makes it easy to transform a city into a smart city, as shown in Fig. 1.

Consider the different entries of the smart system denoted a and b, and these are defined as the following Eq. (1):

$$A^{q}(a,b)=\frac{2*d(a,b|C)}{d(a,a|C)+d(b,b|C)} \tag{1}$$

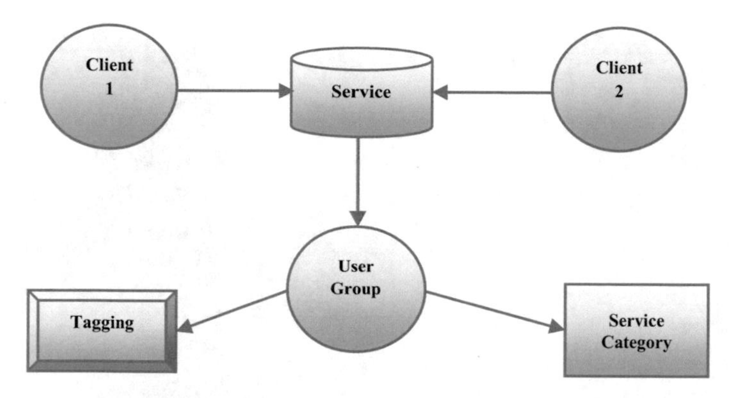

Fig. 1 Proposed model design

where,

C = entries route type.

$d(a, b|C)$ = number of different routes between a and b.

The predefined request from the client 'x' for the service 's' was declared as the following.

Equation (2),

$$m_{x,s} = \sum_{a \in C} \sum_{A=1}^{B} \alpha_A X^A(s, a) \tag{2}$$

where,

$X^A(s, a)$ = the service list values between the services s and a in the route A.

B = no of various Meta track.

α_A = weight of the Meta track.

C = position of the services.

The intelligence derived from the road map of the leading smart cities around the world is summarized into four pillars. Figure 2 shows the sustainable IoT module for smart cities.

It is on these pillars that the foundation of smart cities is laid. Cities can be transformed into smart cities by integrating these facilities into the Smart City project. To transform a city into a smart city, set measurements after its data collection. After that, the local authorities start some brilliant projects and elevate them to higher levels based on their success. Reviewing the experiences of the community helps to make improvements in areas where needed. Figure 3 shows the proposed model service analyzer block and Fig. 4 shows the proposed model service analyzer block.

The proposed steps to transform the ordinary city into a smart city.

Step-1: Gather data about the city and its major development areas.

Step-2: Set the criteria for what it wants to become, along with the availability deadline and budget.

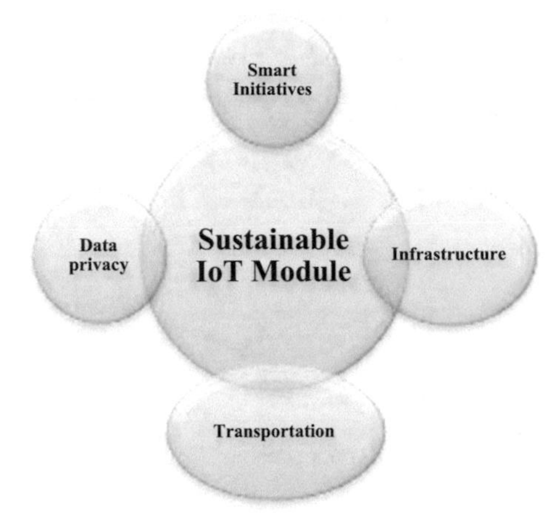

Fig. 2 Sustainable IoT module for smart city

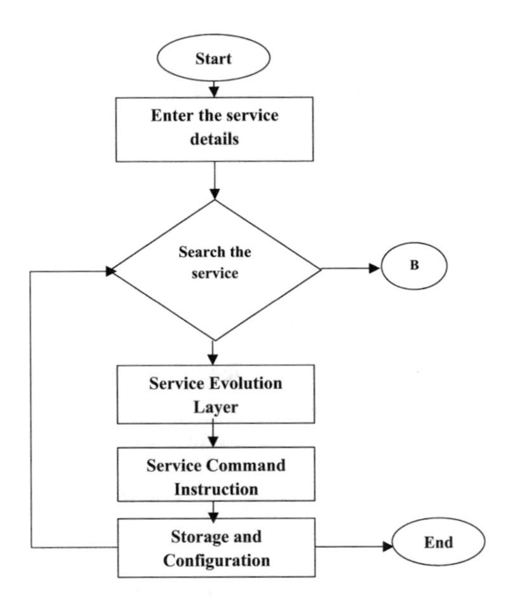

Fig. 3 The proposed model service analyzer block

Step-3: Start with 1–2 small projects and complete them within a specified time. The city's current sustainable operating infrastructure and IT practices may be changed / modified upon completion of the project.

Step-4: Measure all the practices of the city.

Step-5: Review the experience and start working with other projects to make a normal city smarter.

By adopting a step-by-step approach, the authorities can transform ordinary cities into smart cities for a better life.

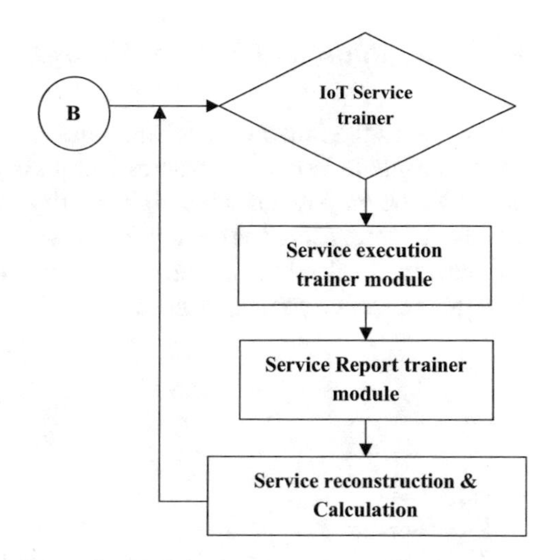

Fig. 4 IoT based training module

Algorithm: Sustainable IoT (SIoT) for Smart cities	
1	*Start*
2	*Enter the service details*
3	*Search the service availability in the record*
4	*Provide the input approval*
5	*If (service = available)*
6	*Then execute the service*
7	*Then transfer the service details to the data centre*
8	*Store the execution details in Data base*
9	*Else go-to step 3*
10	*Inform the service confirmation*
17	*end*

4 Results and Discussion

The proposed Sustainable IoT (SIoT) model was compared with the existing Distributed ITS (DITS), Energy-Aware Smart Connectivity (EASC), Hybrid Model of Machine Learning (HMML) and An IoT Architecture (IoTA). Here the network simulator tool was used to compare the result parameters.

4.1 Calculation of Service Recognition (S_R)

The work on the Smart City IoT module will continue after all the data inside has been thoroughly tested. It separates the quality of the various compounds that enter inside. In the proposed manner high priority and low priority services are calculated and the incoming services are implemented accordingly.

The service selections are performing a random priority module. Hence the service recognition can be computed as per the below Eq. (3).

$$S_R = \sum_{x=1}^{y} P_i \tag{3}$$

where,

S_R = Service Recognition.

P_i = Total number of random services under the active assembly (i, j).

Table 1 and Fig. 5 represents the calculation comparison of service recognition of the existing and proposed models.

4.2 Calculation of Service Rejection (S_{RN})

If the services coming here are of high importance they can be executed immediately. Less important services are also effectively handled in the proposed manner. Thus, the volume of data executed in this mode is high and the error services are exceptionally low shown in Eq. (4).

$$S_{RN} = \frac{\mathrm{P_i} - \mathrm{SR}}{P_j} \tag{4}$$

where,

S_{RN} = Service Rejection.

Table 1 Calculation comparison of Service Recognition

No. of inputs	DITS	EASC	HMML	IoTA	SIoT
1000	65.25	71.62	84.86	78.83	92.84
2000	65.58	73.12	85.45	80.70	93.85
3000	66.92	74.23	86.43	81.53	94.01
4000	68.06	74.61	87.64	82.44	94.97
5000	69.11	75.62	88.78	83.36	94.54
6000	69.82	76.55	89.89	84.69	95.74
7000	71.12	77.55	90.59	85.77	95.90

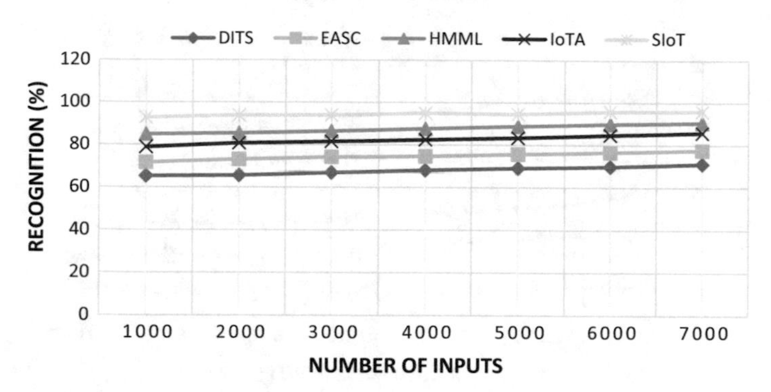

Fig. 5 Comparison of service recognition

Table 2 Calculation comparison of service rejection

No. of inputs	DITS	EASC	HMML	IoTA	SIoT
1000	34.75	28.38	13.41	19.44	5.43
2000	34.42	26.88	12.82	17.57	4.42
3000	33.08	25.77	11.84	16.74	4.26
4000	31.94	25.39	10.63	15.83	3.3
5000	30.89	24.38	9.49	14.91	3.73
6000	30.18	23.45	10.11	15.31	4.26
7000	28.88	22.45	9.41	14.23	4.1

S_R = Service Recognition.

P_i = Total number of random services under the active assembly (i, j).

P_j = Non block random service under the active assembly (i, j).

Table 2 and Fig. 6 represents the calculation comparison of service rejection of the existing and proposed models.

4.3 Calculation of Service Accuracy (S_A)

The service accuracy is the calculation to identify the exact work executed by the system while all the services are in the route. The service accuracy is the ratio between the sum of true positive service and true negative service and the Total number of random services under the active assembly (i, j) shown in Eq. (5)

$$S_A = \frac{TS_P + TS_N}{P_i} \tag{5}$$

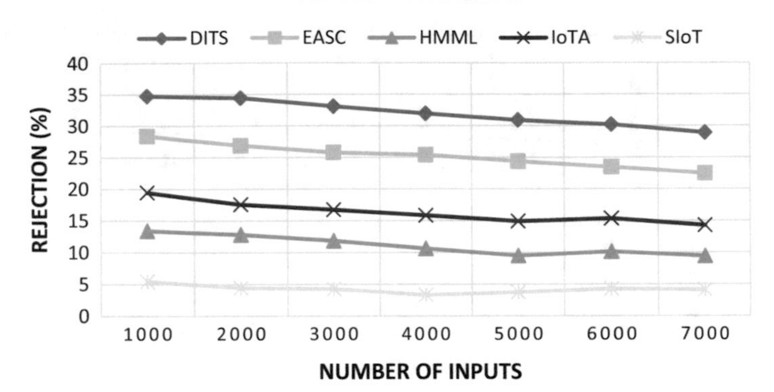

Fig. 6 Comparison of service rejection

Table 3 Calculation comparison of service accuracy

No. of inputs	DITS	EASC	HMML	IoTA	SIoT
1000	62.95	69.32	88.26	81.57	91.93
2000	63.28	70.82	88.85	83.44	92.97
3000	64.62	71.93	89.83	84.27	93.10
4000	65.76	72.31	91.04	85.18	94.06
5000	66.81	73.32	92.18	86.10	93.63
6000	67.52	74.25	93.29	87.43	94.87
7000	68.82	75.25	93.99	88.30	94.98

where,

$S_A =$ Service Accuracy.

$TS_P =$ Positive true service.

$TS_N =$ Negative true service.

$P_i =$ Total number of random services under the active assembly (i, j).

Table 3 and Fig. 7 represents the calculation comparison of service recognition of the existing and proposed models.

4.4 Calculation of Service Precision (S_P)

The service precision is the ratio between the true positive service and the sum of true positive service and the false positive service under the active assembly (i, j) shown in Eq. (6)

$$S_P = \frac{TS_P}{TS_P + FS_P} \tag{6}$$

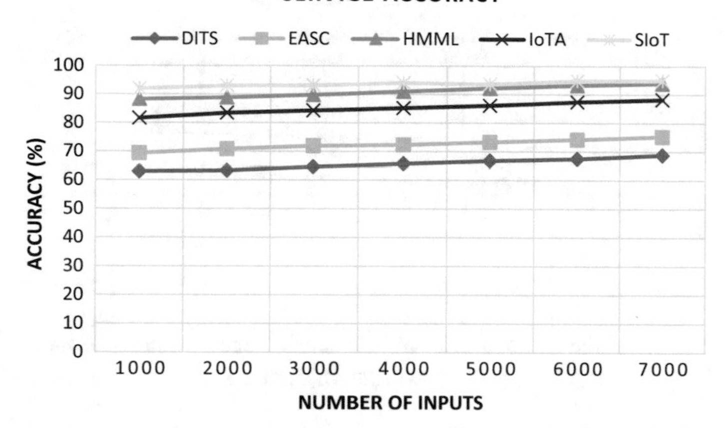

Fig. 7 Comparison of service accuracy

Table 4 Calculation comparison of Service Precision

No. of inputs	DITS	EASC	HMML	IoTA	SIoT
1000	64.21	61.58	80.70	73.13	92.67
2000	65.84	63.32	82.28	74.55	93.96
3000	66.32	65.66	84.48	75.81	94.97
4000	67.61	66.47	86.11	77.80	95.86
5000	69.72	68.76	87.25	80.27	96.23
6000	71.21	70.69	89.45	81.71	97.27
7000	73.02	72.42	90.60	83.43	98.04

where,

S_P = Service Precision.

TS_P = Positive true service.

FS_P = Positive false service.

Table 4 and Fig. 8 represents the calculation comparison of service recognition of the existing and proposed models.

4.5 *Calculation of Service Recall* (S_R)

The service recall is the ratio between the true positive service and the sum of true positive service and the false positive service under the active assembly (i, j) shown in Eq. (7).

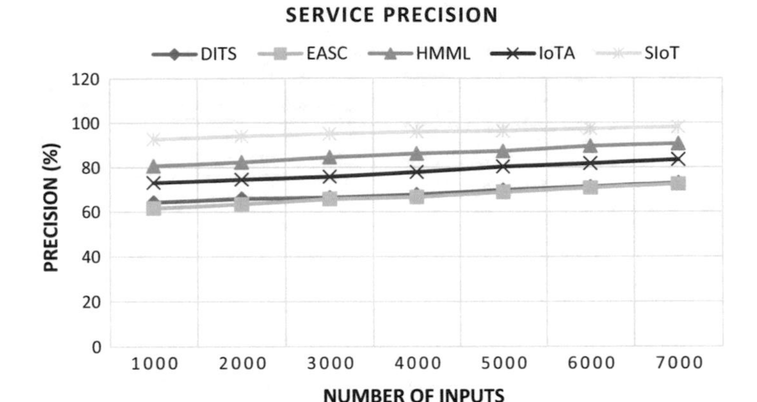

Fig. 8 Comparison of service precision

Table 5 Calculation comparison of service recall

No. of inputs	DITS	EASC	HMML	IoTA	SIoT
1000	54.32	65.68	80.86	74.14	92.67
2000	55.81	67.65	83.28	76.34	92.66
3000	56.61	68.78	83.69	77.14	93.86
4000	58.94	69.97	85.29	77.81	94.34
5000	59.95	70.36	87.61	79.24	95.77
6000	60.59	71.88	88.86	80.33	96.93
7000	61.25	72.12	91.59	80.81	97.70

$$S_P = \frac{TS_P}{TS_P + FS_N} \tag{7}$$

where,

S_P = Service Precision.

TS_P = Positive true service.

FS_N = Negative false service.

Table 5 and Fig. 9 represents the calculation comparison of service recognition of the existing and proposed models.

4.6 Calculation of Service F1-Score ($S_{F1-Score}$)

The service score was calculated by the ratio between the combination of precision and recall under the active assembly (i, j) shown in Eq. (8)

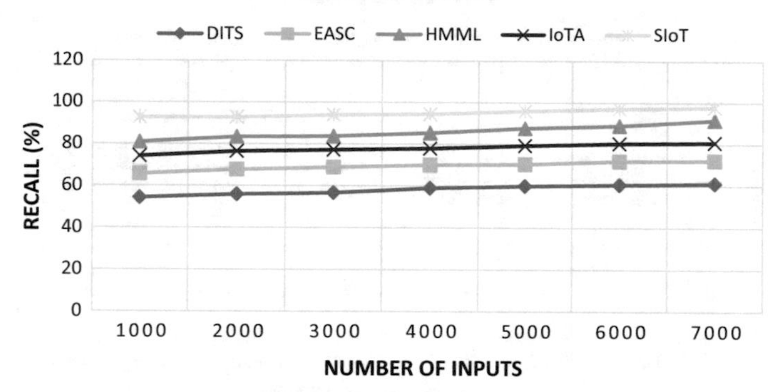

Fig. 9 Comparison of service recall

Table 6 Calculation comparison of service F1-score

No. of inputs	DITS	EASC	HMML	IoTA	SIoT
1000	62.83	62.05	78.35	69.94	92.83
2000	62.72	62.07	78.18	69.67	92.33
3000	62.70	62.95	78.91	69.97	92.45
4000	65.80	65.78	82.25	73.48	95.68
5000	67.00	67.10	82.98	74.80	96.06
6000	67.61	67.93	83.87	75.34	96.63
7000	68.02	68.33	83.95	75.64	96.33

$$S_{F1-Score} = \frac{2 * (S_R * S_P)}{(S_R + S_P)} \tag{8}$$

where,

$S_{F1-Score}$ = Service F1-Score.

S_P = Service Precision.

S_R = Service Recall.

Table 6 and Fig. 10 represents the calculation comparison of service recognition of the existing and proposed models.

5 Conclusion

Finally, all the above elements must occur in an interconnected city. From anywhere, residents can contact the city, get the information or service they value, and provide information about any incident or need. The proposed Sustainable IoT (SIoT) model is compared among the existing Distributed ITS (DITS), Energy-Aware Smart

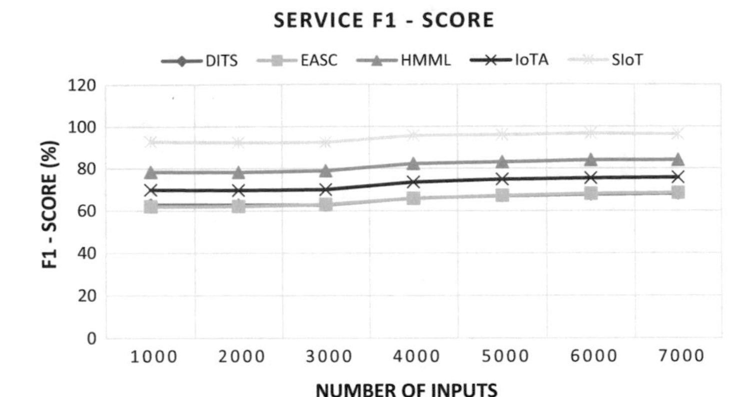

Fig. 10 Comparison of service F1-Score

Connectivity (EASC), Hybrid Model of Machine Learning (HMML), and an IoT Architecture (IoTA). Here, the network simulator tool is used to investigate the result parameters. This connection is only through electronic devices, and that must be an interconnected economy, as this connection is to the environment in such a way that not everyone uses it, creating the full social support that no one leaves out the harmonious way everything is integrated into the real circular economic system. At a cut-off level, the proposed model achieved 94.97% of service recognition, 3.3% of service rejection, 94.06% of service accuracy, 95.86% of service precision, 94.34% of service recall, and 95.68% of F1-Score while compared with the existing models. Thus, the smart city is getting a greater number of quality services in digital, economic, ecological, and social fields. In future, we aim to improve the security using cryptography-based models in smart cities.

References

1. Singh S, Sharma PK, Yoon B, Shojafar M, Cho GH, Ra IH (2020) Convergence of blockchain and artificial intelligence in IoT network for the sustainable smart city. Sustain Cities Soc 63:102364
2. Ullah Z, Al-Turjman F, Mostarda L, Gagliardi R (2020) Applications of artificial intelligence and machine learning in smart cities. Comput Commun, 154, 313-323
3. Allam Z, Dhunny ZA (2019) On big data, artificial intelligence, and smart cities. Cities 89:80–91
4. Chui KT, Lytras MD, Visvizi A (2018) Energy sustainability in smart cities: Artificial intelligence, smart monitoring, and optimization of energy consumption. Energies 11(11):2869
5. Zhang Y, Geng P, Sivaparthipan CB, Muthu BA (2021) Big data and artificial intelligence based early risk warning system of fire hazard for smart cities. Sustainable Energy Technol Assess 45:100986
6. Ahad MA, Paiva S, Tripathi G, Feroz N (2020) Enabling technologies and sustainable smart cities. Sustain Cities Soc 61:102301

7. Li X, Huang K, Xu L (2022) Hybrid model of machine learning refractory data prediction based on IoT smart cities. Wirel Commun Mob Comput
8. Muhammad K, Lloret J, Baik SW (2019) Intelligent and energy-efficient data prioritization in green smart cities: Current challenges and future directions. IEEE Commun Mag 57(2):60–65
9. Huang C, Nazir S (2021) Analyzing and evaluating smart cities for IoT based on use cases using the analytic network process. Mob Inf Syst
10. Ghadami N, Gheibi M, Kian Z, Faramarz MG, Naghedi R, Eftekhari M, Tian G (2021) Implementation of solar energy in smart cities using an integration of artificial neural network, photovoltaic system, and classical Delphi methods. Sustain Cities Soc 74:103149
11. Yigitcanlar T, Desouza KC, Butler L, Roozkhosh F (2020) Contributions and risks of artificial intelligence (AI) in building smarter cities: Insights from a systematic review of the literature. Energies 13(6):1473
12. Mehta S, Bhushan B, Kumar R (2022) Machine learning approaches for smart city applications: emergence, challenges and opportunities. In Recent Advances in Internet of Things and Machine Learning (pp 147–163). Springer, Cham
13. Ghazal TM, Hasan MK, Alshurideh MT, Alzoubi HM, Ahmad M, Akbar SS, Akour IA (2021) IoT for smart cities: Machine learning approaches in smart healthcare—A review. Future Internet 13(8):218
14. Adams D, Novak A, Kliestik T, Potcovaru AM (2021) Sensor-based big data applications and environmentally sustainable urban development in internet of things-enabled smart cities. Geopolit, Hist, Int Relats 13(1):108–118
15. Said O, Tolba A (2021) Accurate performance prediction of IoT communication systems for smart cities: An efficient deep learning-based solution. Sustain Cities Soc 69:102830
16. Jiang L (2021) Evolution of regional economic spatial structure based on IoT and GIS Service. Wirel Commun Mob Comput
17. Rjab AB, Mellouli S (2018, May) Smart cities in the era of artificial intelligence and internet of things: literature review from 1990 to 2017. In Proceedings of the 19th Annual International Conference on Digital Government Research: Governance in the Data Age (pp. 1–10)
18. Zahmatkesh H, Al-Turjman F (2020) Fog computing for sustainable smart cities in the IoT era: Caching techniques and enabling technologies-an overview. Sustain Cities Soc, 59, 102139
19. Mishra KN, Chakraborty C (2020) A novel approach toward enhancing the quality of life in smart cities using clouds and IoT-based technologies. In Digital Twin Technologies and Smart Cities (pp. 19–35). Springer, Cham
20. Sharma I, Garg I, Kiran D (2020) Industry 5.0 and smart cities: A futuristic approach. Eur J Mol & Clin Med, 7(08), 2515–8260
21. Wu J, Zhang J, Xiao Y, Ji Y (2021) Cooperative offloading in D2D-enabled Three-Tier MEC networks for IoT. Wirel Commun Mob Comput
22. Mu J, Tan Y, Xie D, Zhang F, Jing X (2021) CNN and DCGAN for spectrum sensors over rayleigh fading channel. Wirel Commun Mob Comput
23. Sodhro AH, Pirbhulal S, Luo Z, De Albuquerque VHC (2019) Towards an optimal resource management for IoT based Green and sustainable smart cities. J Clean Prod 220:1167–1179
24. Kummar S, Bhushan B, Bhatia S (2022). Blockchain based big data solutions for internet of things (IoT) and smart cities. In New Trends and Applications in Internet of Things (IoT) and Big Data Analytics (pp 225–253). Springer, Cham
25. Geno Peter A, Sherine Y, Teekaraman R, Kuppusamy A Radhakrishnan (Mar. 2022) Histogram shifting-based quick response steganography method for secure communication. Wirel Commun Mob Comput, 2022, pp 1–11, https://doi.org/10.1155/2022/1505133
26. Bhushan B, Kadam K, Parashar R, Kumar S, Thakur AK (2022) Leveraging blockchain technology in sustainable supply chain management and logistics. In Blockchain Technologies for Sustainability (pp 179–196). Springer, Singapore
27. Saxena S, Bhushan B, Ahad MA (2021) Blockchain based solutions to secure IoT: background, integration trends and a way forward. J Netw Comput Appl 181:103050
28. Chenniappan M, Gnanavel D, Gunasekaran KP, Rajalakshmi RR, Ramya AS, Stonier AA, Peter G, Ganji V (2022) Prediction of fault occurrences in smart city water distribution system

using time-series forecasting algorithm. Math Probl Eng 2022:e9678769. https://doi.org/10.1155/2022/9678769
29. Haque AB, Bhushan B, Dhiman G (2022) Conceptualizing smart city applications: Requirements, architecture, security issues, and emerging trends. Expert Syst 39(5):e12753
30. Hoang AT, Nguyen XP (2021) Integrating renewable sources into energy system for smart city as a sagacious strategy towards clean and sustainable process. J Clean Prod 305:127161
31. Lv Z, Chen D, Li J (2021) Novel system design and implementation for the smart city vertical market. IEEE Commun Mag 59(4):126–131
32. Kashef M, Visvizi A, Troisi O (2021) Smart city as a smart service system: Human-computer interaction and smart city surveillance systems. Comput Hum Behav 124:106923
33. Ageed ZS, Zeebaree SR, Sadeeq MM, Kak SF, Rashid ZN, Salih AA, Abdullah WM (2021) A survey of data mining implementation in smart city applications. Qubahan Acad J 1(2):91–99
34. Chu Z, Cheng M, Yu NN (2021) A smart city is a less polluted city. Technol Forecast Soc Chang 172:121037
35. Jasim NA, TH H, Rikabi SA (2021) Design and implementation of smart city applications based on the internet of things. Int J Interact Mob Technol, 15(13)

Empowering Artificial Intelligence of Things (AIoT) Toward Smart Healthcare Systems

Ayasha Malik, Veena Parihar, Bhawna, Bharat Bhushan, and Lamia Karim

Abstract The internet's workings have undergone a transformation from an Internet of Computers (IoC) to an Internet of Things (IoT). Additionally, as a result of the integration of numerous aspects like infrastructure, embedded technology, smart objects, people, and physical environments, massively interconnected systems, often referred to as Cyber-Physical Systems (CPSs), are forming. The next smart revolution may be brought about by IoT and CPS combined with Data Science (DS). The challenge posed by the rapid expansion of data generation is how to effectively handle and process the large volumes of information, given the current limitations in computational capacity. An answer to this issue has shifted the focus of researchers over DS and Artificial Intelligence (AI). IoT with AI can therefore lead to a significant and advanced role in the development of the smart world. This is not just cost-effective, smart work, less human interaction, or following any smart activity for a better and developed world, it is more than that; it is enhancing human lives. But it has been seen that IoT is facing many issues, such as safety and ethical dilemmas. In the end, what matters is how the general public views IoT with AI—whether they see it as a benefit, a burden, or a threat.

Keywords Artificial Intelligence · Healthcare · Internet of Things · Artificial Intelligence of Things

A. Malik (✉)
Delhi Technical Campus (DTC), GGSIPU, Greater Noida, India
e-mail: ayasha07.am@gmail.com

V. Parihar · Bhawna
KIET Group of Institutions, Delhi-NCR, Ghaziabad, India

B. Bhushan
School of Engineering and Technology (SET), Sharda University, Greater Noida, India

L. Karim
Hassan 1St University, National School of Applied Sciences of Berrechid (ENSA), Berrechid, Morocco

B. Bhushan et al. (eds.), *AI Models for Blockchain-Based Intelligent Networks in IoT Systems*, Engineering Cyber-Physical Systems and Critical Infrastructures 6,
https://doi.org/10.1007/978-3-031-31952-5_6

1 Introduction

Despite our fascination with the term "smart," current technology falls short of human intelligence. For instance, a smartphone may be considered smart, but it lacks the ability to automatically adjust settings, such as silencing notifications when the user is driving. To minimize distractions, a connection between the person, their smartphone, and their car is necessary. Similarly, in the event of illness, a smartphone should be able to call an emergency contact or nearby hospital, which would require information about family members and hospitals. We see that for various scenarios, physical objects need to be connected and possess AI to truly be deemed "smart" [1, 2].

The World Health Organization states that an aging population is becoming a widespread issue in both developed and developing countries, causing economic and social difficulties. For instance, in Japan, 27.6% of citizens are 65 or older. In the era of Human-Centred AI (HAI), AI and the IoT have been utilized to make significant advances in healthcare and health monitoring [3]. The combination of these two factors in the Fourth Industrial Revolution has been referred to as Artificial Intelligence of Things (AIoT). By utilizing ubiquitous sensors and advanced signal processing and machine learning techniques, promising achievements have been made [4]. AI and IoT can provide a better quality of life for individuals with chronic illnesses and those who require special care, especially the elderly population. In response to the global COVID-19 pandemic, AIoT has the potential to offer eldercare solutions by utilizing smart home sensor data. The goal of AI is to create computer systems that exhibit human-like behavior and its progression will drive the digital revolution across a range of industries. By connecting all things—be they people, animals, plants, machines, etc.—and enabling them to make smart decisions, we can create a self-sufficient world. For complete self-governance, the system must incorporate machine learning, which imitates human learning, and a data analysis component [5]. Machine Learning (ML) would develop techniques to allow different components and devices within the network to learn and function autonomously. Data Analysis (DA), on the other hand, would examine generated data to determine past patterns and enhance future efficiency. The integration of ML and DA into the sensors and embedded systems of smart systems is becoming more widespread. AI's potential is fascinating and prompts us to reevaluate our perspectives on life and work. As ML and DA continue to drive AI at a rapid pace, it's important to address the growing trends, challenges, and threats [6].

As the world becomes more connected, the idea of an IoT and Internet of Everything (IoE) is becoming increasingly popular, where every living and non-living object is interconnected [7]. This leads to a concept of a CPS, which generates large amounts of data that needs to be processed and analyzed. To handle this data, various fields such as database management, pattern recognition, data mining, machine learning, and big data analytics need to improve their methods and overlap largely in their scope [8–10].

Furthermore, the rest of the paper is organized as follows Sect. 2 discussed the role and impact of the trending technology named AI with its various applications and scope. Additionally, Sect. 3 enlightens the importance of IoT in today's world along with its features and various use cases. Section 4 deliberates the need and benefits of combining AI with IoT to make it AIoT along with various existing examples of it. Moreover, Sect. 5 highlights the various challenges and needs of future research into this growing technology named AIoT. Finally, the paper concludes with Sect. 6.

2 Artificial Intelligence

The birth of AI marked a new era in computer programming, aimed at creating programs that enable computers to learn. Many scientists sought to equip computers with intelligent programs that could replicate, learn and regulate their surroundings, essentially modeling the human brain, imitating human learning, and replicating biological evolution. Despite the fact that computers can only understand binary code, AI has provided a new avenue for advancing the capabilities of these machines [11]. The concept of AI and the pursuit of equipping computers with the ability to learn and imitate human intelligence dates back to the post-World War II era. Alan Mathison Turing is widely considered the pioneer in this field, publishing a seminal article on the topic in 1950 [12]. The article delved into the concept of the Turing Test, a method to determine if a machine has achieved a level of intelligence that can be mistaken for human intelligence. Turing's ideas laid the foundation for the creation of AI systems, starting with the development of a "child machine" that would learn and grow into an "adult machine". Over the years, many more researchers joined the field, contributing to the growth and advancement of AI technology [13]. Furthermore Fig. 1 shows the various boxes of AI.

There are various definitions and theories in psychology and cognitive science. For example, some definitions of intelligence refer to the ability to solve complex problems, adapt to new situations, learn from experience, understand abstract concepts, and use reasoning and logic. In the field of AI, there are several approaches and techniques, such as ML, DL, expert systems, and robotics, which are used to create intelligent machines. Despite the progress that has been made, creating truly intelligent machines that can equal or surpass human intelligence is still a major challenge and a subject of ongoing research and debate [14]. The goal of Artificial Intelligence is to create machines with human-like intelligence, allowing them to comprehend human language, solve problems, and reach objectives. While some believe that the ultimate goal of AI is to replicate human consciousness in machines, intelligence can be defined as the computational aspect of achieving objectives, which exists in different forms and degrees in humans, animals, and machines. The Turing Test is one way to assess the level of machine intelligence [15]. AI can be classified into various subfields, including ML, Computer Vision, Image Processing, DL, Fuzzy Logic, smart agents, etc. as shown in Fig. 2. These branches aim to make machines more intelligent, capable of understanding human language, solving problems, and achieving goals like a human beings. Each of these branches has a different approach and application to achieve AI's overarching goal.

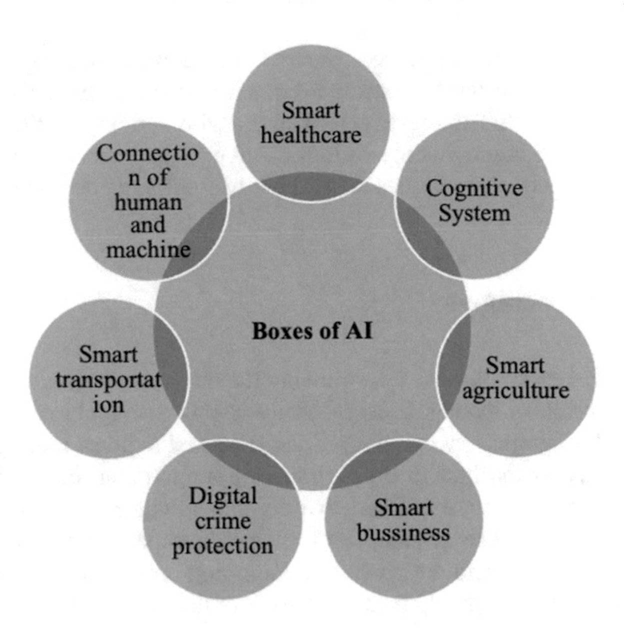

Fig. 1 Boxes of AI

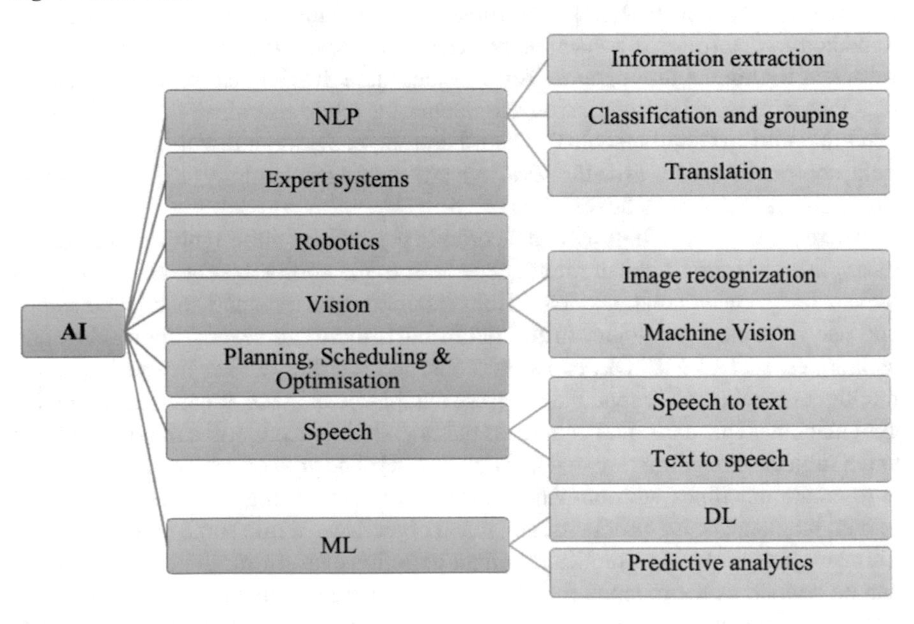

Fig. 2 Branches of AI

The following points state six rules that should be followed when designing any AI application.

- AI needs to be created to help people, in order to protect human employees, it is necessary to respect human autonomy by deploying collaborative robots to carry out hazardous tasks like mining.
- AI must be open and honest, understanding the rules of the technology. Or, in the words of Satya Nadella, "Humans should know about machines, but technology will know things about humans." This would make it possible for people to comprehend how technology perceives and interprets the world because ethics and design must coexist.
- Without undermining human dignity, AI must maximize efficacy. It has to foster diversity while preserving cultural commitments. Because technology shouldn't determine the future's morals or virtues, a wider, deeper, and more diverse commitment to the populace is required [16–18].
- AI needs to be built with intelligent privacy in mind. To ensure personal information is secure and to earn trust, sophisticated protection measures are required.
- AI must be algorithmically responsible so that humans may repair any harm that was not intended. Designing AI technology for both expected and unexpected outcomes is therefore crucial.
- AI must ensure sufficient and representative research to prevent the use of an inaccurate heuristic for discrimination [19–23].

3 Internet of Things

The IoT refers to a network of interconnected devices equipped with sensors, software, and other technologies that allow them to exchange data with other systems over the Internet [24]. There are currently over 7 billion IoT devices, and this number is expected to grow to 10 billion by 2020 and 22 billion by 2025. The emergence of IoT has revolutionized communication, allowing for continuous exchange between people, processes, and things [25]. Devices like household appliances, cars, thermostats, and baby monitors can now be connected to the Internet through embedded devices, thanks to low-cost computing, cloud technology, big data, analytics, and mobile technology. The combination of the physical and digital world has allowed for capturing, observing, and modifying every interaction between connected objects in today's hyper-connected environment. The concept of IoT has existed for a while, but only recent advancements in technology have made it a reality [26].

- **Cost-effective, power-effective sensor**: Affordable and dependable sensors are enabling more manufacturers to adopt IoT technology. The cost-effective and power-efficient sensors are a critical aspect of making IoT a reality.

- **Connection**: The presence of various internet network protocols has made it easier to link sensors to the cloud and other devices, leading to efficient data transmission.
- **Availability of cloud computing**: Advancements in cloud computing have facilitated business growth and made it easier for companies and consumers to access the infrastructure they need, without the burden of managing it themselves. The increased availability of cloud platforms has made it possible to access the resources needed for expansion.
- **Admittance of ML and data analytics**: Admittance of ML and data analytics has greatly impacted the IoT by allowing businesses to quickly gather insights from huge amounts of data, stored over the cloud. These advancements, along with access to the cloud, have enabled businesses to scale up their operations without having to manage the infrastructure themselves. The combination of these technologies is constantly pushing the limits of IoT and feeding its growth.
- **Conversational AI**: The integration of conversational AI has made IoT devices, like digital personal assistants Alexa, Cortana, and Siri, more accessible, affordable, and attractive for household use. This is due to advancements in neural networks, which have enabled NLP to be incorporated into these devices [27, 28].

3.1 Use Cases of IoT

The integration of sensors and computing into everyday objects has led to the creation of the IoT. This network of connected devices enables constant communication and data exchange between physical objects, processes, and people. With the growth of cloud computing and big data analytics, the IoT has become even more powerful as companies and individuals can access vast amounts of data stored in the cloud and gain valuable insights quickly with the help of ML [29].

3.1.1 Smart and Associated Vehicles

Various means, such as infotainment systems or an in-car connected gateway, can connect smart cars to the internet. This link makes it possible to collect a range of data points, including data on the brakes, accelerator, wheels, speedometer, and fuel tank, to keep an eye on both the performance of the driver and the state of the vehicle. This not only improves the driving experience but also makes it easier to proactively identify and fix any potential car problems. The capabilities and features of smart vehicles are anticipated to keep growing as technology progresses [30].

3.1.2 Smart and Linked Homes

Smart home devices have been designed with the prime goal of enhancing the overall efficiency, safety, and connectivity of the household. One example is smart outlets, which monitor electricity usage and help reduce energy waste. Intelligent thermostats, on the other hand, ensure optimal room temperature control by learning and adapting to the homeowner's preferences. IoT sensors can be integrated into hydroponic systems, enabling gardeners to manage their indoor gardens with ease. Additionally, IoT smoke detectors can detect the presence of tobacco smoke and provide an early warning to residents, ensuring their safety. Smart home security systems have revolutionized the way homes are protected from potential threats. These systems, such as door locks, CCTV cameras, and water tank leak detectors, are equipped with advanced features that can detect and prevent dangers and send real-time notifications to homeowners. Additionally, the interconnected devices can be utilized for various other purposes, such as automatically shutting off unused appliances, managing rental properties and locating misplaced articles such as keys, and automating daily tasks like cleaning and coffee-making. These features bring both convenience and peace of mind to homeowners and renters alike [31].

3.1.3 Emerge and Smart Cities

Smart city initiatives aim to leverage IoT to improve the quality of life for citizens, reduce energy consumption, minimize waste, and create more sustainable urban environments. The deployment of smart sensors, wireless networks, and cloud computing technologies in cities enables continuous monitoring, analysis, and control of urban systems, resulting in optimized resource utilization, reduced downtime, and lower costs. Ultimately, the goal of smart cities is to create more livable, sustainable, and resilient urban environments for their citizens. Some of the applications of IoT in these areas include:

- Measuring air quality and radiation levels to monitor the environment and maintain public health.
- Implementing smart lighting systems to reduce energy bills and conserve resources.
- Monitoring the condition of critical infrastructures such as streets, bridges, and pipelines to detect maintenance needs and prevent potential disasters.
- Improving profits through the implementation of efficient parking management systems, reducing costs, and increasing revenue [32, 33].

3.1.4 Connected Street Lights

Connected streetlights are a significant advancement in urban planning and infrastructure. Historically, the primary purpose of street lighting poles in urban areas was to provide visibility and safety at night. However, with the integration of IoT technology, streetlights can serve a much wider range of applications and opportunities. By connecting the lighting systems and poles to the Internet, cities can establish a foundation for smart city initiatives, providing a platform for numerous smart city projects. This innovative technology opens up new possibilities for smart cities, allowing them to leverage connected street lights to enhance their infrastructure and improve their communities [34].

3.1.5 Smart IoT for Research

Smart IoT technology is revolutionizing research and conservation efforts aimed at saving endangered species. Scientists, researchers, and conservationists are using IoT devices to study the behavior and habitat of animals, in an effort to understand the crucial factors that impact their survival. The natural habitats of many species are under threat due to factors such as climate change and disruption in the food chain, leading to declining populations and even extinction. In order to save these species, it is crucial to gain a comprehensive understanding of their behavior and living patterns. By utilizing IoT devices to track the animals' travel habits and food-sourcing locations, scientists can gather crucial information that may help in their conservation efforts and ensure the survival of endangered species [35].

3.1.6 Industrial Use Cases

The utilization of IoT technologies in the manufacturing industry is referred to as the Industrial IoT i.e. IIoT. To understand the application of IoT in a real-world setting, let's consider a smart factory scenario. An industrial company is responsible for manufacturing large machinery equipment at multiple factories located globally. Each day, the factory produces 200 pieces of machinery using a welding machine as a crucial component of the manufacturing process. The integration of IoT technology in this scenario enhances the efficiency and reliability of the manufacturing operations, streamlining the production process and ultimately improving the overall competitiveness of the industrial company [36].

3.1.7 Smart Grid Automation

The integration of IoT technology plays a crucial role in the automation and optimization of smart power grids. The increasing complexities of the grid, such as the integration of renewable energy sources and stringent regulations, demand efficient management and maintenance of smart grid systems. With the ability to detect and respond quickly to faults, the smart grid can ensure a reliable and stable power supply. The variability in renewable energy production and the challenges in integrating it into the microgrid makes real-time control and management of the generated energy essential. To achieve this, a fast and reliable signaling system is necessary between all micro-grid nodes, allowing for seamless communication and control of the grid's energy distribution. IoT technology plays a key role in ensuring the smart grid operates optimally to meet the energy demands of the future.

3.1.8 Collaborative Robots

IoT technology plays a crucial role in the orchestration and coordination of arm and mobile robots, allowing for the autonomous completion of various tasks. These robots require numerous sensors and advanced processors to function independently. However, when several robots must work together to accomplish a complex task, a centralized management system and increased processing power are necessary. It's important to note that incorporating the necessary processing power into these robots can be cost-prohibitive, especially for smaller robots. The use of wired IoT connectivity may also present challenges in mobile environments, such as with Automated Guided Vehicles (AGVs) [37].

4 Combination of AI and IoT

The integration of AI and the IoT has caused a major disruption in the technological landscape, leading to a new era of innovation. Both of these technologies augment each other and fully leverage their data and technical strengths to bring a novel experience to people's daily lives. The IoT, being a massive network of interconnected sensors, actuators, data storage, and processing capabilities, can sense its surroundings, collect and transmit data, store it, process it, and then take actions based on the processed data [38]. Additionally, Fig. 3 shows the various stages of implementing AIoT technology.

The ability of an IoT service to process and take actions based on the gathered data is ultimately what determines how effective the service is. Lacking AI capabilities, an IoT system is constrained in its operation and unable to advance with the data, but with AI integration, it acquires the capacity to automate and adapt. When AI and IoT are combined, an Intelligent IoT is created, which not only allows for data interchange but also uses AI technology to transform raw data into useful information. The impact of AIoT can be seen across various domains, from smart homes to industrial automation [39]. Additionally, Fig. 4 shows the features of AIoT.

4.1 Limitations

- **Privacy and security concerns**: One of the biggest challenges facing AIoT is the security of connected components and the protection of sensitive data. With the

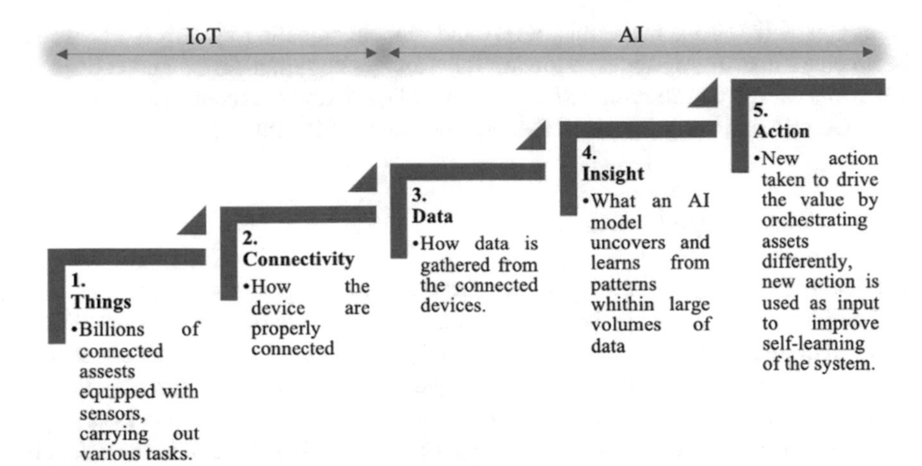

Fig. 3 Stages of AIoT

Fig. 4 Features of AIoT

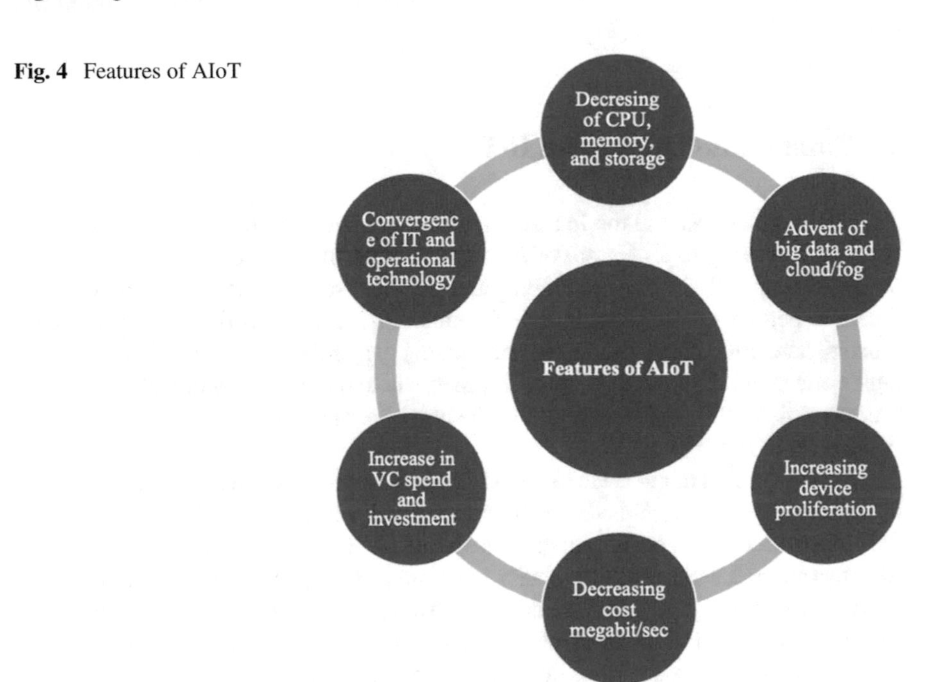

increasing number of connected devices, there is a growing risk of cyber-attacks and data breaches.

- **Interoperability**: Interoperability refers to the ability of different devices and systems to work together seamlessly and effectively. In the context of AIoT, devices from different manufacturers may use varying communication protocols and standards, which poses a challenge to the seamless exchange of data and information between them.

- **Accuracy and reliability**: The accuracy and reliability of AIoT systems are also a concern. Machine learning algorithms may not always produce the correct results, and the performance of AIoT systems can be affected by various factors such as data quality, training data, and hardware limitations.
- **Ethical Concerns**: The use of AI and IoT raises ethical concerns such as bias, accountability, and transparency. The algorithms used in AIoT systems may perpetuate biases that exist in the training data, and it can be difficult to determine who is responsible when things go wrong.

4.2 Perspectives

- **Edge computing**: The future of AIoT will heavily rely on edge computing. Edge computing makes it possible for data processing to take place near the data's origin, cutting down on the quantity of data that must be transferred over the network and lowering latency.
- **5G**: The introduction of 5G technology will bring significant improvements to the AIoT. 5G offers low latency, high bandwidth, and increased reliability, making it possible for AIoT devices to communicate in real time and transmit large amounts of data.
- **Integration with other technologies**: AIoT will continue to integrate with other technologies such as blockchain and augmented reality, creating new opportunities and solutions.
- **Industry applications**: AIoT has a huge potential for use in various industrial sectors, from manufacturing and supply chain management to healthcare and transportation. AIoT-based solutions can help organizations optimize processes, improve efficiency, and make more informed decisions.

4.3 Use Cases of AIoT

A few examples of existing IoT services, powered by AI, have been mentioned in the literature to give an idea of the practical implications of this combination.

4.3.1 Voice Assistants

Voice assistants are virtual personal assistants that can be accessed through cloud-based voice services and are used to perform various tasks with just a voice command. These tabletop assistants may communicate with other nearby smart devices and third-party applications to respond to questions, book reservations at restaurants, make cab calls, play music, turn on/off smart lighting, and do a variety of other tasks.

- Alexa, developed by Amazon, is one of the most well-known voice assistants. It is integrated into popular devices such as Amazon Echo and Amazon Tap and can be personalized through the Alexa Skills Kit (ASK), which is a set of skills that can be updated and modified to improve certain functionalities.
- Apple Inc. has its voice assistant named Siri, which is used in the Apple Homepod. Similar to Alexa, Siri can be used to perform various tasks through voice commands.
- Google Assistant, which is used in Google Home, has the added advantage of recognizing up to six different users and accessing their details to converse with them. This makes it a highly personalized experience for each user [40].

The seamless execution of tasks is made possible through the utilization of various subfields of AI. These processes work in tandem to provide real-time results and a smooth user experience. This not only makes it easier for users to get things done with just their voice but also opens up a world of possibilities for the future of smart homes and personal devices [41].

4.3.2 Human-Like Robots

These most recent developments in robotics have paved the way for the development of more human-like robots that can converse with people and understand and convey emotions. With numerous sensors, actuators, and AI that enables continuous learning and adaptation, these robots are essentially IoT devices.

- Pepper, developed by SoftBank Robotics, is a human-like robot designed as a companion for humans. It uses a combination of facial expressions, body movement, tone of voice, and language to understand human emotions and respond appropriately through movement, touch, words, and displays on its screen. This robot can identify four basic human emotions, such as joy, sadness, anger, and surprise [42].
- Sophia, created by Hanson Robotics, is a highly human-like social humanoid robot that is able to express a wide range of emotions through over 50 different facial expressions. During a conversation, Sophia maintains eye contact with the person she is talking to and has even made appearances in interviews and concerts. This robot is unique in that it is the world's first to receive full citizenship from a country [43].

4.3.3 Smart Devices

In addition to voice assistants and robots, there are also smart devices in the IoT ecosystem that are designed to make everyday tasks more convenient for humans. These devices are often AI-enabled and utilize various AI technologies such as object recognition, facial recognition, voice recognition, speech and expression analysis, deep neural networks, transfer learning, and computer vision.

- One example of a smart device that utilizes AI technology is the June Smart Oven. This oven has an HD camera and a food thermometer, so it can automatically check on the food while it cooks and change the cooking settings as needed. Alexa may be used to manage the oven, and the appliance can even suggest recipes depending on the user's tastes.
- The Honeywell SkyBell HD Wi-Fi doorbell is another illustration. With this gadget, the user can answer the door using their voice assistant or smartphone. A notification and live feed are delivered to the homeowner's phone when the doorbell is pressed, allowing them to converse with the visitor from a remote location. This added security feature helps to deter intruders and burglars [44–46].

4.3.4 Industrial IoT

The integration of IoT and AI has paved the way for a range of innovative solutions that can benefit industries across various sectors. By collecting and analyzing vast amounts of data, these solutions can provide insights that can help companies improve their operations and make informed decisions.

- One such example is Primer, a product from Alluvium. It creates real-time stability score analysis by collecting data from various sources and utilizing sensors in the system. This solution is designed to detect potential issues before they become serious and assist operators in making necessary changes.
- Another example is Plutoshift, an industrial IoT solution that enables companies to monitor the performance of their assets and measure the financial impact of these assets. With this information, companies can make more informed decisions that can help improve their operations and increase their overall efficiency [47].

It is clear that the combination of IoT and AI has tremendous potential, as IoT generates vast amounts of data and AI has the ability to extract valuable insights from this data. In other words, without AI, the data generated by IoT would remain largely useless. The future potential of both IoT and AI is immense and the possibilities are endless when they are integrated [48].

5 AIoT for Healthcare

It is very likely to boost the practical efficacy of AI and IoT combined in the creation of smart healthcare. Tracing, monitoring, controlling, optimizing, and automation is key phases in the execution of advanced and efficient AI technology in IoT devices. Together, they are able to relieve the workload for the patients and the hospital's management or personnel as a whole. In addition, doctors can devote more time and

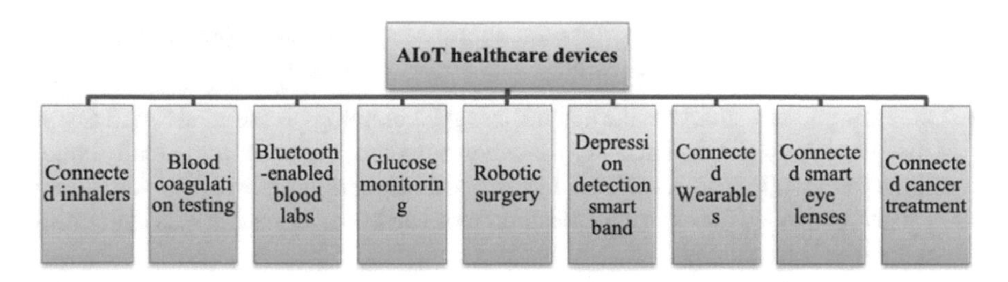

Fig. 5 AIoT-enabled smart healthcare devices

attention to patients with better hospital staff productivity in managing all workflows, resulting in a more patient-centered approach to healthcare delivery.

By giving physical items and devices the ability to see, hear, and think as well as "speak" and share information to express their decisions, AI and IoT connect physical objects and gadgets. By linking previously inanimate things to the internet via a variety of embedded devices, communication protocols, sensor networks, internet protocols, and applications, AIoT technology transforms them into sentient beings. The medical sector uses a number of AIoT-based healthcare services, such as diagnostic, preventative, rehabilitative, and monitoring tools, as well as electronic health and telecare networks. IoT technology includes radio frequency identification systems and wireless body area networks, but they are not necessary. Remote health monitoring is technically possible, according to research in related fields, but the advantages under various conditions are far larger. By monitoring non-critical patients at home instead of at the hospital, remote health monitoring could ease the burden on hospital resources like beds and doctors. Facilitating remote healthcare access might enable older individuals to remain in their homes for longer. In essence, it offers a chance to lessen the burden on healthcare institutions, enhance access to professional treatments, and give people more control over their health. Figure 5 displays the many AIoT-based medical devices available.

The field of Artificial Intelligence of Things (AIoT) is rapidly expanding, with new applications emerging in various domains such as healthcare, entertainment, and daily life. In particular, AIoT is playing an increasingly important role in assistive technologies aimed at enhancing the quality of life of the elderly. This chapter provides an overview of the current state-of-the-art works in the AIoT for elderly-care applications and discusses the current challenges and future perspectives.

5.1 Cutting-Edge Works

The AIoT field for eldercare applications is relatively new, but current studies have shown promising results. In terms of Fall Detection (FD), audio and video-based methods have achieved high FD sensitivity, exceeding 95%. By utilizing advanced signal processing techniques, such as background subtraction algorithms and codebooks, these models can perform well even in different environments. In addition, the use of Wavelet Transform (WT) can aid in extracting features from sensor and radar data, which provides a multi-resolution analysis of non-stationary signals related to falls.

For Activity Recognition (AR), most studies have achieved a Weighted Accuracy Rate (WAR) of over 90% for recognizing the daily activities of elderly individuals. Non-wearable sensors have shown great performance in this area and avoid privacy issues associated with audio and video-based methods, as well as causing minimal inconvenience for elderly users. Human behavior analysis has a significant impact on the improvement of activity recognition models, and factors such as activity encoding, similarity, and context information provide a better representation of activities compared to low-level descriptors [49–51].

5.2 *Limitations and Perspectives*

Despite the numerous advantages and potential solutions offered by AIoT, there are still a number of security challenges and limitations that need to be addressed.

5.2.1 Previous Study

Previous research on the AIoT technology has shown promising results in various fields. However, it is important to consider that these studies were conducted under controlled conditions, often in a laboratory setting. For example, in the FD task, the data collected was performed by actors in a laboratory, which may not reflect real-world situations. Thus, in order to effectively implement these methods in the real world, further research is needed on data collection in real-world environments. Additionally, some studies lack independence in their experiments, which can affect the models' ability to generalize to a larger population. The widely used WAR is also not an appropriate evaluation metric for imbalanced datasets, and unweighted measures should be used instead. These limitations in previous research may result in overoptimistic results and expectations [52].

5.2.2 Improper Examination of Data

One of the major challenges in AIoT for elderly care is the limited examination of data modalities that can be used in these applications. While some studies have explored specific data modalities, there is a need to investigate other sensory inputs that could provide additional benefits [53]. For example, haptics (the sense of touch) could play a crucial role in facilitating daily activities and improving rehabilitation outcomes through the use of AI-powered prostheses. It is also important for AIoT designers to consider factors such as power consumption and computational capacity when developing real-world applications, as these can greatly impact system performance and cost [54].

5.2.3 Data Scarcity

In the realm of AIoT for aged care, notably for fall detection, the lack of data is a recurring problem. More sophisticated technologies must be used, even though data augmentation has shown to be useful in enhancing the generalization of deep learning models. Due to the scarcity of labeled data, techniques including unsupervised learning, semi-supervised learning, active learning, reinforcement learning, and their variations, should be investigated. In addition, despite the rising number of sensors, gathering annotated data from human specialists might still be expensive and difficult [55]. In order to create fresh samples with a comparable distribution to the original dataset and to obtain more reliable high-level representations of the data, generative adversarial networks can be used. In real-world applications, signal augmentation is also essential, especially in smart buildings where IoT-based micro-location is applied. To combat the impacts of wireless device interference, Kalman filtering is a useful technique. Finding more effective data fusion solutions to handle the difficulty of utilizing data from various sensors is made possible by the growing growth of IoT devices [56].

5.2.4 Correct Integration of Various Technologies

When it comes to incorporating ML into AIoT eldercare applications, it is important to consider how to make the models more robust and generalizable. As an illustration, a recent study has demonstrated that a noisy parallel hybrid DL model architecture can result in an accurate and dependable prediction of remaining useful life [57]. Greater consideration should be given to more sophisticated DL architectures given the significance of sociological data in this area. According to research, an encoder-decoder temporal convolutional network model can make predictions that are more accurate and stable when compared to other framewise or sequential models when analyzing electromyographic signals [58]. This finding highlights the importance of the structure of DL models. Temporal information can be used for higher-level tasks like early disease detection and emergency alarms since it contains crucial information about an aged person's behavior and daily routine. Fundamental research on particular tasks is still lacking, nevertheless [59, 60].

6 Conclusion

The current era is characterized by big data and machine intelligence, owing to the advancements in technology, particularly artificial intelligence. People are heavily dependent on the Internet and Internet-enabled devices, producing vast amounts of data every day. However, without effective processing and analysis, the data generated are not put to their full use. The development of AI has made it possible to effectively and promptly analyze large amounts of data, increasing efficiency. The IoT is greatly influenced by globalization and the technological revolution, and its development is closely tied to the advancement of intelligent technology. To keep pace with the rapidly evolving technological landscape, it is important to harness the power of science and technology and stay updated with the latest developments. In conclusion, AIoT is an emerging technology that has the potential to bring significant benefits to various fields, including elderly care. By integrating artificial intelligence and the internet of things, AIoT can help to provide intelligent, data-driven, and automated solutions for elderly care. Despite the potential benefits, some challenges need to be addressed, such as improper examination of data, data scarcity, and the correct integration of various technologies. Nevertheless, these challenges can be overcome by continued research and innovation in the field. The development of AIoT is deeply interconnected with the advancement of the Internet of Things and artificial intelligence technology. As such, it is important for developers to continuously explore new possibilities and solutions in this field and to remain aware of their responsibility to serve the people and society through the application of AIoT.

References

1. Ghosh A, Chakraborty D, Law A (2018) Artificial intelligence in the Internet of things. CAAI Trans Intell Technol 3(4):208–218
2. González García C, Núñez Valdéz ER, García Díaz V, Pelayo García-Bustelo BC, Cueva Lovelle JM (2019) A review of artificial intelligence in the internet of things. Int J Interact Multimed Artif Intell 5
3. Kharya S, Onyema EM, Zafar A, Wajid MA, Afriyie RK, Swarnkar T, Soni S (2022) Weighted Bayesian belief network: a computational intelligence approach for predictive modeling in clinical datasets. Comput Intell Neurosci
4. Mohamed E (2020) The relation of artificial intelligence with internet of things: a survey. J Cybersecur Inf Manage 1(1):30–24
5. Gupta PK, Siddiqui MK, Huang X, Morales-Menendez R, Pawar H, Terashima-Marin H, Wajid MS (2022) COVID-WideNet—a capsule network for COVID-19 detection. Appl Soft Comput 122:108780
6. Wu H, Han H, Wang X, Sun S (2020) Research on artificial intelligence enhancing internet of things security: a survey. IEEE Access 8:153826–153848
7. Yarlagadda RT (2018) Internet of things & artificial intelligence in modern society. Int J Creat Res Thoughts (IJCRT). ISSN 2320-2882
8. Wajid MA, Zafar A (2022) Neutrosophic image segmentation: an approach for the treatment of uncertainty in multimodal information systems. Int J Neutrosophic Sci

9. Zafar A, Wajid MA (2020) A mathematical model to analyze the role of uncertain and indeterminate factors in the spread of pandemics like COVID-19 using neutrosophy: a case study of India, vol 38. Infinite Study
10. Kishor A, Chakraborty C (2022) Artificial intelligence and internet of things based healthcare 4.0 monitoring system. Wirel Personal Commun 127(2):1615–1631.
11. Radanliev P, De Roure D, Nicolescu R, Huth M, Santos O (2021) Artificial intelligence and the Internet of Things in industry 4.0. CCF Trans Pervasive Comput Interact 3:329–338
12. Tien JM (2017) Internet of Things, real-time decision making, and artificial intelligence. Ann Data Sci 4:149–178
13. Malik A, Kumar A (2022) assimilation of blockchain with Internet of Things (IoT) with possible issues and solutions for better connectivity and proper security. In: Sharma R, Sharma D (eds) New trends and applications in Internet of Things (IoT) and big data analytics. Intelligent systems reference library, vol 221. Springer, Cham. https://doi.org/10.1007/978-3-030-99329-0_13
14. Rana AK, Krishna R, Dhwan S, Sharma S, Gupta R (2019) Review on artificial intelligence with internet of things-problems, challenges and opportunities. In: 2019 2nd international conference on power energy, environment and intelligent control (PEEIC). IEEE, pp 383–387
15. Raman DR, Saravanan D, Parthiban R, Palani DU, David DDS, Usharani S, Jayakumar D (2021) A study on application of various artificial intelligence techniques on Internet of Things. Eur J Mol Clini Med 7(9):2531–2557
16. Kuzlu M, Fair C, Guler O (2021) Role of artificial intelligence in the Internet of Things (IoT) cybersecurity. Discov Internet Things 1:1–14
17. Khayyam H, Javadi B, Jalili M, Jazar RN (2020) Artificial intelligence and Internet of Things for autonomous vehicles. Nonlinear Approach Eng Appl Automot Appl Eng Prob 39–68
18. Poniszewska-Maranda A, Kaczmarek D (2015) Selected methods of artificial intelligence for Internet of Things conception. In: 2015 federated conference on computer science and information systems (FedCSIS), pp. 1343–1348. IEEE
19. Malik A, Bhushan B, Kumar A, Chaganti R (2022) Opportunistic Internet of Things (OIoT): elucidating the active opportunities of opportunistic networks on the way to IoT. In: Sharma R, Sharma D (eds) New trends and applications in Internet of Things (IoT) and big data analytics. intelligent systems reference library, vol 221. Springer, Cham. https://doi.org/10.1007/978-3-030-99329-0_14
20. Balas VE, Kumar R, Srivastava R (eds) (2020) Recent trends and advances in artificial intelligence and Internet of Things. Springer International Publishing, Cham
21. Iwendi C, Rehman SU, Javed AR, Khan S, Srivastava G (2021) Sustainable security for the internet of things using artificial intelligence architectures. ACM Trans Internet Technol (TOIT) 21(3):1–22
22. Song H, Bai J, Yi Y, Wu J, Liu L (2020) Artificial intelligence enabled Internet of Things: network architecture and spectrum access. IEEE Comput Intell Mag 15(1):44–51
23. Ghazal TM (2021) Retracted article: Internet of Things with artificial intelligence for health care security. Arab J Sci Eng 1–1
24. Esenogho E, Djouani K, Kurien AM (2022) Integrating artificial intelligence Internet of Things and 5G for next-generation smartgrid: a survey of trends challenges and prospect. IEEE Access 10:4794–4831
25. Subeesh A, Mehta CR (2021) Automation and digitization of agriculture using artificial intelligence and Internet of Things. Artif Intell Agricul 5:278–291
26. Malik A, Bhushan B (2022) Challenges, standards, and solutions for secure and intelligent 5G Internet of Things (IoT) scenarios, smart and sustainable approaches for optimizing performance of wireless networks: real-time applications. https://doi.org/10.1002/9781119682554.ch7
27. Shi Q, Dong B, He T, Sun Z, Zhu J, Zhang Z, Lee C (2020) Progress in wearable electronics/photonics—moving toward the era of artificial intelligence and Internet of Things. InfoMat 2(6):1131–1162

28. Chander B, Pal S, De D, Buyya R (2022) Artificial intelligence-based internet of things for industry 5.0. Artif Intell Based Internet Things Syst 3–45
29. Farrokhi A, Farahbakhsh R, Rezazadeh J, Minerva R (2021) Application of Internet of Things and artificial intelligence for smart fitness: a survey. Comput Netw 189:107859
30. Qian K, Zhang Z, Yamamoto Y, Schuller BW (2021) Artificial intelligence Internet of Things for the elderly: from assisted living to health-care monitoring. IEEE Signal Process Mag 38(4):78–88
31. Shi F, Ning H, Huangfu W, Zhang F, Wei D, Hong T, Daneshmand M (2020) Recent progress on the convergence of the Internet of Things and artificial intelligence. IEEE Netw 34(5):8–15
32. Chen WL, Lin YB, Ng FL, Liu CY, Lin YW (2019) RiceTalk: Rice blast detection using Internet of Things and artificial intelligence technologies. IEEE Internet Things J 7(2):1001–1010
33. Sepasgozar S, Karimi R, Farahzadi L, Moezzi F, Shirowzhan S, Ebrahimzadeh SM, Aye L (2020) A systematic content review of artificial intelligence and the internet of things applications in smart home. Appl Sci 10(9):3074
34. Kumar V, Malik A (2023) Heart disease prediction using machine learning. DTC J Computat Intell 1(2). https://jci.delhitechnicalcampus.ac.in/wp-content/uploads/2022/12/DTCJCI-4.pdf
35. Katare G, Padihar G, Qureshi Z (2018) Challenges in the integration of artificial intelligence and Internet of Things. Int J Syst Softw Eng 6(2):10–15
36. Oniani S, Marques G, Barnovi S, Pires IM, Bhoi AK (2021) Artificial intelligence for internet of things and enhanced medical systems. Bio-Inspired Neurocomput 43–59
37. Zhang L, Liang YC, Niyato D (2019) 6G visions: mobile ultra-broadband, super Internet-of-Things, and artificial intelligence. China Commun 16(8):1–14
38. Seng KP, Ang LM, Ngharamike E (2022) Artificial intelligence Internet of Things: a new paradigm of distributed sensor networks. Int J Distrib Sens Netw 18(3):15501477211062836
39. Gautam S, Malik A, Singh N, Kumar S (2019) Recent advances and countermeasures against various attacks in IoT environment. In: 2019 2nd international conference on signal processing and communication (ICSPC), Coimbatore, India, pp 315–319 https://doi.org/10.1109/ICSPC46172.2019.8976527
40. Mellit A, Kalogirou S (2021) Artificial intelligence and Internet of Things to improve efficacy of diagnosis and remote sensing of solar photovoltaic systems: challenges, recommendations and future directions. Renew Sustain Energy Rev 143:110889
41. Kumar S, Raut RD, Narkhede BE (2020) A proposed collaborative framework by using artificial intelligence-Internet of Things (AI-IoT) in COVID-19 pandemic situation for healthcare workers. Int J Healthc Manage 13(4):337–345
42. Ting DS, Lin H, Ruamviboonsuk P, Wong TY, Sim DA (2020) Artificial intelligence, the Internet of Things, and virtual clinics: ophthalmology at the digital translation forefront. Lancet Digit Health 2(1):e8–e9
43. Zaman S, Alhazmi K, Aseeri MA, Ahmed MR, Khan RT, Kaiser MS, Mahmud M (2021) Security threats and artificial intelligence based countermeasures for Internet of Things networks: a comprehensive survey. IEEE Access 9:94668–94690
44. Dec G, Stadnicka D, Paśko Ł, Mądziel M, Figliè R, Mazzei D, Solé-Beteta X (2022) Role of academics in transferring knowledge and skills on artificial intelligence, Internet of Things and edge computing. Sensors 22(7):2496
45. Tomazzoli C, Scannapieco S, Cristani M (2020) Internet of Things and artificial intelligence enable energy efficiency. J Ambient Intell Human Comput 1–22
46. Ghosh T, Al Banna MH, Rahman MS, Kaiser MS, Mahmud M, Hosen AS, Cho GH (2021) Artificial intelligence and Internet of Things in screening and management of autism spectrum disorder. Sustain Cities Soc 74:103189
47. Puri V, Kataria A, Sharma V (2021) Artificial intelligence powered decentralized framework for Internet of Things in Healthcare 4.0. Trans Emerg Telecommun Technol e4245
48. Arora S, Sharma N, Bhushan B, Kaushik I, Ahmad A (2020) Evolution of 5G wireless network in IoT. In: 2020 IEEE 9th international conference on communication systems and network technologies (CSNT). https://doi.org/10.1109/csnt48778.2020.9115773

49. Stadnicka D, Sęp J, Amadio R, Mazzei D, Tyrovolas M, Stylios C, Navarro J (2022) Industrial needs in the fields of artificial intelligence, Internet of Things and edge computing. Sensors 22(12):4501
50. Mansour RF, El Amraoui A, Nouaouri I, Díaz VG, Gupta D, Kumar S (2021) Artificial intelligence and Internet of Things enabled disease diagnosis model for smart healthcare systems. IEEE Access 9:45137–45146
51. Orchi H, Sadik M, Khaldoun M (2022) On using artificial intelligence and the Internet of Things for crop disease detection: a contemporary survey. Agriculture 12(1):9
52. Goyal S, Sharma N, Kaushik I, Bhushan B, Kumar A (2020) Precedence & issues of IoT based on edge computing. In: 2020 IEEE 9th international conference on communication systems and network technologies (CSNT). https://doi.org/10.1109/csnt48778.2020.9115789
53. Barnawi A, Chhikara P, Tekchandani R, Kumar N, Alzahrani B (2021) Artificial intelligence-enabled Internet of Things-based system for COVID-19 screening using aerial thermal imaging. Futur Gener Comput Syst 124:119–132
54. Gupta BB, Tewari A, Cvitić I, Peraković D, Chang X (2022) Artificial intelligence empowered emails classifier for Internet of Things based systems in industry 4.0. Wirel Netw 28(1):493–503
55. Stracener C, Samelson Q, Mackie J, Ihaza M (2019) The Internet of Things grows artificial intelligence and data sciences. IT Prof 21(3):55–62
56. Sil R, Roy A, Bhushan B, Mazumdar A (2019) Artificial intelligence and machine learning based legal application: the state-of-the-art and future research trends. In: 2019 international conference on computing, communication, and intelligent systems (ICCCIS). https://doi.org/10.1109/icccis48478.2019.8974479
57. Dhar Dwivedi A, Singh R, Kaushik K, Rao Mukkamala R, Alnumay WS (2021) Blockchain and artificial intelligence for 5G-enabled Internet of Things: challenges, opportunities, and solutions. Trans Emerg Telecommun Technol e4329
58. Paśko Ł, Mądziel M, Stadnicka D, Dec G, Carreras-Coch A, Solé-Beteta X, Atzeni D (2022) Plan and develop advanced knowledge and skills for future industrial employees in the field of artificial intelligence, internet of things and edge computing. Sustainability 14(6):3312
59. Mittal V, Tyagi A, Bhushan B (2020) Smart surveillance systems with edge intelligence: convergence of deep learning and edge computing. SSRN Electron J. https://doi.org/10.2139/ssrn.3599865
60. Rjab AB, Mellouli S (2018) Smart cities in the era of artificial intelligence and Internet of Things: literature review from 1990 to 2017. In: Proceedings of the 19th annual international conference on digital government research: governance in the data age, pp 1–10

AI Enabled Internet of Medical Things in Smart Healthcare

S. Jayachitra, A. Prasanth, S. Hariprasath, R. Benazir Begam, and M. Madiajagan

Abstract In recent years, Artificial intelligence (AI) has been burgeoning hastily in various research areas such as healthcare, living assistance, biomedicine, and disease diagnosis. The inception of AI provides enormous amenities to enrich patient monitoring, Clinical outcomes, and limits costs. The Internet of Medical Things (IOMT) is one of the tremendous developments in healthcare. Hence, the integration of AI with IOMT yields machine to machine, human to machine, and human to human communication perhaps completely updated with e-healthcare for the improvement of society. The AI-based IOMT entrust the clinically associated devices and their incorporation enhances the e-healthcare. The application of AI yields abundant growth in e-healthcare from diagnosis to treatment. The utilization of IOMT sensors assists in real-time disease prediction which significantly reduces the mortality rate. Therefore, this chapter discusses the roles of sensors in e-healthcare, smart monitoring, ambient assisted living, smart treatment reminders, security challenges, and opportunities. In addition, the major key issues and challenges concerned with the use of AI in e-healthcare and also delineate the avenues for future research.

S. Jayachitra (✉)
Department of Electronics and Communication Engineering, PSNA College of Engineering and Technology, Dindigul, Tamil Nadu, India
e-mail: Jayachitra0804@gmail.com

A. Prasanth
Department of Electronics and Communication Engineering, Sri Venkateswara College of Engineering, Sriperumbudur, Tamil Nadu, India

S. Hariprasath
Department of Electronics and Communication Engineering, Saranathan College of Engineering, Panjappur, India

R. Benazir Begam
Department of Electronics and Communication Engineering, Rajalakshmi Engineering College, Thandalam, Tamil Nadu, India

M. Madiajagan
School of Computer Science Engineering, Vellore Institute of Technology, Vellore, Tamil Nadu, India

B. Bhushan et al. (eds.), *AI Models for Blockchain-Based Intelligent Networks in IoT Systems*, Engineering Cyber-Physical Systems and Critical Infrastructures 6,
https://doi.org/10.1007/978-3-031-31952-5_7

Keywords Artificial intelligence · Internet of medical things · Smart healthcare · Opportunities · Challenges

1 Introduction

The framework commutes in digital technologies such as conventional healthcare to smart healthcare which is built to invent the smart healthcare around the world. The era of digital technology amalgamate smart healthcare to transit healthcare information, linking resources, individuals, and organization. The advent of smart healthcare connects various stakeholders namely doctors, patients, and service providers. Typically, it was obtained with most emergent technologies like Blockchain, Internet of Things (IoT), Artificial Intelligence (AI), Cyber security, and Internet of Medical Things (IOMT) [1].

AI employs novel algorithms according to man-made hindrance solving skills and thinking expertise provides the platform to obtain innovative solution for intricate problems. The AI-based techniques have been widespread in society which gains space for enforcement in healthcare and various areas [2]. The growth of computer science offers the rapid advancement of algorithms can surpass the speed and ability of healthcare workers. The IoMT in smart healthcare is depicted in Fig. 1. A decade ago, the diagnosis of disease can be done through clinical examination of patients. The technologies such as smart watch, wireless body area network, Bluetooth, and RFID are presently utilized in healthcare environment [3].

A wearable sensor in clothing monitors the signals of EMG, ECG, Temperature, and blood pressure. In addition, the sensed records have been stored in the cloud and it is communicated with hospitals, doctors through Smartphone or any personal digital assistants. Wearable gadgets have become popular due to the various application healthcare monitoring systems that paves the way for the evolution of Internet of medical things (IoMT). The motive of IOMT is to enrich the quality of services in

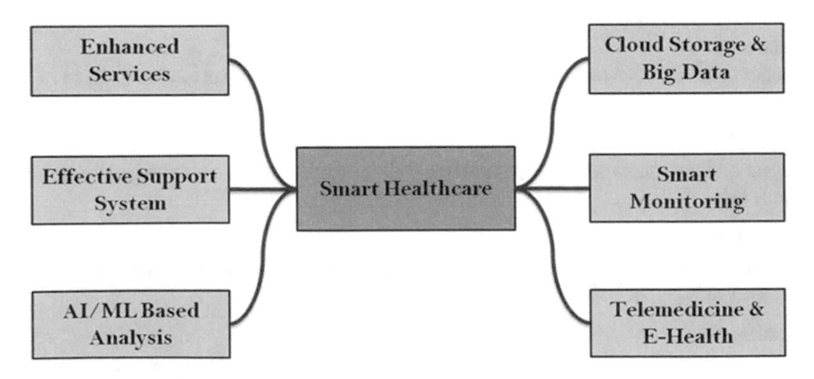

Fig. 1 IoMT in smart healthcare model

healthcare providers, and improve patient satisfaction. The Machine learning techniques reduces the diagnosis time and computational costs. Smart healthcare services are an effective method for reducing the mortality rate, and clinical test costs [4, 5].

1.1 Need for Artificial Intelligence (AI)

The incorporation of automation with AI brings revolution in healthcare industry and healthcare 5.0. The smart device helps us to monitor and predict the disease based on the data extraction through the sensors. The AI in healthcare involved in computer aided automated disease prediction, smart drug development, robotic surgery, and AI smart devices based on sensor. The smart wearable sensor employed in healthcare lags with cognitive facet to supervise the physical emotion of patients, particularly posttraumatic stress disorder patients [6]. The cognitive AI can transmit the healthcare monitoring system in a variety of ways. For example, the integration of cognitive AI with advancement of health monitoring enables to heed the patient mobile calls and evaluate the patient voice to predict the sign of anxiety, and heart disease. In addition, it examines the risk behavior of neurological disorder, and patients with chronic condition are provided with personalized smart healthcare. Henceforth, AI combined with numerous burgeoning technologies that could be used to enhance medical disease prediction, clinical trials, rug development, and robotic surgery [7].

1.2 Role of Artificial Intelligence in Smart Healthcare

AI involved in smart treatment, smart healthcare management, and remote diagnosis. The AI can be efficiently implemented in medical diagnosis. Nevertheless, the user-level system does not have the ability to attain the computations in minimum running time. The processing function can be carried out in cloud platform which is the most prompt solution for high computations. Furthermore, the cloud environment transits the samples to clinicians for better detection of disease and retains the samples for training in neural networks [8]. The downtime in data transmission among cloud servers and healthcare professionals can be mitigated through deploying edge computing with cloud servers. The Artificial intelligence in smart healthcare is demonstrated in Fig. 2.

The AI has sub domains such as Augmented Reality and Virtual Reality in internet of medical things. Nowadays, the clinical application of these methodologies are classified with various areas such as training the data, patient visual impairment, medical assistance, fitness, and disorders. In addition, the sub-domains can be used to training surgery in clinical education. The Virtual reality can be utilized to govern mental illness whereas the augmented reality employed to assist for surgery through

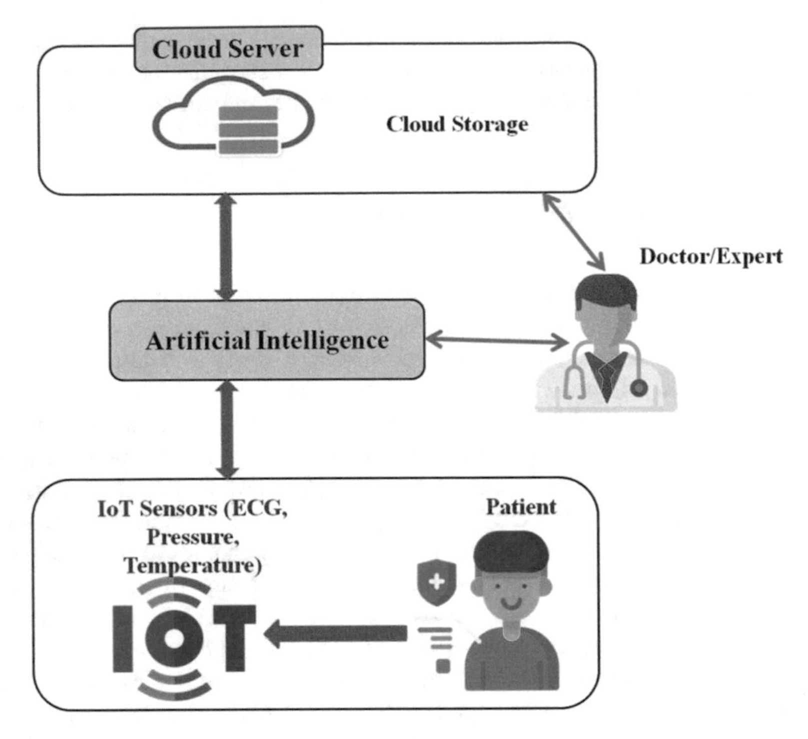

Fig. 2 Artificial intelligence in smart healthcare

robots. The wide variety utilization of augmented and virtual reality validates the security and protection in real world applications [9].

In medical field, AI offer instincts to enhance patient and clinical results which minimize the cost and strengthens the patient health. The AI can be deployed in several healthcare data that incorporates organized and chaotic data. Furthermore, AI has vector machines for organized data whereas the chaotic data can be used in Natural Language Processing [10]. The AI has been implemented for disease diagnosis and treatment in various domains such as cardiology, neurology, pulmonology, and cancer. The literature of AI in healthcare is demonstrated in Table 1.

The AI methodology in radiology incorporates transparency, autonomous, and explicability. The study reveals that AI and radiology integration can create the augmented intelligence that yields the better healthcare system in future. Due to the technological advancement, AI helps to minimize the clinical and pharmaceutical errors. In twentieth century, the healthcare with Clinical Decision System prompts a significant key aspect in healthcare industry [17]. Nevertheless, the General Rule-based system requires principles and rules for making decision in healthcare which are unreliable.

The key contribution of research work is outlined as

Table 1 Illustration of AI in healthcare

References	Year	Topic addressed	Methodologies	Kind of disease	Shortcomings
[11]	2021	General healthcare applications	General rule based clinical decision support system	General support system	Unstable method and lack of interoperability
[12]	2019	AI applications in various domain	Strength weakness opportunities and threats	Radiology system	Chaotic and high dimensional data
[13]	2019	Deployment of machine learning in traditional healthcare and public healthcare records	NA	Heart disease, nervous system disorder	Trust and privacy issue arises
[14]	2019	Implementation of machine learning classifier to predict disease	Decision tree and random forest algorithm	Healthcare support system	Interoperability challenges occurs
[15]	2017	Describes AI based disease prediction	Support vector machine	Cardiac and neurologic disorder	No security and poor performance
[16]	2017	Scarcity of healthcare professionals	Robotic technique	Social utilization	No security and privacy

- The platforms and tools for predicting heart disease is highlighted with different computational techniques.
- The vital responsibility of the proposed technique is to provide the significance of medical things in smart healthcare.
- The illustration of AI in heart Disease Diagnosis prediction is demonstrated and compared with different optimization algorithm.

The rest of this paper adheres as follows: Sect. 2 summarizes the smart healthcare services through sensors. The Internet of Medical Things (IOMT) is represented in Sect. 3. In Sect. 4, AI augmentation for smart healthcare is described briefly. Section 5 corroborates the challenges and future directions. Finally, Sect. 6 concludes the paper.

2 Smart Healthcare Services Through Sensors

The building of smart healthcare systems stipulates various incorporation of developing technologies to offer individualized care, disease prediction, and remote patient monitoring.

2.1 Smart Patient Monitoring and Tracking

The smart healthcare involved in tracking and monitoring the patient remotely through different modes of communication techniques and digital health mechanism. It can be obtained through smart wearable sensors, smart watches, smart phone, and smart intelligent devices. For example, the wearable sensor devices are used to monitor the patient physiological parameters that are needed for any disease prediction. Hence, smart patient monitoring have been fortified in different research areas namely, respiration illness monitoring, cardiovascular disease, blood oxygen saturation, sleep monitoring. The remote patients can be notified while there is any physiological change. The E-clinics can be implemented through digital contact by utilizing the video link, phone call and cloud assisted platforms among patients and doctors for remote consultation with clinicians. In addition, by minimizing the patient direct contact, the E-clinics helps in alleviating highly infectious disease, waiting time and thriving patient healthcare in efficient manner [18].

2.2 Emotional Telemedicine

Telemedicine is the procedure of giving treatments based on the diagnosis of disease through the information acquired through telecommunication methodologies. It has several toolkits which provide cutting edge results to healthcare service. From these, the tele-rehabilitation is a method of post-diagnosis provided to patients those who need rehabilitation for their wellbeing. E-rehabilitation is offered to patients those who suffered with trauma and physical injuries [19]. The proper procedure of training is given to the people in the form of arbitration may not be inflated and for existing users, the technology based involvement have to be feasible to use. The telemedicine have been employed to handle different disease which reduces the burden on healthcare and mitigate poor services. Furthermore, the experimental studies illustrates that technological and individual attributes were detected as significant hindrance for effective deployment of telemedicine. In recent years, the developing countries may depend on infrastructure and funding opportunities whereas several patients may not have computing services and internet facilities to ingress telemedicine services. Nevertheless, the present services lagging in emotional aspects to supervise the state of patients [20].

2.3 Ambient Assisted Living (AAL)

The advent of technological developments like robotic surgery, AI, predicting an earlier treatment for disease, and remote monitoring of patient becomes inventive nowadays. Ambient assisted living aids to supervise the patients of all ages and

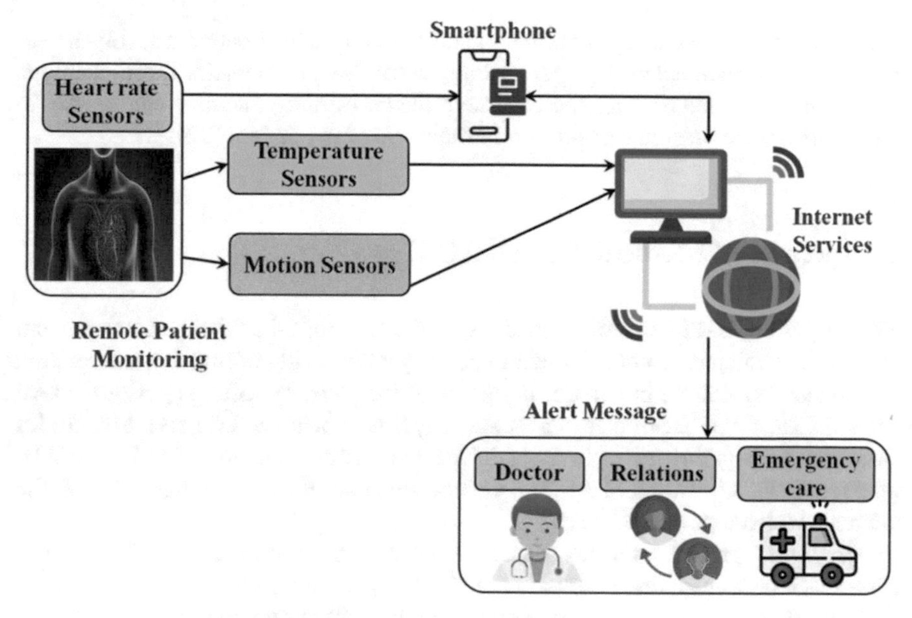

Fig. 3 Scenario of ambient assisted living

monitor their daily activities which empower them to live solitarily irrespective of medical conditions. For example, the utilization of smart healthcare system for monitoring patient health condition and retrieving clinical suggestion from doctor has been deployed efficiently in ambient assisted living. It can be used persistently for autistic patients, aged people, and patients with impairments. This system incorporates wireless body area network, smart devices, wearable sensors which communicate with monitoring information system [21]. To build the day-to-day life simpler, the ambient assisted system has been proposed to automate their tasks and frequently monitor their daily activities that ease their independency. The ambient assisted living in IoT scenario is illustrated in Fig. 3. The implementation of ambient assisted living system manipulated with several information based on sensors, ambient information, and user actions.

2.4 *Smart Treatment Reminders, Compliance, and Adherence*

The IoMT becomes more popular in remote monitoring therapeutics. The wide utilization of AI in robots assists the surgery for patients and implementation of virtual assistants for healthcare services is predicted. The medical robots such as robot-assisted rehabilitation, disease diagnosis, therapeutic treatment, surgical robots are utilized for different applications. In addition, the smart treatment involves in e-healthcare management through prescribing and delivery of medicine through online. If the patient forgets or skips the treatment then a warning message can be sent to the healthcare professionals, and family members of the patient [22].

3 Internet of Medical Things (IOMT)

The advent of IOMT in medical field paves the way for e-healthcare management. Hence, the clinicians is not extended to be busy with storing of massive information of patients data but the internet of things helps the patient to offer superior medical services. The IOMT is an emerging technology in healthcare which gets the amenities from cloud computing by deploying AI. The automatic interaction of patients can be attained by the utilization of AI. The Chatbots enables the communication with the clients which are powered by AI.

The IOMT provides various services such as data transfer for efficient healthcare, machine-to-machine interaction, and interoperability. The sensors collect the data, examine the original data, and negotiate the requirement to maintain raw data. The IOMT system gathers information of illness and transit the information to the doctors for real-time monitoring where the smart sensors send sends alert message to individuals [23]. The e-health monitoring through connected devices can save the life of human being during medical emergency like heart attacks, diabetes, etc. The healthcare mobility innovate the healthcare facilities for the patient and reinforces the medical practitioners through offering timely treatment.

Nowadays, the IoT is elevated through incorporating smart chips for healthcare monitoring. The most challenges in IOMT are privacy, interoperability, and security. The connected model provides routing the data, monitoring the healthcare, sending alarms, and delivering drugs. The e-healthcare in IOMT utilizes wearable sensors for patient monitoring, temperature, pressure, ECG monitoring that helps to improve accuracy, nurture efficacy, minimize the costs, flatter the standards, and ensure the safety of the patients. In recent years, the IOMT postulates strong requirements for healthcare applications. It monitors and controls the patient whether the patients have taken the medication properly, and on time due to continuing complications at hospitals and healthcare facilities [24]. The wireless body area network assesses the power consumption in IOMT healthcare system. The e-healthcare system has been engrossed on resolving the hindrance of chronic illness. The IoMT can be widely used in variety of applications such as sleep monitoring, cardiovascular disease, mental illness monitoring, etc. [25].

3.1 *Technologies of IOMT*

3.1.1 Radio Frequency Identification (RFID)

The RFID is the emerging methodologies in IoMT that yields automatic authentication. This employs radio frequency technique to estimate the targeted value through reading, writing the data and detecting model can estimate the detection without any contact during data communication. This offers several amenities such as distance detection, non-hindrance, and no human intervention [26]. The RFID model encompasses E-Tag, data reader, and information framework model. The main objective of E-Tag is to save the data in targeted value and it disseminates the value to the reader.

The RFID is essential for monitoring the patient's breath, temperature, blood pressure, EEG, and ECG monitoring. The RFID have multi objective detection and objection moving detection that is utilized in managing healthcare system, tracking and controlling real-time data, and personal recognition. The inkjet RFID have the benefits of patient medication dosage, time consuming, enormous resources, and lower costs. However, it has relentless loss of information; minimize the accuracy in detection, signal loss, channel contention, and deterioration of tag [27]. To resolve this constraint, a meager exemplifying classification methodology based on dictionary partitioning is employed in medical support system.

3.1.2 Wireless Sensor Network

The wireless sensor incorporates internet of things, information processing, and communication development to perform the operation of acquisition of data, information processing, and transferring of data to the cloud. It has massive amount of sensor nodes with computing abilities and telecommunication that are chaotically employed around the monitoring area via multi-hop communication [28]. This can gather and control the information received from various framework in real-time and convey the processed data through wireless to the clients. The wireless sensor network plays a significant role in healthcare due to its stability, secure data access, and reliable data transfer characteristics. The application involves monitoring the physiological parameters of patient, emergency care in hospital.

A wearable body area network or grapheme based wearable sensor that yields higher sensitivity and stability which are utilized to enhance the patient's clinical parameters. A novel sensor employed to monitor the pressure via the human temperature, patient respiration, and their activity speed which diminish the venture of health. Nowadays, the optimization in wireless sensor network has been focused by the researchers in E-healthcare. In addition, the security hindrance arises in wireless network [29]. To alleviate this constraint, low energy dissipation methodology has been proposed in smart grid monitoring. Alternatively, balanced weighted routing mechanism has been proposed to minimize the energy dissipation through efficiently shrinking the heavy load transfer on sensor nodes. However, the principal component analysis deployed along with optimal clustering mechanism minimizes the energy consumption in sensor nodes. Henceforth, the wireless networks have extreme advancement in healthcare support system.

3.1.3 Application Programming Interface (API)

The API plays significant contribution in IoMT for managing healthcare support system. The API is positioned among back-end and front-end that performs as arbitral in management system. This demands the requirements of variety of applications, assist prevalent computing, and sensor system. The general platform creates the interface in several environments for communication of information among application devices. This records the information from the sensor and gathers the data from the sensor in real-time environment that performs proofreading and processing of data. It transfers the processed data to RFID reader to analyze the information association among back-end and RFID reader system [30].

3.2 Applications of IOMT

The internet of medical things finds better medication for the patient and managing the people and devices which is particularly utilized to minimize the cost as well as assure the patient health. The object recognition system plays a major role in healthcare because the patient information has high sensitivity. The solitary system recognition has low privacy and does not authorize the extreme growth of information in healthcare [31]. Furthermore, the multi-system biometric recognition holds the incorporation of fingervein, face, and thumbprint which yields higher diagnosis rate, greater security and privacy that brings pivotal growth in healthcare field. Besides, the vital sign monitoring brings gradual modification in individualized healthcare treatment. The patient real-time collection of data identifies the biological condition of patients which analyze the prediction of disease and provide earlier treatment to save the life of human being. The patients are not necessary to visit the doctors frequently to get the treatments. The smart phone based integrated system, fog-based computing system, internet of things have create more augmentation, ameliorate the procedural system in hospital, enhance the efficacy and exploitation of resources that directs the employment of IOMT in current scenario [32].

The vital sign monitoring system connected with wearable devices can be realized through telemedicine. The smart wearable devices such as smart watch, smart band forms the interrelation among medical practitioner and patients. Nevertheless, the disruptive transformation in healthcare, the stability, flexibility, and linking of multiple wearable system has to be enhanced. Due to the vast development of healthcare, the utilization of wearable system should be esteemed in a greater way [33].

3.3 Platforms and Tools for IoMT

The Apple has proposed the environment such as ResearchKit and CareKit to gather the healthcare information that was perceived in Apple systems. The tool Researchkit is an open-source tool which can be employed to accumulate the medical data for analysis through Smartwatches and iPhone [34]. The ResearchKit assets the medical data from confined Health application. The Glucosuccess App would be implemented to enhance the quality of patient life through changing diet in app environment. Furthermore, the Asthma Health App bewitched on monitoring the patient data about the asthma disorders and their sickness. The shortcoming of Researchkit limited set of aspiring patients minimized to proprietors of Apple Commodities and also deficiency of details about the privacy of patients information.

The Microsoft introduced a tool named Health Vault to provide computerized chatting amenities which incorporates AI to operate communication with the people through Healthcare Bot. The patient can able to communicate with the bot regarding the healthcare services and facilities. In addition, the Microsoft Immunomics employed to predict the disease according to the patient data acquired on antigen outline. The Antigen outline interprets a key aspect in constructing the outline of T-cells to antigens [35]. The Ambient medical intelligent solution was contributed by Microsoft which infers that doctor is deliberating more on patients and medical authentication. The Environment that support Medical Data from IoMT are depicted in Table 2.

Table 2 Environment that support medical data from IoMT

Tool	Design	Information resources	Transaction procedure	Information privacy	Incorporated machine learning
Researchkit	Disseminated	Healthkit	Customized	Access prone by patient	No
Microsoft E-Health Management System	Cloud	Middleware for RHIR	RHIR	Access prone by patient	Azure environment
1up.Health	Cloud	Electronic health record system	RHIR	Hospitals are the proprietors for medical information	No
onRHIR.io	Cloud	RHIR	RHIR	Access grant once by patient	No
Bridgera Monitoring	Cloud	Bridgera myhealth	Customized	Access grant once by patient	No
Docboxmed	Deficit of data	Customized IoMT sensors	Customized	Depends upon hospital authorization	Deficit of data

The Microsoft provides Azure environment middleware for Rapid Healthcare Interoperability Resource (RHIR). This is one of the major possibilities of integration of medical healthcare records using HL7. It formulates the RHIR connector to gather medical information and store the data in RHIR Server. The environment 1up.health promotes the sharing of patient medical information and incorporates the data sources like Meditech, Athenahealth, Nextgen, etc. The medical data gathered from 1uphealth middleware is feasible for medical practitioners, clinicians, and Patients [36].

The tool on RHIR.io maintains separate repository for storing and receiving the Medical Health Records. The amalgamation controls the procedures like Healthkit, Middleware FitBit, and Googlefit. The procedure and rules are made feasible in software enterprise. The platform Bridgera Monitoring can be implemented to monitor the patient's real-time data. The data can be recorded through mobile application or cloud platform. The developer acquires permission to ingress the data through conventional protocol like JSON and CSV.

DocBoxMed is an emerging tool utilized to monitor the patients those who are under Intensive Care Units. This model is imposed to encompass the medical equipments with patients to predict the disease through AI and Machine learning technique formed through healthcare providers. According to the data acquired from Docboxmed, it sends warning alerts to the clinical staff to take care of the patients. The prevalent environment collects data from various platforms and the information can be shared through RHIR. Hence, a new system will gather the data from several IoMT systems in real-time and the healthcare system can be maintained with better privacy and security [37].

4 AI Augmentation for Smart Healthcare

The conventional healthcare has been characterized with higher costs, prediction error, maintenance complexity, and incompetence workflow. In the year 2011, the United States healthcare system cost was computed with 18% of gross product of their nation. The predicted development of 6% per annum yields the healthcare GDP around 20% in 2020. From the observation, the rising price was intricate through the data by 1 in 20 Unites States grown-ups endured with prediction error. Thereby, one-half people may suffer due to severe harmful of error [38]. The scarcity creates the innovation in healthcare and the AI development integrates collectively to examine the key factor in enhancing the healthcare delivery and reduce the cost of healthcare supporting systems.

4.1 Artificial Intelligence in Data Extraction

The implementation of AI in electronic health record is to computerize the record maintenance of patient details that presently elevate considerable proportionality of contributor time and instigate higher rates of exhaustion among physicians. The patient variation of health record is probable source of harm. The electronic data extraction and representation from healthcare records would intensify the efficacy of healthcare along with research efforts. For example, the detection and rationalizing the clinician medication is a typical application of AI.

4.2 Healthcare Disease Predictions

The capability to detect clinical results can aid in better enlighten medical trajectory and diagnosis. For instance, the precise computation of health risk of re-admittance in hospital can be avoided risky discharge or assure the resources and collateral services have been placed to reinforce the patients while discharged from hospitals [39]. The AI acts a major role in medical decisions and concerns about the stringency of sickness, diagnosis, and clinical risk of impairment. The developments in computation power and information storage in cloud have to be analyzed, the researchers computing the diagnosis disease in clinical model endures to extend at remarkable rate.

4.3 Heart Disease Diagnosis

In recent years, the machine learning algorithm interprets key aspects in heart disease detection. The supervised machine learning technique predicts the abnormality of a patient and classifies the disease based on prediction. To mitigate the complexity associated with disease a new attention mechanism is necessary for healthcare system. The IoMT framework has been incorporated along with machine learning classifier to detect the heart disease with high precise accuracy. The attributes such as blood pressure, age, cholesterol, and sugar level from medical dataset are used to examine the heart disease [40, 41].

In [42], the wavelet oriented principal component analysis has been proposed to pre-process the raw data and extract the relevant features. The extracted features are sent to back-propagation neural network to detect the heart disease according to the acquired data. Alternatively, the IoT with machine learning algorithm employed to detect heart disease from cloud data [43]. The information is gathered from patients through medical sensors, where the UCI repository dataset are utilized to analyze the method. The J48 classifier outperforms than other machine learning classifier and obtains the accuracy of 91.48%. Table 3 illustrates the heart disease diagnosis systems, the utilized methodologies for detection, and the function of each developed model in AI environment.

Table 3 Illustration of AI in heart disease diagnosis

References	Year	Methodologies	Topic addressed	Function
[40]	2019	SVM classifier	Classification and regression analysis	Pattern analysis and non-linear regression
[41]	2018	Probabilistic classifier	Naïve Baye's classifier	Sentiment analysis, fake news detection
[44]	2019	Data clustering analysis	Classifying the samples into different groups	Employed KNN to compare with NICV
[45]	2019	CNN	Examine visual imagery	Classify abnormality through ventricular dysfunction
[46]	2017	Deep learning	Prediction model	Non linear technique to process the data
[43]	2018	RFNN	Neural network	Implement DPSS to protect healthcare and ensure security in medical data
[47]	2019	ReliefF and feature selection	Incorporated feature selection for heart disease prediction	Data mining is used in pharmacology and medical applications
[42]	2019	Multilayer perceptron	Combination of SLP and ANN	Offering continuous functions to resolve real-time problems

4.3.1 Detecting Methods

Neural network is utilized for casualty patient to enhance multi-dimensional detection. The outcome of neural network system test depicts that it have proven efficacy for non-linear aspects in respiratory casualty patients. Presently, the IoMT gathers medical information from wearable sensor and images like CT, MRI, and Ultrasonic Images. Hence, the Convolution Neural Network (CNN) evaluates the detection of disease and their conditions through collected data from sensors [45]. There are several methodologies such as deep belief neural networks and recurrent neural networks are also employed to predict disease. The machine learning approaches utilize non-invasive methodologies to lay hypothesis among biomarker dialysis process. The various detecting methods supported by IoMT are demonstrated in Table 4.

The fog-IoT model for diabetes detection integrates smart sensors to examine the glucose level of patients. The medical data is transferred into cloud environment to predict the abnormalities through machine learning algorithm. The researchers have incorporated various AI methods to improve the life of patients. The computer aided diagnosis and IoT based system can gather medical data from patients [49].

The pulmonary cancer can be detected through computer aided diagnosis which employs deep learning methods. The CNN and DCNN shows that the detection

Table 4 Summary of detection methods supported by IoMT

References	Year	Methodologies	Outcomes	Application
[47]	2019	Backpropagation neural network	Promote the non-linear relationship among casualty patients	Diagnose the respiratory illness
[45]	2019	CNN	Developed RNN and DBNN	Diagnosing the chemical composition using deep learning
[41]	2018	KNN, Naïve Baye's, SVM, and random forest algorithm	The method is a lay of biomarker dialysis derived in probabilistic way	Machine learning with IoT for threat detection
[48]	2018	Linear regression and neural network	Rule based clustering in classifier for detecting the disease severity	Medical disease detection through fuzzy neural network classifier
[49]	2019	Case based reasoning	Implemented fuzzy set theory	Data privacy and security
[46]	2018	Deep fuzzy neural network and CNN	Aids radiologist to detect cancer nodule prediction accuracy	Cancer prediction

accuracy of 77.6 and 84.58%. A novel cryptography system is needed to develop medical data security. It is used to accumulate the patient record into cloud server with high security. The IoMT provides better healthcare services with high security of data [48]. The hybrid optimization, particle swarm optimization, and grasshopper optimization applied to encrypt and decrypt the data. This can be carried out through handling big-data with high security through appropriate cryptographic methods.

4.3.2 Robotic Surgery

In recent years, robotic surgery plays a significant role in healthcare industry. The surgical robots are frequently employed in tremor filtration and instrument stability. The utilization of robots in surgery have amenities like quicker recovery time, low blood loss, less infection risk, and better visualization. The contemporary experiment depicts endo-wristed equipment with unimpaired abdominal wall yields better results for patients as doctors.

Transoral robotic surgical method combines retractors as an efficient model for tissue supervision and schematic representation in larynx. Further, the medrobotics have been implemented for surgical cases for 37 patients with various diseases. The factual outcomes assure that transoral robots can be applied to dysplasia, leukoplakias, and papillomas. The adoption of robots in medicine paves the way to enhance complex surgical techniques like coronary surgery, mitral valve, and atrial fibrillation surgery. Furthermore, the hospitals, clinics have employed surgical robots due to their proven records which results with reduced costs, ensure the safety of patient,

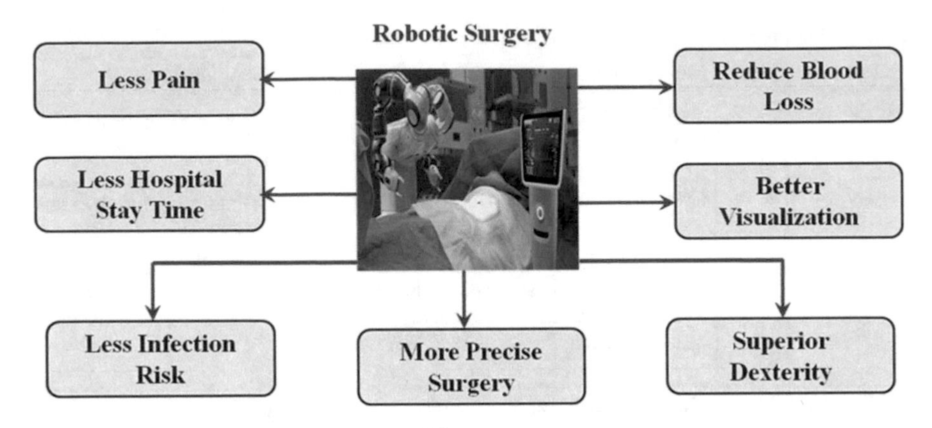

Fig. 4 AI augmented robotic surgery

and clinical efficacy. However, the surgical robots face challenges in cardio bypass surgery, and myocardial invulnerability have to be resolved. The AI augmented Robotic Surgery is shown in Fig. 4.

The Robotic surgery employed in oropharyngeal cancer surgery particularly transoral robots have downcast the risk during surgery. The various surgical robots are deployed to reinforce endoscopy methods according to AI and machine learning algorithms [50]. The endoscopy robots exploited for transnasal skull based surgery in which the accuracy obtained is highly notable. In addition, the invasive surgical methods with prominent safety and necessitate the utilization of robots as support for highly complicated surgery. This methodology motivates the enhancement in precision tool to promote the capability of surgeons. The smart robots in medication offer challenging possibilities in medical images, data collection, and data visualization.

The IoMT with Robots has been explored to provide high safety, and accuracy of surgery is maintained in hospitals. This variety of model can invade tangled pathology and acquire medical details with micron scale. These methodology have an integrate information communication configuration to offer cyber security for big data in medical applications. The invasive models envelope energy reservoir regulator and volatile tele-operation function. A new smart tissue autonomous robot has been exploited for intestine surgery [51]. The experimental outcome shows that autonomous robot outperforms more accurately with safety intervention than clinical surgeons. Nevertheless, the autonomous robot requires evaluation for assisting the surgeons. The development of autonomous robots authorizes surgical robots for governing tissue surgical robots through extent monitoring system. The robot and sensor integrated computer aided surgery focused on neurosurgery and manifests to improve robotic model with highly intelligent robotic controllers to assist clinical surgeons. The robot with sensor gathers patient data during surgery executes diagnosis and notify the clinical surgeons through real-time monitoring system.

5 Challenges and Future Directions

The various key concerns of IoMT and AI-based health monitoring system are node communication, interoperability of sensors, utilization of AI, data management, security and privacy. In recent healthcare sector, the IoMT devices can be employed to gather the patent's information which is further used to recognize and classify the disease. Typically, a large-scale healthcare sector will operate only if it has better sensing ability along with the processing capacity to generate the optimal information. The entities in IoMT should manipulate with compatible data technique and knowledge illustration method. The IoMT offers international partnership for distributing the private medical data that will endorse remote telemedicine and the establishment of finest medical care. There is a rapid growth in the usage of IoMT device for several smart applications. However, the variety of issue is arisen while collecting the sensed data from heterogeneous sensors. These issues include limited batteries, hardware malfunctions, and connectivity limitations [52–54].

IoMT deeply depends on data gathering from distinct devices and sensors associated to human's bodies. The traditional healthcare model can get unreliable data due to occurrence of various constraints. This incomplete data may affect the remote monitoring system. The patients will tolerate several devices for prolonged time period. When designing the IoMT sensors, the major challenge is to incessantly supervise the human vigorous signs without tumbling user coziness.

Another vital challenge in the medical sensors is correlated to babies. The sensors generally placed closer to the baby's body in order to gather the data. Some of the deployed sensors may impend the baby with a sharp hazard. Wearable sensors are deployed in elderly as well as children to track the body movement where the tracked information will be reported to doctor for appropriate action. In practice, the wearable sensors are furnished with batteries, communication module, and other comport materials and they were intended to be committed to human skin. Radio frequencies of IoMT sensors might have a consequence on interpretation areas and observers might provide incorrect readings. This kind of redundant data will collapse the smart healthcare systems.

The conventional AI-based healthcare models will not increase satisfactoriness to the clinicians. Hence, explicable AI-based model can be established where the clinicians can envision the identification or categorization of diseases. The effectual distribution and employment of data aggregation in practice will permit for more consistent measurement and assessment of physical actions. Alternatively, AI models require numerous training samples to acquire superior performance. AI is an innovative area which enhances the patient's trust through taking people into the loop and eradicating clarity anxieties. Furthermore, AI will have a tough time while identifying intermittent categories of cyber-attacks. Contempt challenges in utilizing AI and blockchain, they are ahead position in IoMT models [55, 56].

According to the aforesaid challenges, IoMT and AI-based health monitoring system still have a long way to obtain better performance. In particular, some challenges are more significant than others because it is very difficult task to tackle them.

For instance, emerging sensors with superior energy efficacy and enlarging security of patients' record appear to be listed over the other concerns. At the same time, imminent research should be directed on the AI systems to evaluate their efficiency and direct them more valuable for many real-time applications. The hosting the e-health record in a public library with suitable protections and investigating present data aggregation approaches using such public e-health record will be a critical possible route for future research.

6 Conclusion

The AI model and machine learning algorithm in healthcare system needs tools and environment to assure high quality medical data, validating the methods, and their outcomes. Further, the integrated methodologies yields better services, secure data ingress to the patients, clinician, and hospitals. The AI can extend their revolution in healthcare industry. It prevents and promotes the earlier diagnosis and treatment where the AI can be employed to enhance clinical decisions, personalized healthcare services, and public health results. Specifically, AI provides numerous services with reduced costs, therapeutic care yield its diagnosis accuracy of efficient outbreaks, and capability to improve the individual and global health. The utilization of IoMT, smart sensors, the integration of IoT and cloud computing in healthcare provides better security and data privacy in the network.

References

1. Wazid M, Singh J (2022) ASCP-IoMT: AI-enabled lightweight secure communication protocol for internet of medical things. IEEE Access 10:57990–58004
2. Okolo CT (2022) Optimizing human-centered AI for healthcare in the Global South. Patterns 3(100421):1–15
3. Ghosh A, Saha R, Misra S (2022) Persistent service provisioning framework for IoMT based emergency mobile healthcare units. IEEE J Biomed Health Inform 1–12
4. Radoglou-Grammatikis P et al (2022) Modeling, detecting, and mitigating threats against industrial healthcare systems: a combined software defined networking and reinforcement learning approach. IEEE Trans Industr Inf 18:2041–2052
5. Idrees AK, Idrees SK (2022) An edge-fog computing-enabled lossless EEG data compression with epileptic seizure detection in IoMT networks. IEEE Internet Things 9:13327–13337
6. Tabari P (2022) The role of artificial intelligence in human-computer interaction: using a smart topic extraction system. In: IEEE Symposium on visual languages and human-centric computing, pp 1–3
7. Murtaza M, Ahmed Y, Usman M (2022) AI-based personalized E-learning systems: issues, challenges, and solutions. IEEE Acccess 10:81323–81342
8. Panagopoulos A, Minssen T (2022) Incentivizing the sharing of healthcare data in the AI era. Comput Law Secur Rev 45(105670):1–9
9. Young AS (2022) AI in healthcare startups and special challenges, intelligence-based medicine. IEEE Access 6(100050):1–10

10. Nazar M, Alam MM, Yafi E (2021) A systematic review of human-computer interaction and explainable artificial intelligence in healthcare with artificial intelligence techniques. IEEE Access 9:153316–153348
11. Pawar U, O'Reilly R (2020) Explainable AI in healthcare. In: Proceedings in international conference cyber situational awareness, data analytics assessment, pp 1–5
12. Mahajan A, Vaidya T, Gupta A (2019) Artificial intelligence in healthcare in developing nations: the beginning of a transformative journey. Cancer Res Statist Treat 2:182–187
13. Noorbakhsh-Sabet N, Zand R (2019) Artificial intelligence transforms the future of health care. Am J Med 132:795–801
14. Fritchman K, Saminathan K (2018) Privacy-preserving scoring of tree ensembles: a novel framework for AI in healthcare. In: Proceeding in IEEE international conference big data, pp 2413–2422
15. Jiang F, Jiang Y, Zhi H (2017) Artificial intelligence in healthcare: past, present and future. Stroke Vasc Neurol 2:230–243
16. Clarke I (2017) State of the art: a study of human-robot interaction in healthcare. Int J Inf Eng Electron Bus 9:43–55
17. Durán JM (2021) Dissecting scientific explanation in AI (sXAI): a case for medicine and healthcare. Artif Intell 297(103498):1–12
18. Jayachitra S, Prasanth A (2022) An efficient clinical support system for heart disease prediction using TANFIS classifier. Comput Intell 38:610–640
19. Zhou C, Wang Z, Jiang Z (2021) Interactive interface design for telemedicine and the emotional needs of patients. In: Proceedings in 16th international conference on computer science & education, pp 554–559
20. Yu H, Zhou Z (2021) Optimization of IoT-based artificial intelligence assisted telemedicine health analysis system. IEEE Access 9:85034–85048
21. Koren A, Šimunić D (2016) Requirements and challenges in wireless network's performance evaluation in ambient assisted living environments. In: Proceedings in international convention on information and communication technology, electronics and microelectronics, pp 624–627
22. Daniel A, Lattanzi G (2022) Medical devices, smart drug delivery, wearables and technology for the treatment of diabetes mellitus. Adv Drug Deliv Rev 185(114280):1–11
23. Ghubaish A, Salman T (2021) Recent advances in the internet-of-medical-things (IoMT) systems security. IEEE Internet Things J 8:8707–8718
24. Yuldashev Z, Sergeev A (2021) IoMT technology as the basis of wearable online monitors for space distributed monitoring systems for pregnant women. In: Proceedings in wave electronics and application in information and telecommunication systems, pp 1–4
25. Prasanth A (2020) Implementation of efficient intra- and interzone routing for extending network consistency in wireless sensor networks. J Circuits Syst Comput 29
26. Prasanth A (2021) Certain investigations on energy-efficient fault detection and recovery management in underwater wireless sensor networks. J Circuits Syst Comput 30:1–11
27. Wang X, Mao S (2019) On remote temperature sensing using commercial UHF RFID tags. IEEE Internet Things J 6:10715–10727
28. Prasanth A (2020) A novel multi-objective optimization strategy for enhancing quality of service in IoT enabled WSN applications. Peer Peer Netw Appl 13:1–11
29. Prasanth A (2021) A Tuned classification approach for efficient heterogeneous fault diagnosis in IoT-enabled WSN applications. Measurement 183:1–12
30. Abdulhadi AE, Denidni TA (2017) Self-powered multi-port UHF RFID tag-based-sensor. IEEE J Radio Freq Identif 1:115–123
31. Jayachitra S (2021) Multi-feature analysis for automated brain stroke classification using weighted Gaussian Naïve Baye's classifier. J Circuits Syst Comput 30:1–21
32. Dwivedi R, Mehrotra D, Chandra S (2022) Potential of internet of medical things (IoMT) applications in building a smart healthcare system: a systematic review. J Oral Biol Craniofac Res 12:302–318
33. Dang LM, Han D (2019) A survey on internet of things and cloud computing for healthcare. Electronics 8:1–12

34. Quynh Pham CL (2020) A ResearchKit app to deliver paediatric electronic consent: protocol of an observational study in adolescents with arthritis. Contemp Clin Trials Commun 17(100525):1–12
35. Pirbhulal S, Shang P (2018) Fuzzy vault-based biometric security method for tele-health monitoring systems. Comput Electr Eng 71:546–557
36. Ford JP (2016) Root aggregated prioritized information display: a single screen display for efficient digital triaging of medical reports. J Biomed Inform 61:214–223
37. Çiçek E, Gören S (2022) Physical activity forecasting with time series data using Android smartphone. Pervas Mobile Comput 82(101567):1–12
38. Singh H, Meyer A (2014) The frequency of diagnostic errors in outpatient care: estimations from three large observational studies involving US adult populations. BMJ Qual Saf J 23(727131):1–11
39. Kaur P, Kumar R (2019) A healthcare monitoring system using random forest and internet of things (IoT). Multimed Tools Appl 78:19905–19916
40. Kapa ZI, Lopez-Jimenez S (2019) Screening for cardiac contractile dysfunction using an artificial intelligence-enabled electrocardiogram. Nat Med 25:70–74
41. Putte V, Boumans R (2019) A social robot for autonomous health data acquisition among hospitalized patients: an exploratory field study. In: International conference on human-robot interaction, pp 1–12
42. Heidari A (2019) An efficient hybrid multilayer perceptron neural network with grasshopper optimization. Soft Comput 23:7941–7958
43. Jayachitra S (2021) A novel eye cataract diagnosis and classification using deep neural network. J Phys: Conf Ser 1–8
44. Kumar PM, Lokesh S (2018) Cloud and IoT based disease prediction and diagnosis system for healthcare using Fuzzy neural classifier. Future Gener Comput Syst 86:527–534
45. Fki Z (2018) Machine learning with internet of things data for risk prediction: application in ESRD. In: Proceedings in international conference on research challenges in information science, pp 1–6
46. Yao C (2019) A deep learning model for predicting chemical composition of gallstones with big data in medical internet of things. Future Gener Syst 94:140–147
47. Li B, Zhou B (2019) Power system transient stability prediction algorithm based on reliefF and LSTM. Artif Intell Secur 74–84
48. Masood A (2018) Computer-assisted decision support system in pulmonary cancer detection and stage classification on CT images. J Biomed Inform 79:117–128
49. Sangaiah AK (2019) Hybrid reasoning-based privacy-aware disease prediction support system. Comput Electr Eng 73:114–127
50. Park AD (2019) Comparative safety and effectiveness of transoral robotic surgery versus open surgery for oropharyngeal cancer: a systematic review and meta-analysis. Eur J Surg Oncol 1–12
51. Albahri AS (2021) IoT-based telemedicine for disease prevention and health promotion: state-of-the-Art. J Netw Comput Appl 173:102873
52. Arun N (2021) Assessing the trustworthiness of saliency maps for localizing abnormalities in medical imaging. Radiol Artif Intell 3:e200267
53. Bahalul Haque AKM, Bhushan B, Nawar A, Talha KR, Ayesha SJ (2022) Attacks and countermeasures in IoT based smart healthcare applications. In: Balas VE, Solanki VK, Kumar R (eds) Recent advances in internet of things and machine learning. Intelligent systems reference library, vol 215. Springer, Cham
54. Goyal S, Sharma N, Bhushan B, Shankar A, Sagayam M (2021) IoT enabled technology in secured healthcare: applications, challenges and future directions. In: Hassanien AE, Khamparia A, Gupta D, Shankar K, Slowik A (eds) Cognitive internet of medical things for smart healthcare. Studies in systems, decision and control, vol 311. Springer, Cham. https://doi.org/10.1007/978-3-030-55833-8_2

55. Hameed K, Bajwa IS, Sarwar N, Anwar W, Mushtaq Z, Rashid T (2021) Integration of 5G and block-chain technologies in smart telemedicine using IoT. J Healthc Eng 8814364
56. Swain S, Bhushan B, Dhiman G et al (2022) Appositeness of optimized and reliable machine learning for healthcare: a survey. Arch Computat Methods Eng 29:3981–4003. https://doi.org/10.1007/s11831-022-09733-8

AI Model for Blockchain Based Industrial Application in Healthcare IoT

Vimal Bibhu, Lipsa Das, Ajay Rana, Silky Sharma, and Shallaja Salagrama

Abstract Artificial intelligence is one of the prominent areas that automates the functional system to reduce human efforts and cost incurs by healthcare industries. Blockchain, the Internet of Things, and Artificial Intelligence can be combined to model an emerging application for the healthcare industry. This chapter will address a secured industrial application of IoT that is based on artificial intelligence and blockchain to automate patient service in the healthcare industry. The application processes all the conditions of patients and automatically sends the patient health information to the healthcare professional in the healthcare industry. We are using the body area network with IoT devices to get the patient information and an AI-based designed algorithm processes the real-time status of health on different parameters and creates the block to add to the blockchain. The healthcare professional of the healthcare industry is notified by the entry of a block in the blockchain and access to the patient health status to provide the monitoring activity remotely. This model application is so scalable that it can run over different types of platforms and hardware. The automation with the help of an AI algorithm eliminates the delay between the healthcare professional of the healthcare industry and the patient. Blockchain technology is immutable and temper proof so that the application working environment for the healthcare system would be robust and secure. Healthcare IoT for healthcare industries acts smartly by AI algorithms under the secured environment

V. Bibhu (✉) · L. Das · A. Rana · S. Sharma
Amity University, Greater Noida (An Atulyam Campus), Greater Noida, Uttar Pradesh, India
e-mail: vimalbibhu@gmail.com

L. Das
e-mail: lipsaentc9@gmail.com

A. Rana
e-mail: ajay_rana@amity.edu

S. Sharma
e-mail: ssharma4@gn.amity.edu

S. Salagrama
Information Technology, University of the Cumberlands, Williamsburg, KY, USA
e-mail: Shaila25@me.com

B. Bhushan et al. (eds.), *AI Models for Blockchain-Based Intelligent Networks in IoT Systems*, Engineering Cyber-Physical Systems and Critical Infrastructures 6,
https://doi.org/10.1007/978-3-031-31952-5_8

of blockchain. AI and blockchain create a smart interface between the patient and healthcare system where IoT enables regular monitoring and data sharing to the gateway from where the patient information is accessed and processed with accurate recommendations.

Keywords Blockchain · Healthcare · Artificial intelligence · Electronic Health Records · Distributed ledger · Internet of Things

1 Introduction

Artificial Intelligence is one of the common topics in the area of smart technologies. This has brought automation into the field of industry and many other applications. At present, almost all the engineering and science domains are adopting artificial intelligence to integrate it to provide intelligent behavior in the current and future systems. This induced intelligent behavior of the system especially, in the field of medical healthcare provides a bundle of opportunities to serve the patients and facilitates the medical services perfectly to save lives. It reduces the time and cost so that the treatment is started early which saves human lives in case of critical diseases even if the patients are in a remote area.

Blockchain is a distributed shared ledger that introduces the temper proof and immutable properties with the data which is stored in the block. These two properties of blockchain provide a high level of security with the data stored in the distributed shared ledger. The artificial intelligence model for blockchain advances the automation of healthcare industrial applications by incorporating security. The Healthcare Internet of Things requires a robust security infrastructure to protect sensitive data related to patients, prescriptions, healthcare professionals, and others. Blockchain-based artificial intelligence model induces the required high-level security framework for the required data. Healthcare Industrial IoT is based on the wireless network and Internet to transfer the sensor data to a centralized processing system, therefore, it is obvious that blockchain inserts the required security to data to not be hacked by any means.

Since the emergence of Bitcoin, blockchain technology has taken the foundation of cryptocurrency [1]. Besides this, blockchain technology has expanded for use in various categories of applications. The two most important applications of blockchain-based AI models are IoT and healthcare. Blockchain is used to exchange data between IoT devices and a centralized system to provide automated healthcare services to users. It is also true that security is one of the benefits availed by employing the blockchain in the field of healthcare and aligned IoT. Security of data is a very crucial component for the management of data at any stage such as data at rest, data at processing, and data in transit so the blockchain integration has ample capacity to provide security to all stages of data related to healthcare and IoT. In the field of healthcare, losing the patient's unique identity prevents messaging and makes the system inoperable [2, 3].

As it is considered blockchain technology, there is no accepted definition for it. Many professionals state that blockchain is a distributed digital ledger [4, 5]. Some of the authors stated that blockchain is a data structure used for the transaction management system. In the field of healthcare, blockchain technology offers a bundle of the benefits such as secured medical data records, Authorized access to patient medical records, instant communication from an authorized entity, tracking of the logs of medical details of a patient, and many more. Artificial Intelligence driven application under the plethora of blockchain is considered the most suitable for an industrial healthcare system with the Internet of Things. This is because AI provides automation, blockchain provides security and IoT can capture the current scenario of the patient's condition. This integration of AI, Blockchain, and IoT under a single entity for the healthcare industry makes the smart healthcare system the capability to prevent fatality through instant medical provision to the patients associated with the healthcare industry.

Electronic Health Records (EHR) are a primary ingredient in the healthcare system which contain patient records with various details in electronic format. This centralized framework of EHR enables the complex computation process with a huge amount of patient data with the help of AI algorithms and in secured states with blockchain technology. This tremendous capability of AI brings the healthcare industries to adopt to provide smart healthcare services to people around the globe. The blockchain and AI integration is shown in Fig. 1 [6].

These days, the healthcare system processes electronic data and reports of patients. The data about the patients is private and sensitive so it requires secured storage and processing by employing the blockchain. The assurance of data and identity related to the patient are very crucial aspects in front of industrial healthcare [7]. Therefore, the AI model for blockchain automates the processing and storing of patient data intelligently without any manual interaction with the computer system. In the context of healthcare, there are many designed systems are already available. These are HER,

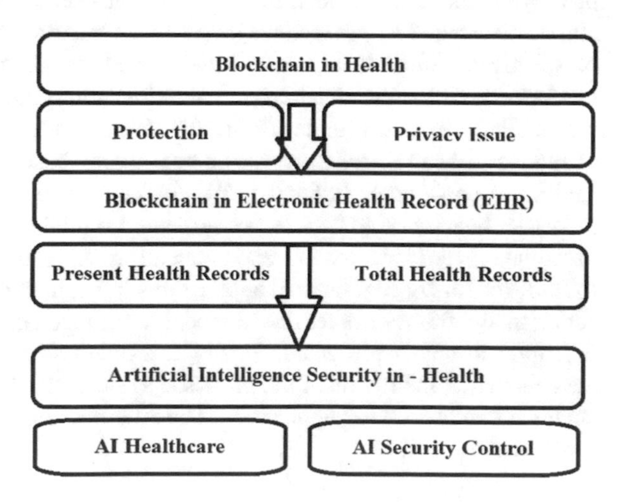

Fig. 1 Blockchain and AI integration for healthcare system [5]

Computerized Provider Order Entry (CPOE), Electronic Medical Record (EMR), Electronic Result Report (ERR), Clinical Decision Support (CDR), Personal Health Record (PHR), Telemedicine, and Health Management Information System (HMIS) [8, 9]. These all systems are independent and need a manual process to perform the task. AI Model for blockchain integrates all healthcare applications into a single entity to automate the functions of each in industrial healthcare IoT [10]. Further, IoT provides the information of patients instantly monitored by healthcare professionals from the healthcare industry.

In this paper, we have discussed and explored AI, blockchain, and IoT for the healthcare industries. The applications and other aspects of AI, blockchain, and IoT are deeply explored for better and smart healthcare services by the healthcare industries. The advantages of AI in the field of healthcare industries with their smart health services are detailed and also the risks associated with AI in healthcare industries are accounted for. The basics and application scenario of blockchain for the healthcare IoT and its feasibility with a suitable algorithm are being proposed. The challenges of IoT in healthcare industries are detailed with suitable examples.

This paper comprises various sections and sub-sections, in the introduction section we elaborate on the scopes of AI models based on blockchain for the applications of healthcare IoT. Discussed blockchain technologies, IoT, and their application scenarios for healthcare IoT systems. In Sect. 2, we elaborate on the AI and healthcare industries with detailed scopes of AI in healthcare, its industrial use for healthcare management, and diagnosis of health ailments. Advantages such as enhancement of accessibility of health facilities, prior diagnosis, cost reduction and faster healthcare processes, assistance in surgery, and mental health support are detailed based on AI and IoT applications. Further, we discussed the challenges of AI in healthcare industries. The challenges such as shortage of AI skill sets in healthcare, digitization, and consolidations of data, legislation, and regulatory challenges. Next, we elaborated risks of AI in healthcare industries. The risks such as injuries and errors, and privacy issues of health data. In Sect. 3, we elaborated and detailed the blockchain and its distributed ledger features such as the structure of blockchain, cryptographic perspectives of blockchain, network nodes and blockchain, and transaction process of blockchains like authentication, authorization, and proof of work. Further, in Sect. 4, we detailed the industrial healthcare IoT. We include the applications perspectives of industrial healthcare, smart health emergency based on IoT, IoT, and industrial healthcare challenges, In Sect. 5. We detailed the AI model for blockchain-based industrial healthcare IoT. Here, we presented a model for AI based on blockchain-based healthcare industry IoT and application. Also, an algorithm for an AI model for blockchain-based industrial applications in healthcare IoT is given. A communication model framework for the AI model for blockchain-based industrial healthcare IoT applications is presented. Finally, the result analysis is performed where we have tabulated the merits of the proposed system of AI model of blockchain-based healthcare industrial IoT application. The efficiency, accuracy, and supportability are summarized.

2 AI and Healthcare Industry

AI technology is a boon for the healthcare industry with its intelligent incorporation of functional provisions in the process of electronic healthcare based on information systems. A reliable artificial intelligent model in the healthcare industry encompasses fruitful results by enhancing the effectiveness of the services and availability of quick information related to patients. The healthcare professionals registered with the healthcare industry access the patient's medical records and artificial intelligence uses the proposed algorithm to decide to quickly prescribe the remedies. Service efficiency, time, and reduction of cost are ensured by the AI model in healthcare industries.

In the last decade, the use of AI in healthcare industries is increased for providing better medical services to people at a reduced cost. Almost all private healthcare industries are employing AI to model the automated version of the healthcare system to reduce cost and increase functional efficiency. Healthcare records in electronic form are the primary ingredient for the AI-based algorithm and application process and take the decision efficiently. Predictive analysis of health information with available healthcare records is one of the wonders in the field of healthcare industries. This predictive analysis provides prior information to the patients or people to follow the preventive guideline to reduce the chance to become a victim of a disease. This predictive analysis depends on the AI algorithms and data availability. A healthy lifestyle can only be recommended by healthcare industries when data and information aggregation through AI-based machine learning tools are employed.

AI is not just used for digital transformation in healthcare industries, but also it provides medical innovation to get a real-time automated system for better healthcare management. For example, virtual health assistants are an AI-powered technology that provides much support to patients in the healthcare industry. This virtual assistant provides facilities such as customer service representatives and tools for problem diagnosis and therapy. In the case of precision medicine, medical imaging, the discovery of the drug, and genomics the real power of AI is to reduce the failure rate by applying highly efficient automated processes. The AI-based programs and applications analyze a huge number of pathology images in case of cancer with high accuracy and predict the best possible anti-cancer drug combinations. In medical imaging, AI provides diagnostics for the spot details of the eye even spot analysis is not possible without AI-based systems. In the field of drug manufacturing, pharmaceutical campiness are using AI to reduce the development cycle. This is a boon for drug discovery and development for critical diseases by cost effort reduction.

2.1 Advantages of AI in Healthcare Industries

There is a bundle of advantages of AI in healthcare industries. The advantages are related to both patients and the healthcare industries.

2.1.1 Enhancement of Accessibility

Following World Health Organization (WHO), an 18.1-year gap in the expectancy of human life is recorded between poor and rich countries around the globe [11]. This is because of the zero-healthcare accessibility by the poor nation's inhabitants. With the innovation of AI, poor countries also can get the accessibility to healthcare and a better ecosystem of healthcare be facilitated. The AI and IoT-based digital platforms of the healthcare system can provide diagnosis and treatment to people in the poorest countries around the globe. AI-based dedicated applications have been developed to render help to national and international healthcare organizations to come together and provide the necessary assistance and help for healthcare to the people who require it. For example, during the global covid 19 pandemic, AI-based applications provided huge support to people around the globe to make self-assessments.

2.1.2 Prior Diagnosis

Disease diagnosis is a major challenge to healthcare professionals. If the actual cause of healthcare issues is not determined in a timely then the stages of some of the diseases advanced to create the complexity of treatment and cure. Tools and technology driven by AI rely upon the existing data of the previous diagnosis which provides a quick indication of findings of the issues by comparing the current with existing. The disease details samples are huge because these are taken from a huge number of people and stored electronically in the database system. AI-driven applications mine the relevant samples and predict the disease similarity by comparing and finding the rationales from the database [12]. This assists the healthcare professionals of healthcare industries to find out the patient disease timely and accurately. Additionally, the prediction for future health issues will also be taken by healthcare professionals for a patient on the ground of comparing the available symptoms and issues with a huge set of stored data in this category. For example, Verily is a google AI-driven application that forecasts genetic diseases.

2.1.3 Cost Reduction and Faster Healthcare Processes

AI applications and algorithms make healthcare processes faster with reduced costs. From tests to diagnostics of healthcare issues of a patient is faster with reduced cost. Biomarkers can easily be identified by AI, which can accurately and efficiently provide a suggestion about the disease in the human body. The manual process which takes time is avoided by AI algorithms to specify the biomarkers to identify the disease in the body. AI has the capability of massive automation which can save more lives as works faster than the manual processes of biomarker specification. The AI-based algorithm is cost-effective in comparison to the traditional methods of the processes of diagnosis and treatment. AI reduces the number of trips to patients to laboratories to evaluate the various health parameters as AI provides the prediction-based solution to the health issues of the patients [13, 14]. These advantages of AI-based algorithms in the healthcare system insists that 88% more healthcare industries get digitally transformed on AI-based platform to provide the more appropriate services of healthcare around the globe.

2.1.4 Surgery Assistance

The robotic system is derived from intelligence through AI-driven applications. In surgery, the machine learning implementation is derived from AI algorithms. In the field of surgery, there is a dedicated AI surgical system that executes with the tiniest movement having an accuracy rate of 100%. Therefore, it can be used in complex operations efficiently having low risks, side effects, low blood loss, and low pain. Also, recovery is faster and very easier after the surgery.

2.1.5 Mental Health Support and Enhanced Human Abilities

AI-based robots assist both medical professionals and patients by providing the ability for paralyzed ones to gain mobility without or with little help from the caretaker. These days smart AI-backed prostheses are suitably fitted with sensors that give more reactive limbs than the traditional methodology of medical processes. Also, the service robots having the machine learning deployment handle the regular tasks of the patients either on the healthcare premise or at the home of the patients. A dedicated companion robot is designed in such a way by use of AI and machine learning to provide the necessary support like tests and checks of blood glucose levels, systolic and diastolic blood pressure levels, temperature management, and pill takings prescribed by healthcare professionals [15]. Some of the robots have built analytical capabilities by AI and machine learning to help the depressed and make feelings positive by helping [16].

2.2 Challenges of AI in Healthcare Industries

There are many challenges yet to be with AI implementation in healthcare industries. These challenges are directly associated with the data involved with AI applications for analysis and prediction.

2.2.1 Shortage of AI Skilled Talent

AI skill shortage around the globe is a big issue in the healthcare industry's model and development the AI-based healthcare smart applications. The gap between the skill set requirement and the need for skilled AI professionals in healthcare industries is a bottleneck and it needs to be sorted out in near future to boost the healthcare services to people. Many AI projects are based on the garbage in and garbage out philosophy, without feeding the massive amount of data in AI systems [17]. Due to this fact, the result or output is erroneous. The non-availability of correct skill set professionals of AI leads the project to be built with a non-relevant set of data which outcomes go in the wrong result.

2.2.2 Data Digitization and Data Consolidation

Data digitization is a process that comes under digital transformation. Adopting digital transformation in healthcare industries needs data and skilled professionals in the information technology field. High-quality sourced data is always a tricky approach to determine which is very much essential to the healthcare industry. The fragmented, and versatility of data such as texts, images, and other multimedia types of data are not organized by a manual system of healthcare industries so organizing them in a suitable form requires a massive amount of time and effort along with the cost to the healthcare industries. Further, frequent changes in insurance companies by patients for healthcare insurance again insist the fragmented data to the healthcare and information technology professional to transform digitally [17, 18]. So, due to the mentioned fact, data acquisition is a major problem in the healthcare industry.

2.2.3 Legislative Regulations Update

Security and privacy are the most problematic domain in the information and digital era. These two issues bring the legislative eyes to the framework and deploy strict regulations for cyber security and information privacy. These types of regulations are more strictly employed over medical records globally. Hence, sharing the data with AI systems is considered a legal violation as per current regulations [18]. In some nations it is legal, but patient consent is mandatory before use for the defined purpose. Therefore, it is required to be a flexible rules and regulations for the acquisition of medical data with identity protection. Also, strict compliance must be employed by healthcare industries with accountability for using the patient health data for acquisition and use.

2.3 *Risks of AI in Healthcare Industries*

There are also risks associated with the use of AI in healthcare systems. These risks are not only from the AI technology side but also from the healthcare industry side. Some of the common and most important risks of AI in healthcare industries are to be assessed and evaluated to reduce the impacts.

2.3.1 Injuries and Errors

Systems are machines that are prone to provide an erroneous outcome. AI system is not different in this context, and due to this fact leads to the injury of a patient or any other significant issue of patient health [18]. For example, an AI-based system can recommend the wrong drug to the patient which leads the fatality for the patient. Similarly, AI-inspired radiological scanning may fail to detect the tumour.

2.3.2 Data Privacy Issue

Privacy of patient data is a bigger and more serious concern behind the acquisition. The risk of data hacking and leaking may bring damage the patient's image and trust in the healthcare industry or many other associated entities. Hackers always try to get private and sensitive data at any cost, so it is a potential risk with the data of healthcare industries [18]. This potential risk can be abated by protecting the data by employing the tools and technologies such as cryptographic security tools, and physical security measures. If the AI system is fed with inadequate data then the prediction will also be wrong which leads to the issue of diagnosis.

3 Blockchain and Its Distributed Ledger Features

Blockchain is defined as the chain of blocks containing some of the specific information. It is a ledger which constant grows and keeps the transactions permanently. In 1991, Stuart Haber and W. Scott Stornetta introduced the blockchain technology. They developed the computer based solution to timestamp the digital documents which could not be tampered or misdated. In 1992, Merkle trees formed a legal corporation on the system developed by Stuart Haber and W. Scott Stornetta. Due to this Merkle trees blockchain became more efficient to store and collect the documents more secured. System of digital cash was introduced in 2004 by cryptographic expert Hal Finney. This digital cash was the game changer for the blockchain and cryptography. This proposed digital cash was also able to solve the double spending problem through keeping ownership token registered on trusted server. In 2008, the distributed blockchain was introduced by modifying the Merkle tree. This distributed blockchain became more secure and capable to store history of exchange of data. This system worked on peer to peer network of time stamping and became the backbone of cryptography. In 2014, various digital currencies were developed on blockchain technology and this technology became popular in the field financial sectors. Afterwhile, blockchain are used to developed various applications such as smart contract, Ethereum, Litecoin etc.

Basically, blockchain is a tamperproof and immutable storage and transaction services to the data [19]. It means that once the block in the blockchain is added then it cannot be altered and deleted by any means. The accessibility of the block

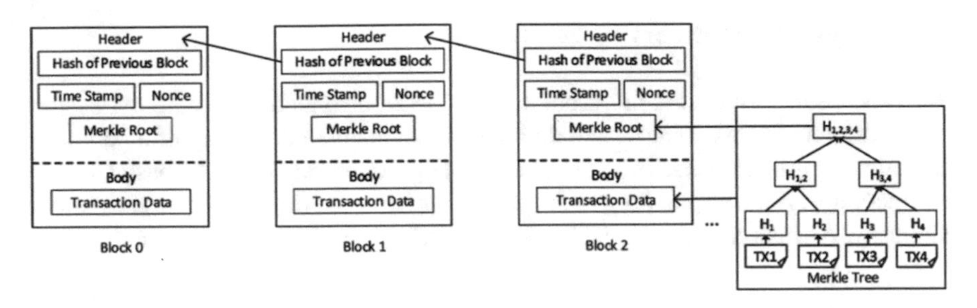

Fig. 2 Structure of blocks in blockchain [19]

information is also depending on the authorization. Therefore, only trusted and authorized entities can access the block of the blockchain. These properties of blockchain provide abundant security and privacy for data stored in a block of the blockchain. The healthcare IoT system mostly relies on Internet communication so, security and privacy become very important in the context of data abuse and misuse by cybercriminals. Blockchain solves the issues of security and privacy of healthcare data through its elegant features of accessibility, tamperproof nature, and immutability.

3.1 Blockchain Structure

The chain of the blockchain is created by the addition of blocks one by one. There is a header, hash value of the previous block, a time stamp, nonce, and Merkle root, and the body having transactional data is being stored [19]. The structure of blocks in the blockchain is stated in Fig. 2.

Each of the successive blocks as per Fig. 1, is linked with just its previous block. The body of each of the blocks in the blockchain stores the different values because it is a data placeholder in the blockchain to store transactional and other types of data. Also, the time stamp, nonce, and hash values for each of the blocks in the blockchain are different. The role played by timestamp and nonce to make the block unique in the blockchain.

3.2 Blockchain and Cryptographic Perspectives

Blockchain is a distributed shared ledger following the peer-to-peer network having a chain of blocks. Block holds the data value and the chain links the blocks concerning time. The linking of blocks uses the cryptographic solution by both asymmetric key cryptography and hashing. A pair of keys such as the private and public keys of the user is used to encrypt and decrypt the block data and Secured Hashing Algorithm 256 (SHA-256) is used to create the hash value of store in the successive block under

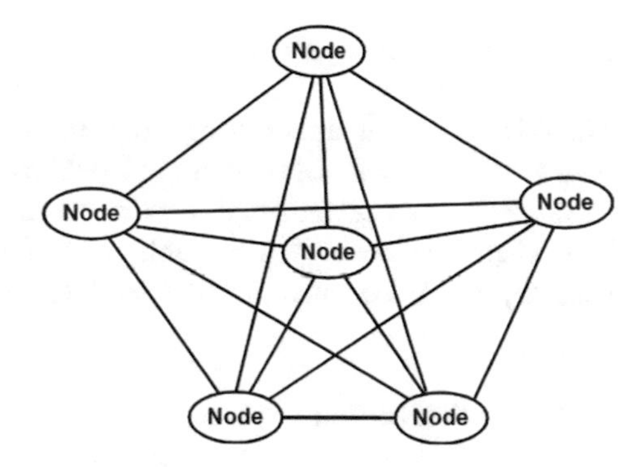

Fig. 3 Blockchain nodes connectivity

the blockchain. Other, benefits of hashing to the blockchain are providing the effect of avalanche, quickness, determinism, and uniqueness [19]. Further, if the block data is changed then the hash value automatically changed as SHA-256 uses the block data to create a hash value for linking, but this is not possible as once the block is added then it becomes tamperproof.

3.3 Blockchain and Network Nodes

A node in terms of networking is considered and known as a computer having processing, storage, and communication capabilities. In the case of blockchain, a node is connected to the blockchain network. The blockchain network is a fully connected network where all the nodes are connected to nodes of the blockchain [20]. The fully connected nodes of the blockchain network are shown in Fig. 3.

Once the node connects with the blockchain the data of the blockchain is downloaded to the node which fully synchronizes with the latest block of the blockchain. Basically, as per Fig. 3, blockchain follows the mess network topology where all the nodes connect with all other nodes in the network.

3.4 Blockchain Transaction Process

There are many pathways for a transaction to process with the blockchain. Authentication for a transaction to proceed with blockchain is performed by use of cryptography and authorization by use of cryptographic proof of work [21].

3.4.1 Authentication Process

There is no central authority of blockchain as it is distributed over the authorized nodes of the client. Even though blockchain is a decentralized technology, the transactions are authenticated by using cryptographic keys. Here each client of the blockchain has its key pair of public and private keys. These two keys of users are used to provide the authentication services in the blockchain.

3.4.2 Authorization

The authorization process of a transaction is used for the blockchain to approve the transaction of different users. In the case of the public blockchain, authorization is required to be approved then transaction to the blockchain takes place by adding the block in a chain of blockchain. Authorization and its approval are mandatory in blockchain to prevent unwanted and unauthorized access and use.

3.4.3 Proof of Work

To add the block to the blockchain proof of work is mandatory. The proof of work is created by solving a complex mathematical problem. This is done by the blockchain owner client first to add the block in a chain of the blockchain. This complex mathematical problem-solving process is termed mining.

4 Industrial Healthcare IoT

The Internet of Things (IoT), is a revolution in the area of healthcare industries. IoT has great potential to provide multiple applications such as remote monitoring and integration of medical devices. Specifically, in the field of healthcare IoT is used to provide monitoring of the patient, a sensor to get the health data of a patient, a detector to capture the patient's real-time health data, and store it in the centralized server to be analysed to make healthcare service better [22].

Smart sensors, healthcare wearable devices health monitoring systems are now IoT enabled so that these smart devices can be used to enhance the fruitfulness of the healthcare of people by the healthcare industries. Real-time efficient tracking of health information improves the overall process of the healthcare system by enabling doctors to take quick decisions about the prescription of the remedy. The errors taken by healthcare professionals due to manual diagnostics through the reports are minimized through the help of an IoT system [22, 23].

Wearable IoT devices such as glucose level detector, blood pressure monitor, heart rate indicator, oxygen level detector, and others provide the patient with real-time health-related data and shares the same with smartphone applications [23]. The smartphone further sends these data to a central service via the internet and that data is accessed by a healthcare industry professional to analyze and take the decision to improve the treatment.

An electronic Health Record (EHR) is a paperless health record of the patients. This paperless EHR provides the smart healthcare industry by centralizing the patient data which eases the accessibility to analysis and management [23]. This smart system enables the healthcare industries to facilitate emerging services such as engagement of the patient, streamlining communication between healthcare professionals and patients, tracking of the assets of healthcare industries, and optimization of workflow [24].

Due to IoT and mobile services, the face-to-face mode of the healthcare system with patients and healthcare professionals is rapidly changing [24, 25]. These technologies provide a different mode for healthcare professionals to care for their patients. For example, telehealth, remote monitoring of patient health, and telemedicine are saving patient's lives in case of health emergencies such as in case of heart attack, diabetes, or asthma attack.

4.1 IoT Application Perspective in Industrial Healthcare

There are many supportive and crucial roles played by IoT in healthcare industries. Some of the common roles are life-supporting implants, preventive healthcare measures, Drug support, and care, and monitoring of the health of a diseased patient, A survey says that 89% of older people around the globe like to stay at home alone and it will go to 2 billion till 2050 [26]. The healthcare facilities by healthcare industries can only be formulated through the IoT when a person would like to stay at home with monitoring and treatment. IoT wearable devices are equipped with smart sensors that can accurately get the health status and send the same to healthcare professionals of healthcare industries to evaluate the values to prescribe the solution remotely [26].

4.2 Smart Health Emergency System Based on IoT

The smart emergency application demands health information sharing among the different healthcare industries to evaluate the best possible supportive solution. A healthcare emergency requires the association of information availability and support to the patients by smart healthcare-enabled transportation, real-time constant monitoring, observation of the healthcare professionals, and many more to prevent fatality [26, 27]. IoT-enabled healthcare solutions can share information quickly with all the participants who are required in this case of emergency. Automated health information dissipation, status monitoring remotely, and advice prescription during transportation enable the proper care and prevention of the death of the patient health data communication provide efficient medical support on ad hoc basis to server the patient by doctor even the patient is remote.

4.3 IoT and Industrial Healthcare Challenges

IoT is no doughty a boon for the healthcare of people around the globe. Its massive applications in the various requirements of healthcare industries reduce the cost and increase the success of treatment. But, there are many challenges with IoT in the healthcare system. These challenges are the security and privacy of healthcare data, communication systems, and ownership management [28]. These days, cybersecurity criminals are so advanced with the latest hacking mechanism to break the security barriers of the deployed security frameworks of information and communication systems that are used with IoT to get sensitive information. This leads to serious harm to both patients and healthcare industries as they abuse these personal and sensitive healthcare data for defined goals set [29]. Therefore, IoT and healthcare industries need robust security technologies to secure the privacy and ownership of patients and their healthcare data in electronic form.

5 AI Model for Blockchain-Based Industrial Healthcare IoT

The AI model of the blockchain-based healthcare industry IoT application scenario is the integration of AI, blockchain, IoT, and healthcare systems having HER and other electronic data under the healthcare industries. All healthcare experts, doctors, and patients in-premise and out the premise of healthcare industries work on the virtual system defined through AI algorithms, IoT, and Blockchain systems [30]. A proposed AI Model for blockchain-based applications for the healthcare industry IoT is given in Fig. 4 [30].

The model given in Fig. 4 represents the complete integration of AI, blockchain, and healthcare IoT sub-systems into a single entity to create the smart healthcare industry. The interface for the healthcare experts and patients should have to be so intelligent that it auto corrects the misspelled entry to make the mining and other AI algorithms search and predicts the results for treatment. Support Vector Machine (SVM) provides the support to find out the appropriate ailments and diagnoses as per the current health status of the patient provided by IoT devices working either with remote or in-premise patients. Optimization of the score of finding by AI algorithms is matched by applying the deductive and predictive analysis of the patient current health information and symptoms by the healthcare professional experts and doctors [30]. Three step algorithmic process for the AI model for blockchain-based industrial application in healthcare IoT follows in the next subsection.

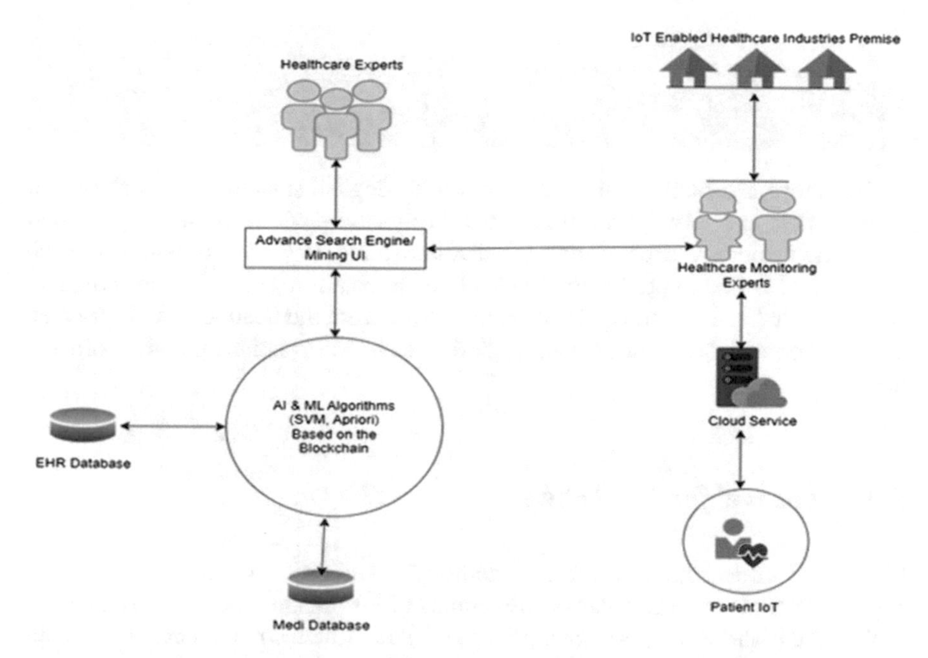

Fig. 4 An AI model of blockchain for application of healthcare industry IoT [30]

5.1 *Algorithm for AI Model for Blockchain-Based Industrial Application in Healthcare IoT*

The algorithmic steps to process the system for AI model for blockchain-based industrial applications in healthcare IoT are formulated as follows.

Step 1. IoT devices capture the health information of the patient (Remote/In-premise Patients)
Step 2. Generate random values from captured health data
Step 3. Create a Key by using a random value
Step 4. Create Block for captured health data
Step 5. Create a hash value by using a random value
Step 6. Add a block in the blockchain, with hash and random value
Step 7. Extract block by healthcare professional
Step 8. Compute hash and match with extracted
Step 9. If Same then
    Execute AI Algorithms to match the symptoms
    Predict and decide by Apriori AI Algorithm
    Extracted the resulting symptoms and medicines
    Analyze the results

Offer better treatment
Step 10. Discard the block

The above algorithm is straightforward with English sentences. This algorithm processes the IoT data from different IoT healthcare devices to the central cloud where the healthcare experts from the healthcare industry access them to run the proposed AI model to predict the issue related to health to prescribe medicine and other remedial action quickly. Healthcare experts from the healthcare industry also monitor the health of remote patients by getting health information through healthcare IoT.

5.2 *AI Model for Blockchain*

Blockchain is a decentralized ledger technology that offers the peer to peer computing between patients and healthcare professionals of healthcare industries. Blockchain enables the high-level security and privacy of the IoT data communication between the IoT devices to the cloud and healthcare professional systems.

The blocks of each of the healthcare data of a particular time interval with the patient are added into the chain which is downloaded through the cloud service to the healthcare professional [31]. Similarly, when a doctor prescribes remedial actions and medicine to the patients is also added to the chain of the block by creating a block by the AI model. The process of blockchain addition and extraction is only taken by authorized entities of healthcare industries and patients.

Patient:-

Addition:

Step 1. Gather IoT data from patient
Step 2. Create block
Step 3. Generate hash value, Nonce, and Timestamp
Step 4. Add the block to the chain
Step 5. Add hash value, Nonce, and Timestamp to block

Access:

Step 1. Solve complex mathematical problem as comes
Step 2. Select the chain

Step 3. Download the block
Step 4. Extract the data from the block

Healthcare Professional:-

Access:

Step 1. Solve the complex mathematical problem as asked
Step 2. Go through the chain of blockchain
Step 3. Download the block
Step 4. Extract data from the block
Step 5. Apply AI Model Algorithm
Step 6. Do symptom and prediction analysis
Step 7. Generate the prescription and remedial action

Addition:

Step 1. Solve complex mathematical problem as comes
Step 2. Select the chain
Step 3. Generate hash value, nonce, and timestamp
Step 4. Add the hash with successive block
Step 5. Add time stamp, nonce, and data block in the chain

5.3 *Communication Model of the Healthcare Industry*

The communication model is the important thing of the blockchain-based AI model for the application healthcare industry IoT. The communication Model framework for the whole system is presented in Fig. 5 [31].

Concerning Fig. 5, a Cloud-based IoT system communicates with healthcare professionals, and it gets the health status and data from various sensors and devices set up with the healthcare IoT devices [31]. Alerts and real-time health status of a patient are communicated to the healthcare professional and monitoring panels. AI and ML algorithms are employed with the cloud interface to analyze data to diagnose and predict the health status of a patient.

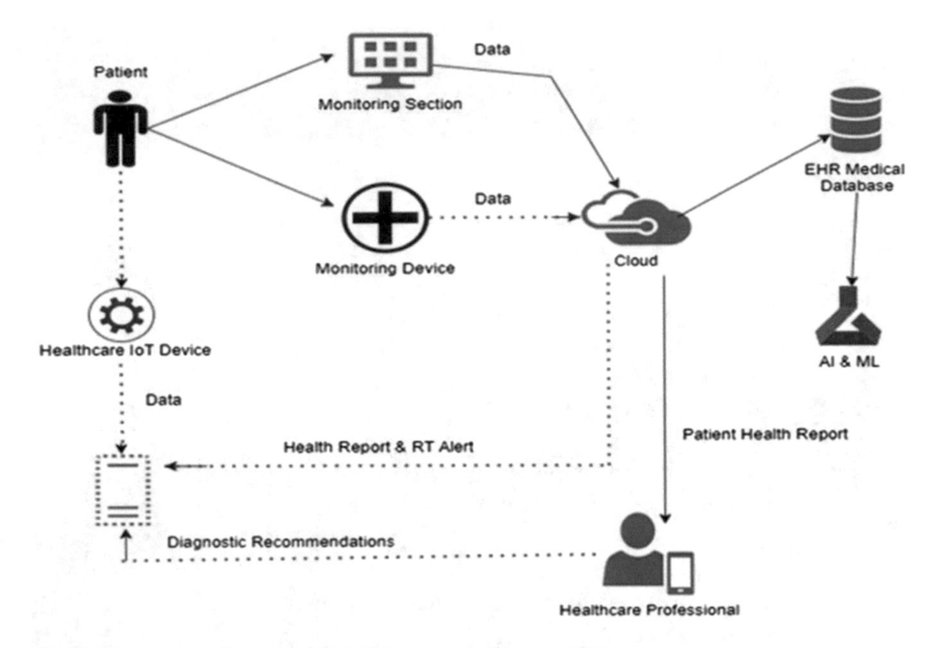

Fig. 5 The communication model of the proposed system [31]

5.4 Result Analysis

The proposed AI model for the blockchain-based application of the healthcare industry IoT is a primary system for the healthcare industry. This provides healthcare support and management to the people either in the case of in-premise patients also to remote patients [31, 32]. The functional aspects of the proposed system are highly secured due to the integration of blockchain for any type of transaction by healthcare professionals and patients for healthcare services. Again, IoT provides the data feeding to the system on which the healthcare professionals from healthcare industries apply the SVM and Apriori algorithm to diagnose the causes and predict the disease to fasten the treatment. The overall merit of the system based on time and accuracy for diagnosis, prediction, and treatment suitability to the patients is presented in Table 1.

Table 1 Diagnosis and treatment time delay for in and remote patients

Patient type	Healthcare data availability delay (S)	Diagnosis time delay (s)	Prediction time delay (s)	Treatment start time delay (s)
In premise	0	100	160	220
Remote	5	105	165	240

Table 1 shows the approximated delay as per the assumption of the functional system of the proposed model of industrial healthcare. This is not actual, and it may be a 10% fluctuation in the real-time scenario. Therefore, the conclusion of the system functional delay for the diagnostics, disease prediction by analysis, and discovery and treatment time delay is approximated in Fig. 6.

The efficiency of the system is assumed by considering the comparison values between the traditional healthcare application and this AI model for blockchain-based applications for the healthcare industry IoT [33]. The approximated value for efficiency in percentage for both systems of healthcare is summarized in Table 2.

The efficiency, accuracy, and supportability parameters approximated for both traditional healthcare and proposed systems are summarized in above Table 1. According to the approximated values for both systems, the proposed system value in percentage is higher in each of the cases such as in efficiency, accuracy, and supportability. This indicates that the proposed system is better than the traditional

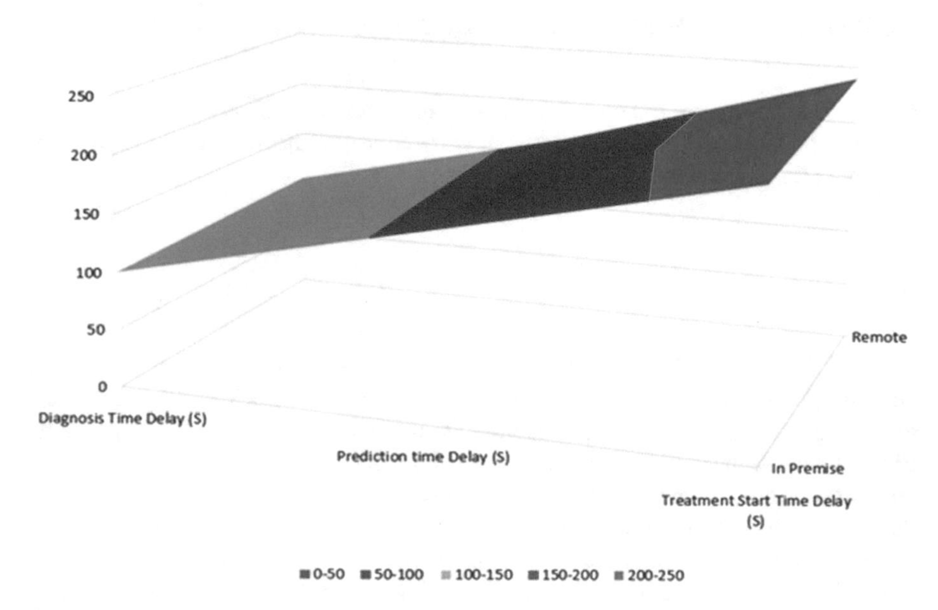

Fig. 6 Time delay by proposed system in its functional aspects

Table 2 Comparison efficiency, accuracy, and supportability

System type	Efficiency (%)	Accuracy (%)	Supportability (%)
Traditional healthcare system	60	70	40
AI-based proposed system	90	95	100

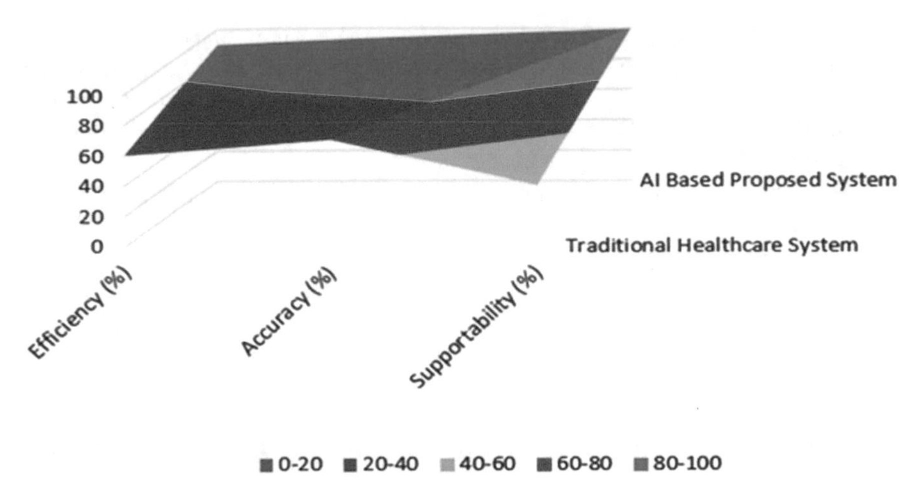

Fig. 7 Efficiency, accuracy and supportability result of traditional and AI based healthcare services

face-to-face healthcare system. The technologies such as AI, Blockchain, and IoT advance the functional aspects of the healthcare industry. The result is systematically presented through a graph in Fig. 7.

6 Conclusion

AI model for blockchain-based industrial applications of healthcare IoT is an emerging technological integration. The healthcare system is faster adopting the technologies such as AI, blockchain, and IoT to provide accurate and timely healthcare services to the people. IoT is one of the known technologies that enable things able to send and receive information via the network. Therefore, IoT-based healthcare devices are most useful for the monitoring and management of healthcare services to patients. AI provides automated services like decision making, diagnosis of disease, prediction of remedies, and much more very quickly by producing the intelligence by the algorithm to the system. Blockchain is distributed ledger which provides traceability and a high level of information security and privacy so that the requirement of medical data security is being fulfilled by employing the system of healthcare. In this chapter, we have proposed a novel AI model for blockchain-based industrial applications for healthcare IoT. All the assumptions and other requirements are modeled and shaped and results are presented by approximation. This chapter starts with an introduction and basic details of AI, Healthcare IoT, and blockchain are presented serially. In the last section, the design and model of AI for healthcare industry application is presented and the diagrams are represented through images. At, last we have assumed the comparison between the traditional and this proposed AI Model

Healthcare system for healthcare industrial applications. The results are tabulated and graphically represented by the use of tables and figures.

References

1. Bhushan B, Sahoo C, Sinha P, Khamparia A (2020) Unification of Blockchain and Internet of Things (BIoT): requirements, working model, challenges and future directions. Wirel Netw. https://doi.org/10.1007/s11276-020-02445-6
2. Kohli R (2016) Electronic health records: how can IS researchers contribute to transforming healthcare? MIS Q 40(3):553–573. https://doi.org/10.25300/MISQ/2016/40.3.02
3. Rabah K (2018) Convergence of AI, IoT, big data and blockchain: a review. Lake Inst J 1(1):1–18
4. Khan MA, Salah K (2018) IoT security: review, blockchain solutions, and open challenges. Future Generat Comput Syst 82:395–411. https://doi.org/10.1016/j.future.2017.11.022
5. Saxena S, Bhushan B, Ahad MA (2021) Blockchain based solutions to secure Iot: background, integration trends and a way forward. J Netw Comput Appl 103050. https://doi.org/10.1016/j.jnca.2021.103050
6. Vyas S, Shabaz M, Pandit P, Rama Parvathy L, Ofori I (2022) Integration of artificial intelligence and blockchain technology in healthcare and agriculture. J Food Qual, Article ID 4228448, 11 pp. https://doi.org/10.1155/2022/4228448
7. Bhushan B, Sinha P, Sagayam KM, Andrew J (2021) Untangling blockchain technology: a survey on state of the art, security threats, privacy services, applications and future research directions. Comput Electr Eng 90:106897. https://doi.org/10.1016/j.compeleceng.2020.106897
8. Edeh MO, Dalal S, Obagbuwa IC, Prasad BS, Ninoria SZ, Wajid MA, Adesina AO (2022) Bootstrapping random forest and CHAID for prediction of white spot disease among shrimp farmers. Sci Rep 12(1):20876
9. Ghouali S, Onyema EM, Guellil MS, Wajid MA, Clare O, Cherifi W, Feham M (2022) Artificial intelligence-based teleopthalmology application for diagnosis of diabetics retinopathy. IEEE Open J Eng Med Biol 3:124–133
10. Sethi R, Bhushan B, Sharma N, Kumar R, Kaushik I (2020) Applicability of industrial IoT in diversified sectors: evolution, applications and challenges. Studies in big data multimedia technologies in the internet of things environment, pp 45–67. https://doi.org/10.1007/978-981-15-7965-3_4
11. de Souza F, Rêgo L (2017) Life expectancy and healthy life expectancy changes between 2000 and 2015: an analysis of 183 World Health Organization member states. J Public Health 26(3):261–269
12. Liu C, Jiao D, Liu Z (2020) Artificial intelligence (AI)-aided disease prediction. BIO Integr 1(3):130–136
13. Curtis C, Gillespie N, Lockey S (2022) AI-deploying organizations are key to addressing 'perfect storm' of AI risks. AI Ethics. https://doi.org/10.1007/s43681-022-00163-7
14. Ball H (2021) Improving healthcare cost, quality, and access through artificial intelligence and machine learning applications. J Healthc Manag 66(4):271–279. https://doi.org/10.1097/jhm-d-21-00149
15. Bickman L (2020) Improving mental health services: a 50-year journey from randomized experiments to artificial intelligence and precision mental health. Adm Policy Ment Health Ment Health Serv Res 47(5):795–843. https://doi.org/10.1007/s10488-020-01065-8
16. Lal A, Pinevich Y, Gajic O, Herasevich V, Pickering B (2020) Artificial intelligence and computer simulation models in critical illness. World J Crit Care Med 9(2):13–19
17. Huang B, Philp M (2020) When AI-based services fail: examining the effect of the self-AI connection on willingness to share negative word-of-mouth after service failures. Serv Ind J 41(13–14):877–899. https://doi.org/10.1080/02642069.2020.1748014.

18. Flavián C, Casaló L (2021) Artificial intelligence in services: current trends, benefits and challenges. Serv Ind J 41(13–14):853–859. https://doi.org/10.1080/02642069.2021.1989177
19. Malik A, Gautam S, Abidin S, Bhushan B (2019) Blockchain technology-future of IoT: including structure, limitations and various possible attacks. In: 2019 2nd International conference on intelligent computing, instrumentation and control technologies (ICICICT). https://doi.org/10.1109/icicict46008.2019.8993144
20. Kaur M, Gupta S (2021) Blockchain technology for convergence: an overview, applications, and challenges. https://doi.org/10.4018/978-1-7998-6694-7.ch001
21. Glaser F (2017) Pervasive decentralization of digital infrastructures: a framework for blockchain enabled system and use case analysis. In: Conference on system sciences HICSS, Waikoloa, Hawaii, USA, pp 1–14. https://doi.org/10.24251/HICSS.2017.186
22. Islam SR, Kwak D, Kabir MH, Hossain M, Kwak KS (2015) The internet of things for health care: a comprehensive survey. IEEE Access 3:678–708
23. Bahalul Haque AKM, Bhushan B, Nawar A, Talha KR, Ayesha SJ (2022) Attacks and countermeasures in IoT based smart healthcare applications. In: Balas VE, Solanki VK, Kumar R (eds) Recent advances in internet of things and machine learning. Intelligent systems reference library, vol 215. Springer, Cham. https://doi.org/10.1007/978-3-030-90119-6_6
24. Onyema EM, Dalal S, Romero CAT, Seth B, Young P, Wajid MA (2022) Design of intrusion detection system based on cyborg intelligence for security of cloud network traffic of smart cities. J Cloud Comput 11(1):1–20
25. Pareta D, Verma IN, Lohani BP, Kushwaha PK, Bibhu V (2022) In: 2022 2nd International conference on innovative practices in technology and management (ICIPTM). IEEE, pp 369–373
26. Zafar A, Wajid MA (2020) A mathematical model to analyze the role of uncertain and indeterminate factors in the spread of pandemics like COVID-19 using neutrosophy: a case study of India, vol 38. Infinite Study
27. Akhtar SM, Nazir M, Saleem K, Haque HMU, Hussain I (2020) An ontology-driven IoT based healthcare formalism. Int J Adv Comput Sci Appl 11(2):479–486
28. Ali Z, Hossain MS, Muhammad G, Sangaiah AK (2018) An intelligent healthcare system for detection and classification to discriminate vocal fold disorders. Future Gener Comput Syst 85:19–28
29. Dash SP (2020) The impact of IoT in healthcare: global technological change & the roadmap to a networked architecture in India. J Indian Inst Sci 100:773–785. https://doi.org/10.1007/s41745-020-00208-y
30. Chamola V, Goyal A, Sharma P et al (2022) Artificial intelligence-assisted blockchain-based framework for smart and secure EMR management. Neural Comput Appl. https://doi.org/10.1007/s00521-022-07087-7
31. Sun J, Yao X, Wang S, Wu Y (2020) Blockchain-based secure storage and access scheme for electronic medical records in IPFS. IEEE Access 8:59389–59401
32. Lauritsen SM, Kristensen M, Olsen MV et al (2020) Explainable artificial intelligence model to predict acute critical illness from electronic health records. Nat Commun 11
33. Jabbar R, Fetais N, Krichen M, Barkaroui K (2020) Blockchain technology for healthcare: enhancing shared electronic health record interoperability and integrity. In: Proceedings of the 2020 IEEE international conference on informatics, IoT, and enabling technologies (ICIoT), Doha, Qatar, February 2020

Formation of a Recurrent Neural Network for the Description of IoMT Processes in Restorative Medicine for Post-stroke Patients

A. N. Trunov, I. M. Dronyuk, V. S. Martynenko, S. I. Maltsev, I. V. Skopenko, and M. Yu. Skoroid

Abstract The successful extension of the growth of 5G net and Wi-Fi devices applied in the healthcare technologies are considered innovative possibilities for the patients, especially moving from the local clinics to rehabilitation in their own home in family doctor technology. Based on Markov chains, improvement of the modeling tool is proposed as a solution to the collective problems of the description. It is shown that the estimate as the square of the ratio of the output to the input vector is determined by the ratio of the next to the prior probability of the states of the process for an arbitrary step of Markov chains. During modeling, a parameter change was ensured within 80% of changes about the average value. The isotropic properties of the transition matrix and the suitability for controlling the parameters were demonstrated during the numerical simulation. The indicator expands the possibilities of controlling the parameters of the simulation model as a whole. In addition, it is shown that due to the coordinated application of the recurrent network, opportunities have been created to improve the modeling algorithms of the structure of the model of recovery procedures for post-stroke patients. It is shown that the proposed calibration of the matrix elements in combination with the condition of normalization as a conditional probability for a set of simultaneous events will ensure further efficiency of the recurrent network in the operation of the complex model.

Keywords Internet of medical things · Wi-Fi devices · Markov chains · Neural networks · Calibration · Estimation instrument

A. N. Trunov · V. S. Martynenko · S. I. Maltsev · I. V. Skopenko · M. Yu. Skoroid
Computer Science Faculty Petro Mohyla Black Sea National University, Mykolaiv 54058, Ukraine

I. M. Dronyuk (✉)
Artificial Intelligence Department, Lviv Polytechnic National University, Lviv 79000, Ukraine
e-mail: ivanna.m.droniuk@lpnu.ua

B. Bhushan et al. (eds.), *AI Models for Blockchain-Based Intelligent Networks in IoT Systems*, Engineering Cyber-Physical Systems and Critical Infrastructures 6,
https://doi.org/10.1007/978-3-031-31952-5_9

1 Introduction

The successful expansion of the growth of 5G network and Wi-Fi devices used in healthcare technology is increasingly seen as innovative opportunities for patients, especially in the transition from inpatient rehabilitation in local clinics to recovery in their own homes using technologies of the family doctor [1]. A computing architecture based on single-board Wi-Fi controllers simplifies data monitoring and collection [2–4], as well as their use in the Internet of Medical Things (IoMT) [5, 6]. In addition, the prevalence of such gadgets as doctors' notebooks, nurses' smartphones, patient monitoring sensors, and other medical technologies accompany their widespread use [7, 8].

However, the further spread of such systems is restrained by a list of problems. Part of them is due to the comfortable fitting of auxiliary elements based on a wireless sensor network for complete long-term monitoring of health and activity [9]. Another is the sensors' suitability for unattended bio-medical non-invasive research [10]. In addition, the readiness of information technologies for data analysis for home medical care of the patient in the role of an expert system [11]. The suitability of hardware, algorithmic, and software of the network and its elements, for recently cheap and common details based on Arduino Uno, are increasingly used [12]. However, the description of the course of restorative procedures and reactions to them for a post-stroke patient during a random process remains such that it complicates the analysis and perception in the existing system of information tools.

The main investigation aim is to build a combined simulation model, with Markov chains and recurrent neural networks for the introduction of Wi-Fi network devices for implementation in rehabilitating medicine.

The article is organized in such a way: In Sect. 2, the brief literature review is presented; In Sect. 3 mathematical model is developed; part 3.1 is devoted to Markov chain modeling, and part 3.2 contains the neuron network implementation; In Sect. 4 the join model containing Markov chain and recurrent neuron network is constructed; Sect. 5 shows the simulation results for post-stroke patients and discussion; Sect. 6 contains the conclusion and the plan of some future work.

2 Analysis of Recent Publications and Identification of Unsolved Problems

A brief review of the problem state is presented. The main goals of publication are defined.

The successful integration of video streams from two or more cameras and time series formed at the output of Wi-Fi devices requires the solution of queue management problems, including recurrent networks, especially those suitable for operation in different modes [13]. As shown in work [14], the network's uninterrupted operation, especially IoMT, is significantly determined by access to the typical physical environment. The direct access protocol is carrier-specified multiple access with collision avoidance (CSMA/CA), efficiently distributing the packet queue between workstations. A network state transition diagram based on Markov chain transition diagrams and CSMA/CA schemes is considered to build a system of differential equations. Based on the solution of the system of differential equations, analytical expressions were constructed for the probabilities of the network station being in one of the possible states [14]. However, in the search for ways to compress streams without loss, the formation of a probabilistic model describing the course of procedures of a post-stroke patient did not ensure their practical use in the network of Wi-Fi devices [1]. Regardless of the functional significance and necessity, the task of improving the model due to the combined synthesis of the properties of Markov chains and artificial neural networks with the properties of express estimation of synaptic weights has not exhausted its hidden properties. The scientific basis of forming a structure suitable for recurrent reconfiguration and increasing the accuracy of the calibration of an artificial neural network is included with the analytical determination of the coefficients of synaptic weights and with recurrent reconfiguration [13, 14]. However, today there are unknown examples of such a complex neural network application that improves Markov chain models [15]. The proposed construction of the matrix of transitions by an expert method improves the properties of the model but does not verify its characteristic features. As a result, the obtained and researched example of the model built based on the hypothesis of the Markov properties of complex exchange of random flows improved the model by combining and implementing an expert matrix of transitions but still did not solve the problem of controlling the parameters of Markov chains and the simulation model. In addition, since a characteristic feature of models for network Wi-Fi devices of restorative medicine is the unification of various stream formats, the work deserves attention [16]. Thus, it demonstrated that at the current stage of the development of photogrammetry, there are methods for determining elements of internal orientation that improve accuracy and informativeness. The latter is carried out due to the calibration proposed in work even for non-metric cameras and systems [16]. However, the formation of a unified flow to expand the spatial perception of movements for medical recovery systems needs further development, according to the authors. The information technology based on Markov chains for calculating unique metrics of the real-time cyber-physical systems and can be used in telemedicine is presented in [17, 18]. A small data set is critical in medicine for insufficient to realize effective intellectual analysis [19, 20]. Mathematical modeling is useful in many medical problems [21, 22].

Thus, the main unsolved problem that will ensure the introduction of Wi-Fi network devices for rehabilitation medicine is the lack of methods for creating a combined simulation express model, which includes the simplicity of Markov chains and suitability for calibration and control of recurrent neural networks.

The goal is to expand the possibilities of controlling the parameters of Markov chains and the simulation model as a whole due to the coordinated use of a recurrent network to improve the modeling algorithms of the structure of the recovery procedures model for post-stroke patients.

To achieve the goal, the following tasks were formulated:

- to form a model for describing the activities of restorative procedures;
- establish a relationship between the ratio of the next to the previous probability of process states for an arbitrary step;
- to create conditions that will ensure the complex application of recurrent neural networks for probabilistic modeling under calibration conditions.

3 Features of Describing the Activities of Recovery Procedures by Markov Chains Using a Recurrent Neural Network

Markov chains develop modeling of the recovery procedures process for post-stroke patients.

3.1 Analysis of Features of the Applicability of Markov Chains for Describing the Activity of Restorative Procedures

It was considered, in the course of the measurement process according to the standard method prescribed by the Ministry of Health, sets of parameters of the patient's condition were recorded for this recovery procedure [1]. If the conditions of synchronicity of fixation as a measurement at one moment of time are fulfilled, then in cases the independence of the conditional distribution of subsequent states from all previous ones is observed, and the processes are Markovian with continuous time. Under such conditions, the initial probability distribution for n steps of a finite-dimensional discrete-time process completely determines and describes the start-up vector:

$$\overline{p} = [p_1, p_2, \dots p_i, \dots p_n]^T; \quad p_i = P(\overline{X}_0 = i); \quad i = \overline{1, n} \tag{1}$$

and the matrix, the elements of which are the values of transition probabilities:

$$\|P(\Delta t)\| = P_{ij}(\Delta t) = P(X_{\Delta t} = j | X_0 = i). \tag{2}$$

We substantiate the connection between the matrix of transition probabilities (2) and the probabilities of two transitions. We denote the probability at the initial moment of time as $P(t)$. In the next step through the time interval Δt, denote the probability, $P(t + \Delta t)$, then under the conditions of continuity and differentiability of the probability functions we write:

$$P(t + \Delta t) = P(t) + P'(t)\Delta t. \tag{3}$$

We will estimate the probability of transition at the next moment in time (3). Based on a well-known geometric inequality, we present the sum of arbitrary positive terms:

$$P(t+\Delta t) = P(t) + P'(t)\Delta t \geq 2\,[P(t)]^{1/2}\left[P'(t)\Delta t\right]^{1/2}. \tag{4}$$

Assuming that the function satisfies the above-mentioned differentiability property and the second factor allows the expansion of the root as a Newton binomial, we will find valid values for the upper and lower bounds of (4):

$$[P(t)][\delta(t)-2] \leq 2\,[P(t)]^{1/2}\left[P'(t)\Delta t\right]^{1/2} \leq [P(t)][\delta(t)+2].$$

In the latter, the weight factor is indicated:

$$\delta(t) = \frac{P(t+\Delta t)}{P(t)}$$

the ratio of the next to the previous probability of the states of the process, then under the conditions $\delta(t) \geq 2$ or the condition of non-negativity of the subroot expression $\delta(t) - 1 \geq 0$, the specified evaluation is performed. In other cases, based on the physical meaning of the definition of probability and its range of values from zero to one, we have:

$$0 \leq 2\,[P(t)]^{1/2}\left[P'(t)\Delta t\right]^{1/2} \leq [P(t)][\delta(t)+2]$$

Based on the obtained estimates, we write down to estimate the upper limit of the prognostic state of the probability of the next successive step of the Markov chain process:

$$P(t+\Delta t) \leq [P(t)][\delta(t)+2]. \tag{5}$$

The last condition (5) generalizes the Kolmogorov-Chapman equation to cases of arbitrary changes in the probability of the next successive step of the Markov chain process. In the absence of applicability conditions for Newton's binomial, we also write, if the coefficient on the right-hand side is one or less than it:

$$P(t+\Delta t) \leq 2[P(t)]\sqrt{\delta(t)+1}.$$

At the same time, it should be noted that there is a second approach to simplification. So, since the state of the Markov chains in the next step (the left part (4) evaluates the vector), we determine the square of its modulus:

$$P(t+\Delta t)^2 = \left[P(t) + P'(t)\Delta t\right]^2 \geq 4\,[P(t)]^2[\delta(t)-1].$$

Whence the upper limit of the weight coefficients determines the ratio of the modules of the state vectors of two consecutive steps:

$$\delta(t)|_{\max} \leq 1 + \frac{P(t + \Delta t)^2}{4\,[P(t)]^2} \tag{6}$$

and listed after each simulation step. Such new possibilities open the following steps for the structure of modeling algorithms.

Let us consider the daily activities of an imaginary patient undergoing rehabilitation post-stroke therapy and demonstrate the simplicity of finding the probability components of the condition in the next step.

Suppose that every day the patient can perform 3 actions, i.e. go through three possible states:

- the patient does not perform the procedure on this day (A);
- the patient performs the procedure, but does not perform the entire list of them in full (B);
- the patient does the procedure (D). Thus, we formed such a space of states, which is represented by a vector:

$$\overline{R} = [A, B, D]^T .$$

Also suppose that on the first day ($n = 0$, the original numerical sample distribution is known), the patient has a probability of 0 to start the procedure, and to attempt the procedure on the first day has a probability of 0.5, while to perform the entire procedure has a probability of 0.5. Then, numerically, the vector describing the initial state has the following form:

$$\overline{R} = \left[0;\ 0.5;\ 0.5\right]^T$$

Also, suppose that the following probabilities are observed:

- when the patient does not take the procedures one day, there is a probability of 0.20 not to undergo the procedure the next day, and the probability of just trying is estimated at 0.5, while visiting and undergoing the procedure is 0.30;
- when the patient does not take procedures one day, there is a probability of 0.5 to undergo the procedure again the next day not in full, and also a probability of 0.5 to visit and undergo the procedure in full;
- when a patient does not take a procedure one day, there is a probability of 0.32 that he will not take it the next day, and the probability of only trying but not performing is estimated at 0.32, and fully performing the procedures is estimated at a probability of 0.36.

The specified information about the course of probability values $\|p\|$ will be presented as a matrix for the selected numerical example:

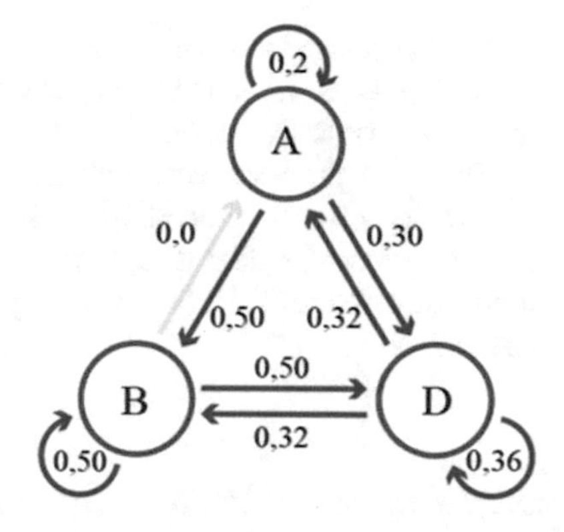

Fig. 1 Schematic representation of an example of procedures, displaying Markov chains as graphs

$$\|p\| = \begin{bmatrix} 0.20;\ 0.50;\ 0.30 \\ 0.00;\ 0.50;\ 0.50 \\ 0.32;\ 0.32;\ 0.36 \end{bmatrix}$$

The probability of each state at the next step ($n = 1$, i.e. for the next procedure) will be calculated:

$$\overline{L} = \overline{R}^T \|p\| = \begin{bmatrix} 0;\ 0.5;\ 0.5 \end{bmatrix} \begin{bmatrix} 0.20;\ 0.50;\ 0.30 \\ 0.00;\ 0.50;\ 0.50 \\ 0.32;\ 0.32;\ 0.36 \end{bmatrix} = \begin{bmatrix} 0.16;\ 0.41;\ 0.43 \end{bmatrix}.$$

The dynamics of an example of the patient's proposed actions described by Markov chains can be graphically represented in the manner shown in Fig. 1.

Such a graphic representation of a probable imaginary post-stroke patient with unusual behavior in the form of graphs that display Markov chains and on the edges of which the values of the probability of transitions are shown, gives a qualitative-quantitative idea of the nature of the course of procedures and reflects discipline in following the recommendation. The Fig. 1 represents a graphic-numerical solution of a direct problem with a known vector of initial states and a matrix of transition probabilities. However, no less important are the tasks of establishing such probabilities of the matrix of transitions with known initial vectors and forming predictive estimates for actions when coordinating the course of procedures. A set of conditional (for a given input vector) samples makes it possible to choose personal recommendations and to define the order and set of prescription parameters for it. Let's highlight the aspects of this process.

The obtained probability vector of each state at the next step ($n = 1$) is equal to $|\overline{L}| = 0.6153$ under the conditions of the value of the modulus of the vector of initial states $= 0.7071$. The latter allows you to calculate the limit value of the weight coefficient according to (6):

$$\delta(t)|_{\max} = 1.1893.$$

Further analysis of the values of the vector modules and the range of values of the weight coefficient and its upper limit shows that their deviations are beyond the limits of permissible errors.

Thus, the process needs repeated refinements. However, the question remains how to organize the process of calculating the weight coefficient and what ensures its convergence? Under what conditions does the process allow convergence and is it convergent at all? And even if the process converges, how to estimate its error at each step?

3.2 Analytical Presentation of a Recurrent Neural Network for the Description of the Course of Procedures in Recovery Devices That Implement the IoMT Principle Without an Operator

The essence and consequences of the stroke itself, as a result of previous changes in the human body or extreme loads and influences, creates unpredictability of t transition matrix elements when describing the process. Chains of transitions, and micro ruptures of capillaries and blood vessels are accompanied by hemorrhage and disruption of metabolic processes. They are usually the cause of unpredictable functional disorders of the nervous system of mental activity and partial motor activity of individual muscles in the form of spasms or weakness. Attempts of tomographic methods do not yet provide opportunities to distinguish differential connections between the specific causes of disorders of impulses and hemorrhages and the consequences—spasms and exhaustion. In this regard, the search for methods of description that take into account the nature of features of the random-regular essence and manifestations of regular-random consequences should be foreseen. The description of the neurosynaptic connections of the brain, including that of the brain damaged by micro-ruptures and clamps of hemorrhages and their consequences, is obviously more logical to be carried out by artificial network methods. However, the ability to adjust and rebuild the recurrent network, especially in the presence of a method of analytically determining the coefficients of synaptic weights and calibration, revealed innovative properties of recurrent networks [13].

To study the effectiveness of this approach, it was assumed that there is an intelligent generator of three values: X_n is a vector of independent value of the argument, the output value of the sensor signal $L(X_n)$, reference signal (law) $L_s(X_n)$ [13]. Also, assuming the existence of such a law of density distribution with given mathematical expectation and root mean square deviation, there is a mode of automated formation of the output of the reference device and fixation of its value in the system memory. Under these conditions, it is advisable to also record the error value of reference displays and Markov chains calculated by step and argument. Also, as in the artificial recurrent network [13] with the ability to calibrate, the calibration range and the calibration factor are determined and stored in the database as a function of the independent value of the argument X_n:

$$k(X_n) = L_s(X_n)/L(X_n).$$

The calibration mode provided in the network functions is ensured by the formation of the physical value of the X_n. However, since all processes are synchronized with the internal countdown of the timer, it is time-dependent. Under these conditions, a stable time step was imposed Δ and the selected set of condition-action rules. Stable trigger time step Δ, provides an unequivocal transition from the waiting phase $[t = t(n-1)]$ to the triggering phase $[t = t(n-1+\Delta)]$. The jump in the output of the regulator with the onset of the triggering phase causes a jump in the input signal.

The calibration mode provided in the network functions provides due to the formation of the physical quantity X_n. However, since all processes are synchronized with the internal timer, it is time-dependent. Under these conditions, the need for a stable time step Δ was assumed, and a set of "condition-action" rules was chosen. A stable time step of activation Δ provides an unambiguous transition from the waiting phase $[t = t(n-1)]$ to the activation phase $[t = t(n-1+\Delta)]$. A jump in the regulator output with the onset of the activation phase causes a jump in the input signal:

$$X_n = X_{n-1} + \delta(n)a_1 + \delta(n-1)a_2,$$

which over time forms new output signals $L_s(X_n)$ and $L(X_n)$.

Thus, if you choose a random number generator generated from a series of natural numbers, it is possible to combine Markov chains and recurrent networks. The implementation of the above forms a system, that allows you to choose, form, and refine the law of density distribution under the features of the random process:

$$\begin{cases} p_{ij} = \int\limits_{1}^{\infty} \frac{1}{\sigma_{ij}\sqrt{2\pi}} e^{-\frac{(x-a_{ij})^2}{2\sigma_{ij}^2}} dx; \\ \frac{1}{a_i} \sum\limits_{i=n,1}^{j=m} p_{ij} = 1;\ i = \overline{1,n}\ j = \overline{1,m} \end{cases}. \tag{7}$$

For this, it is enough to form the elements of the row of the transition matrix using the Monte Carlo method followed by normalization by all their values, which guarantees to find the constant a_i from the second equation of the system for each row of the transition matrix. The formed systems (7) and the proposed calibration of the matrix elements and the condition of normalization as a conditional probability for a set of simultaneous events will ensure further efficiency of the recurrent network.

4 Modeling and Formation of a Probabilistic Model of the Patient of Markov Chains Using a Recurrent Neural Network

A probabilistic modeling with recurrent neural network is calculated. The calculated cases prove the effectiveness of the proposed model.

To carry out the simulation, first, the influence of the value of the central element of the transition matrix on the upper limit of the deviation coefficient was studied. The calculations are given for the example of the initial vector R and the transition matrix:

$$\|p\| = \begin{bmatrix} 0.20;\ 0.50;\ 0.30 \\ 0.00;\ 0.50;\ 0.50 \\ 0.00;\ 0.50;\ 0.50 \end{bmatrix}.$$

Modeling of the influence of Markov chain parameters was carried out taking into account normalization conditions for each matrix term, and its numerical results are presented in Table 1. Analysis of the data in Table 1 shows that the module of the output vector |L| is not sensitive to changes in the P_{22} element of the transition matrix, the value of which is determined by the parameters and the probability density law. Thus, its change by 80% causes a maximum change in the weight coefficient according to (6) by only 3.2% (see Table 1).

It should be noted that the influence of changes in the transition matrix element P_{32} is presented in Table 2. As evidenced by the calculations when the normalization conditions are applied, the results of the calculations of the output vector and the weight coefficient at a fixed value of $P_{22} = 0.5$ are the same.

Thus, regardless of the independence of the rows of the transition matrix in Markov chain models, the isotropic effects on the module of the output vector and the weight coefficient (see Table 2) under normalization conditions are observed even in one dimension. The introduced coefficient of estimation indicator according to (6) based on a geometric inequality clearly demonstrates the model's state. Using other tools in the quantitative assessment of functional analysis operators opens wide opportunities for further refinement and narrowing.

Table 1 Analysis of the model parameters influence on the output vector and coefficient of deviations (case 1)

№	P_{22}	|R|	|L|	$\delta(t)\|_{max}$
1	0.5	0.707107	0.707107	1.25
2	0.6	0.707107	0.710634	1.2525
3	0.7	0.707107	0.72111	1.26
4	0.8	0.707107	0.738241	1.2725
5	0.9	0.707107	0.761577	1.29

Table 2 Analysis of the influence of model parameters on the output vector and coefficient of deviations (case 2)

№	P_{32}	|R|	|L|	$\delta(t)\|_{max}$
1	0.5	0.707107	0.707107	1.25
2	0.6	0.707107	0.710634	1.2525
3	0.7	0.707107	0.72111	1.26
4	0.8	0.707107	0.738241	1.2725
5	0.9	0.707107	0.761577	1.29

In addition, the formed systems (7) make it possible to choose, form and refine the law of density distribution in accordance with the features of the random process. Also, the proposed calibration of the matrix elements in combination with the normalization condition, as a conditional probability for a set of simultaneous events, will ensure further efficiency of the recurrent network.

In the second modeling stage, attention was focused on studying the peculiarities of solving the problem of forming a model describing the activities of rehabilitative procedures and finding a relationship between the ratio of the next to the previous probability of process states for an arbitrary step. The number of states considered was limited. Three states of two modules define possible variants of stages of functioning, as procedures that, by the standards of the Ministry of Health, are performed by the automated control system (ACS) of post-infarction and post-stroke remote recovery A, B, D. Thus, event A is the state of the system in which the procedure of electrical stimulation of muscles that have experienced spasms. Event B is a state of vacuum massage, during which the formation of shear waves intensifies lympho-blood circulation. Event D is a state of periodic contractions of a separate volume of muscles, which is performed by the grip of the manipulator with a controlled amount of contact pressure.

At the third modeling stage, ACS was considered for conducting remote recovery procedures. Unlike the existing system of recovery modules, it involves the integration of modern computer-integrated technologies, modern research and recovery devices [1, 12], and IoT into the treatment system—the family doctor [1]. The block diagram showing the set of connections and the interaction of the main elements is presented in Fig. 2.

Unlike existing systems, monitoring and rehabilitative modules provide controlled recovery at home for post-infarction and post-stroke patients and military personnel who are ill or injured. In the conditions of many broken Ukrainian hospitals, ensuring a controlled recovery becomes an urgent task since the number of places in hospitals has significantly decreased. Unlike analogs, the system of modules under development implements control and logging during procedures. It differs from SaeboMAS Mini, SaeboFlex, and SaeboReach, systems of Continua Health Alliance, PhysioDroid, Forastiere, or from analogs of the research project of the European Union AMON.

The advantages will be:

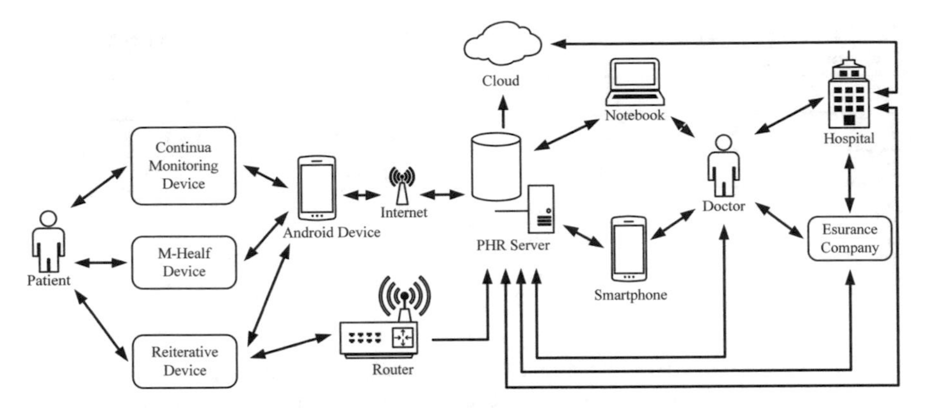

Fig. 2 Schematic representation of the structure and connections of the remote rehabilitative ACS, which is integrated into the system of the family doctor

- in the increasing number of procedures, as they go from one type of procedure (in existing samples) to 5, their list, sequence, and parameters are chosen according to the patient's condition, and the number of which can increase;
- network integration and service by external personnel, which ensures the requirements of accessibility, openness, and provability;
- provision of electrical stimulation cycles that form a special course, adjustable tracking frequency, and layering in the range from 0.25 mm to 0.8 mm, which improves micro lymphatic and micro blood flow and together with the creation of vacuum waves, relieves spasms and edema;
- conducting recovery under supervision at any time through remote access to complete data on the patient's condition and reaction to the course of procedures.

5 Simulation Results

The numerical simulation for proposed model are realizing.

It was proposed to divide the parameter control of the process into two groups to study the features of the optimal models of functioning and search for a set of transitions, the formation of recommended algorithms and parameters of the course of procedures, and the selection of their sequence. The classification was carried out according to the sign of temporal applicability: continuous monitoring and measurement on demand. As indicated in Fig. 2, permanent control, according to the recommendations of the Ministry of Health, is carried out periodically by a group of devices and devices from the continuous monitoring unit Continua Monitoring Device. On-demand or selective or special is carried out as needed and at the request of the ACS by a group of devices and devices from the block of the M-Healf Device type. While examining the patient, these two groups of devices and devices interact with the ACS via a Wi-Fi channel using Android smartphone devices in the data transmission and

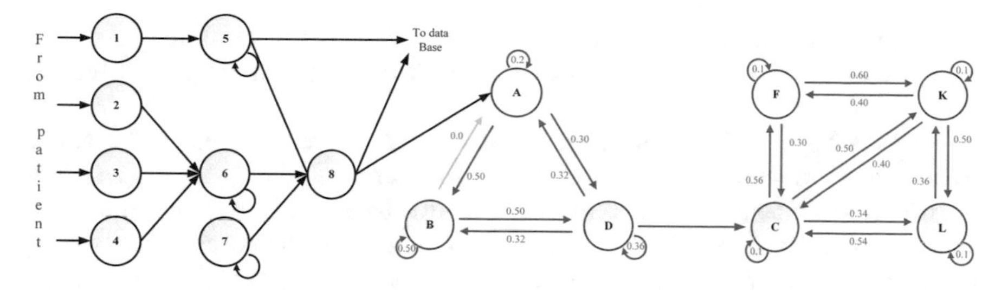

Fig. 3 Schematic representation of the combination of neural recurrent network and Markov chains

storage chain: Android Device—Internet—PHR Server—Cloud. The family doctor, the hospital administration, and the insurance company have access to three of the four elements of the specified communication chain. For effective communication, the system provides connection via a PC or laptop and an Android smartphone, which is regulated by administrative law and internal instructional and regulatory documents of actions the institutions and persons are involved in [1].

To carry out simulation modeling of processes with the help of a Markov chain model of two modules combined in ACS (Fig. 2), it was considered a combined system of a recurrent network and Markov chains.

A schematic representation of the transitions and connections of such a system is shown in Fig. 3.

Here, color marking was used for the transparency of the scheme of presentation. Nodes of the recurrent network, synapses, and connections are marked in black, and the states of the Markov chain system are tentatively marked in red, while the transitions and their probabilities are marked in blue. For the recurrent network, circles with numbers conventionally mark devices that determine the time course of the patient's condition. The input arrow indicates the process of attracting input information to measure the parameters that determine the condition, and the output arrows indicate the data characterizing the patient's condition. We will provide them with the appropriate set of input vectors and output vectors. All outputs of eight nodes of the recurrent network are directly connected to the database (see Fig. 3); these connections are conditionally not shown. The latter ensures orderly data transfer. Thus, conditional node 1 represents a mobile electrocardiograph equipped with a wired (optional connection) and Wi-Fi communication channel, which provides the electrocardiographic diagnostic system with data. Laptop diagnostic software and an expert family doctor assess patients' fitness for procedures with physical loads. Conditional node 2 represents equipment suitable for determining the upper and lower pressure levels, which is also equipped with dual communication channels. Similarly, node 3 provides information about the time course of the pulse. Node 4, unlike the first three, provides qualitative information about paramedical indicators using a survey using a separate system of questions and counting. Nodes 5 and 6 of the synapse of the second layer of the recurrent neural network, which form information about the patient's condition, which is recorded and which is activated by synapse

8 in the presence of signs of relevant information from the thermal imaging study, initiate action A and forms an initial vector that describes the initial probabilities of states.

States A, B, and D retain the original content, while states C, F, K, and L have a range of new events. Thus, event C—determines the state of spasmodic or painful examination based on the quantitative data of infrared and qualitative data of paramedical examination. Event F—defines articulation exercises for leg muscles. Event K—defines articulation exercises for the muscles of the hands. Event L—defines exercises containing the cognitive load.

For modeling the third stage, we will use the output vector of the system of the first stage:

$$\overline{R} = \left[0.16;\ 0.41;\ 0.43\right]^T$$

And form the corresponding square matrix 3 × 3 matrix of transitions:

$$\|p\| = \begin{bmatrix} 0.10;\ 0.56;\ 0.60 \\ 0.10;\ 0.34;\ 0.36 \\ 0.10;\ 0.00;\ 0.50 \end{bmatrix}. \tag{8}$$

Let us define the process of considering three events as the transition from state C to state K. Let's define the output vector of the event K. The probability of each state at the next step (n = 1, i.e. for the next procedure) will be calculated:

$$\overline{L} = (L_1, L_2, L_3) = \overline{R}^T \|p\| = \left[0.16;\ 0.41;\ 0.43\right] \begin{bmatrix} 0.10;\ 0.56;\ 0.60 \\ 0.10;\ 0.34;\ 0.36 \\ 0.10;\ 0.00;\ 0.50 \end{bmatrix} =$$

$$= \left[0.1;\ 0.229;\ 0.4586\right].$$

We will demonstrate by modeling how the patient's condition affects the choice and sequence of procedures. The dynamics of the modeling example and the actions offered to the patient, described by Markov chains, are graphically presented in Fig. 4.

Thus, the influence of the parameters of the input vector and step n on the output probability vector is presented on the graph. Its components L1, L2, and L3 respectively indicate the probability of choosing each type of procedure: electrical stimulation, vacuum massage, and compression with the controlled force of the muscles that have experienced spasms.

The dynamics of the simulation of the second group of states C, F, K, and L for actions and options for possible transitions with the same input vector to state A at each of the n steps are graphically presented separately in Fig. 5. The influence of the parameters of the input vector and step n on the output probability vector is presented here. Its components are similarly labeled L1, L2, and L3, respectively, indicating the probability of choosing a transition from procedure C to event K, which determines articulation exercises for the muscles of the hands. Note that its

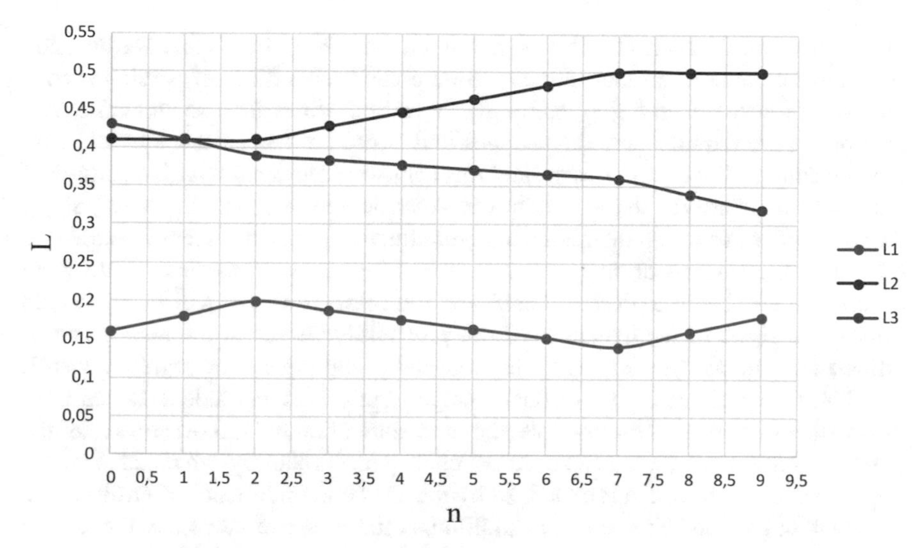

Fig. 4 Graphic representation of the probability of states A, B, and D stepwise changes in the components of the output vector L1, L2, and L3, respectively

achievement is assumed by three separate transitions, the probabilities of which are given by the rows of the transition matrix (8).

The dynamics of the probability, firstly, proves the variability of the preference of the procedures. Secondly, it shows at which step, according to the minimum limit, it is useful to use them only with a lower sequence or at what limit value of the probability to reduce the sequence or not to implement the procedure at all. The minimum

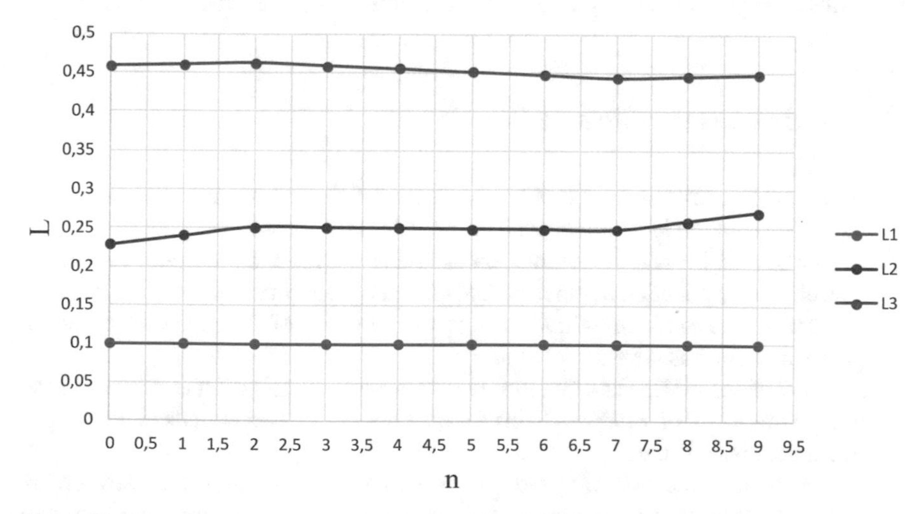

Fig. 5 Graphic representation of the probability of transitions from state C through states F, and L to state K, stepwise changes of components of the output vector L1, L2, and L3, respectively

probability limit is determined by the maximum number of inputs and is calculated from the condition of normalization of conditional probabilities of parallel events. Analysis of curves (see Fig. 4) shows that electrical stimulation as an anticonvulsant procedure requires periodic changes and is not the first priority. In addition, the procedures of vacuum massage and mechanical compressions with a controlled amount of force are equally important in the considered segment of recovery steps. Although after increasing their number, their effectiveness will decrease in anticipation. With the progress of the number of procedures received, mechanical massage procedures will begin to decrease, and vacuum massage waves will begin to gain priority. Analysis of the dynamics of the probability curves of the next group of articulating exercise procedures for leg muscles F and arms K or cognitive load L (see Fig. 5) represents both a gradual change in priorities and independence from the number of steps. So, for example, the procedure of articulation exercises for the muscles of the legs F (curve L1) does not change its probability, which confirms its importance as such, which is explained by the need and importance of ensuring the quality of large blood circulation. In addition, in the course of modeling, it becomes obvious the need to introduce tools that would allow representing individual states as a set of additional interconnected states of Markov chains or fragments of integrated networks that allow calibration.

Thus, the coordinated application of the recurrent network improves and expands the capabilities of the modeling algorithms of the model structure for the selection of individual recovery procedures for post-infarction and post-stroke patients. Possibilities for choosing the priority and sequence of procedures are opened thanks to the formed model describing the dynamics of the probabilities of restorative procedures and transparent visualization. The tool for selecting the weight coefficient of Markov chains according to (6) was formed, along with the calibration tool and system (7) for further analysis and setting of the problem of the minimal structure of ACS.

6 Conclusions and Future Works

1. Based on Markov chains, a tool has been created that allows for evaluation, and its properties as an indicator expand the possibilities of controlling the parameters of Markov chains and the simulation model as a whole. In addition, due to the coordinated application of the recurrent network, opportunities have been created to improve the modeling algorithms of the structure of the model of recovery procedures for post-stroke patients.
2. A model of description as a quantitative assessment of the activities of restoration procedures based on Markov chains for the upper and lower bounds of the weight coefficient was formed.
3. The estimate of the ratio of squares (6) is defined as the relationship between the ratio of the next to the previous probability of process states for an arbitrary step of Markov chains. During modeling, a change in parameters was ensured within 80% of changes in relation to the average value. The isotropic properties of the

transition matrix and the suitability for parameter control were demonstrated during the numerical simulation.

4. To create conditions to ensure the complex application of recurrent neural networks for probabilistic modeling under calibration conditions. It is shown that the proposed calibration of the matrix elements in combination with the condition of normalization as a conditional probability for a set of simultaneous events (7) will ensure further efficiency of the recurrent network in the operation of the complex model.
5. Future works is planning for extend the possibilities of telemedicine. The extension proposed model for children rehabilitation is planning. Next one is the extention for using proposed model in medicine for recovering patient after myocardial infarction diseases.

References

1. Trunov A, Beglytsia V, Gryshchenko G, Ziuzin V, Koshovyi V (2021) Methods and tools of formation of general indexes for automation of devices in rehabilitative medicine for poststroke patients. Eastern-Eur J Enterp Technol 4(2–112):35–46. https://doi.org/10.15587/1729-4061.2021.239288
2. Gay V, Leijdekkers P (2007) A health monitoring system using smart phones and wearable sensors. Int J ARM 8(2):29–35
3. Otto CA, Jovanov E, Milenkovic A (2006) A WBAN-based system for health monitoring at home. In: Proceeding of 3rd IEEE/EMBS international summer school on medical devices and biosensors, pp 20–23
4. Baraka A, Shokry A, Omar I, Kamel S, Fouad T, El-Nasr MA, Shaban H (2012) A WBAN for human movement kinematics and ECG measurements. E-Health Telecommun Syst Netw 1(02):19
5. Roşu M, Paşca S (2013) A WBAN-ECG approach for real-time long-term monitoring. In: Proceedings of 8th international symposium on advanced topics in electrical engineering (ATEE). IEEE, pp 1–6
6. Al RM, Lee BH, Sudarsono A (2015) Implementation of body temperature and pulseoximeter sensors for wireless body area network. Sens Mater 27(8):727–732
7. Kommey B, Kotey SD, Opoku D (2018) Patient medical emergency alert system
8. Saha R, Biswas S, Analytical study on data transmission in WBAN with user mobility support. In: Proceedings of international conference on wireless communications, 2018, signal processing and networking (WiSPNET), pp 1–5
9. Lee YD, Chung WY (2009) Wireless sensor network based wearable smart shirt for ubiquitous health and activity monitoring. Sens Actuators, B Chem 140(2):390–395
10. Burns A, Greene BR, McGrath MJ, O'Shea TJ, Kuris B, Ayer SM, Cionca V (2010) SHIMMER™—a wireless sensor platform for noninvasive biomedical research. IEEE Sens J 10(9):1527–1534
11. Chen CM (2011) Web-based remote human pulse monitoring system with intelligent data analysis for home health care. Expert Syst Appl 38(3):2011–2019
12. Pawar PA, Heart rate monitoring system using IR base sensor and Arduino Uno. In: 2014 conference on IT in business, industry and government (CSIBIG), IEEE, pp 1–3
13. Trunov A (2016) Recurrent approximation as the tool for expansion of functions and modes of operation of neural network. Eastern-Eur J Enterprise Tech 5/4(83):41–48. https://doi.org/10.15587/1729-4061.2016.81298

14. Auzinger W, Obelovska K, Dronyuk I, Pelekh K, Stolyarchuk R (2022) A continuous model for states in CSMA/CA-based wireless local networks derived from state transition diagrams. Lecture notes in networks and systems, vol 287. Springer, Singapore. https://doi.org/10.1007/978-981-16-5348-3_45
15. Vlasenko OV, Lebed VV, Gogunsky VD (2012) Markov models of communication processes in international projects. Management of development of difficult systems. Kyiv, Ukraine, KNUCA, pp 35–39 (In Ukrainian)
16. Pashchetnyk OD, Glotov VM (2009) Determination of the necessary accuracy of the focal length of digital non-metric cameras, Geodesy, cartography and aerial photography: interv science and technology coll, I:231–237 (In Ukrainian)
17. Kovtun V, Izonin I, Gregus M (2022) The functional safety assessment of cyber-physical system operation process described by Markov chain. Sci Rep 12(1). https://doi.org/10.1038/s41598-022-11193-w
18. Izonin I, Tkachenko R, Shakhovska N, Ilchyshyn B, Singh KK (2022) A two-step data normalization approach for improving classification accuracy in the medical diagnosis domain. Mathematics 10(11):1942. https://doi.org/10.3390/math10111942
19. Kumar P, Kumar Y, Tawhid MA (eds) (2021) Machine learning, big data, and IoT for medical informatics; intelligent data centric systems. Academic Press, Cambridge, MA, USA. ISBN 978-0-12-821777-1
20. Izonin I, Tkachenko R (2022) Universal intraensemble method using nonlinear AI techniques for regression modeling of small medical data sets. In: Cognitive and soft computing techniques for the analysis of healthcare data. Elsevier, Amsterdam, The Netherlands, pp 123–150. ISBN 978-0-323-85751-2
21. Domanski A, Domanska J, Filus K, Szyguła J, Czachórski T (2020) Self-similar Markovian sources. Appl Sci (Switzerland) 10(11). https://doi.org/10.3390/app10113727
22. Sheketa V, Pasieka M, Chupakhina S, Pasieka N, Ketsyk-Zinchenko U, Romanyshyn Y, Yanyshyn O (2021) Information system for screening and automation of document management in oncological clinics. Open Bioinf J 14(1):39–50. https://doi.org/10.2174/1875036202114010039

AI and Blockchain for Cyber Security in Cyber-Physical System

Manpreet Kaur Aiden, Shweta Mayor Sabharwal, Sonia Chhabra, and Mustafa Al-Asadi

Abstract The convergence of communication networks among devices and systems is the focus of much recent technological innovation and development. CPS has a wide range of applications, such as in industrial control systems and important industries including healthcare and electricity production. As CPS with internet networks become more integrated, security breaches and vulnerabilities arise. Blockchain technology and artificial intelligence are expanding quickly in both business and daily uses. The real-time processing of data across several information systems is a bottleneck in centralised systems and applications, such as healthcare. Blockchain's decentralized database management system, protect data storage, data interchange, and authentication would be able to address this issue. In addition to the blockchain, AI may offer perceptions from the collective data that is acquired and applied to foresee the future. Blockchain is indeed a cutting-edge security technology which, through mutual agreements between nodes, creates chains that chronologically link the new block to the previously stored blocks in nodes. The rise of numerous businesses, including banking, finance, cybersecurity, forecasting, healthcare, cryptocurrencies, etc., is accelerated by technology convergence. The threat of hacking such systems increases as more digital technologies are used and services are offered by these companies. Artificial intelligence and blockchain technology working together can create a robust safeguard opposing such attack and protection risks. This chapter will explore how blockchain technology and artificial intelligence are collaborating to enhance cybersecurity. We'll go into more detail about how they help secure cyber-physical systems.

M. K. Aiden (✉) · S. Chhabra
Maharishi University of Information Technology, Noida, India
e-mail: preetniet@gmail.com

S. M. Sabharwal
Galgotias University, Greater Noida, India

M. Al-Asadi
Faculty of Engineering and Natural Sciences, Department of Computer Engineering, KT KTO Karatay University, Konya, Turkey
e-mail: ahmedj.aljanaby@uokufa.edu.iq

B. Bhushan et al. (eds.), *AI Models for Blockchain-Based Intelligent Networks in IoT Systems*, Engineering Cyber-Physical Systems and Critical Infrastructures 6,
https://doi.org/10.1007/978-3-031-31952-5_10

Keywords Cyber security · Artificial intelligence (AI) · Cyber-Physical Systems (CPS) · Security breaches · Vulnerability · Blockchain

1 Introduction

Today's world has undergone significant change, necessitating our reliance on technology in order to advance. Nowadays, everything is connected, unlike ten years ago. We discuss about the industry revolution that transformed the industrial world when we talk about history. There was a period when technology was so primitive that it hardly ever required planning. For instance, take the beginning of the telephone. Once it was developed, it was exclusively used by the very wealthy, who could only communicate with one another. For example we can take the development of the computer. The machines were so large that moving them was nearly impossible. These days, we have quite modern desktop and even laptop computers that are very portable [1].

In contrast hand, the development of earlier phones has led to the development of foldable mobile devices. With all the technical developments that have been achieved, the world has become a small village, and it is now simple to speak with anyone who has access to the Internet.

Improvements haven't just been in the computer and phone industries, either. The entire dynamic of our existence has shifted in the era we currently live in [2]. Almost all tasks that a human could perform are now handled by machines. Businesses have transitioned into the digital era, where computers can now perform some of the tasks once performed by humans, making the workforce much more flexible.

The Internet of Things (IoT) and Cyber-Physical Systems are an important development in Internet and technology. While technology has greatly improved, the number of devices has also increased. For this reason, the majority of businesses have started enabling their gadgets to link to different devices [3]. The gadgets' connectivity and accessibility are intended to be improved by this. The technology world has undergone a huge change and development with the introduction of devices like Alexa, which can even talk with us. In the beginning, mobile phones might be connected to our laptops, but today, the majority of the electronic equipment we have at home is interconnected [4]. A more thorough analysis of the trends is much appreciated and will contribute to numerous improvements in how people live. A significant breakthrough is having a fridge that, thanks to connectivity, can recognize what is running low and needs to be replaced, among other things. Modern technology has a huge impact on every part of our life. Consider the fact that some cars, like the Tesla, are driverless and can connect to other technologies, including our phones [5].

Even if our modern devices may communicate with one another, if the proper security precautions are not taken, it is easy for them to be accessed by outsiders due to their separate existence. Every advancement in technology is accompanied by an increase in cybersecurity. Cybersecurity refers to defending against external attacks on computers, electronics, and related technological equipment. The devices are vulnerable to hostile assaults if they are not adequately safeguarded because the data and information stored there are crucial [6].

The cybersecurity risks spread while technology experts keep developing new innovations.

Because of this, cybersecurity needs to grow along with technological development [7]. The majority of technological developments include upgrades for already-existing equipment, which, if poorly overwritten and covered, might result in the cracks that adversaries are waiting to exploit. As a result, the cybersecurity industry should receive the same or even greater attention than we do to advancements.

Every day, we need enhancements to existing technology, yet These Technologies work with data as well as access some areas that call for privacy and security. Nearly every sector of the economy, including the companies we operate for, the financial industry, as well as the healthcare system, use new technology [8]. They need data, and if it ends ending up in the wrong hands, it may be really harmful to people. Because of this, cybersecurity is essential for any aspect whose operation involves the use of any technology.

One of the most intriguing technologies is blockchain, which is rapidly gaining acceptance as a transversal technology that is frequently applied in several fields [9]. The utilization of blockchain innovation gives an exceptionally powerful answer for get rid of the requirement for a focal framework to manage and approve interchanges and moves among various members. Every exchange in a blockchain should be gotten and checked by all mining hubs that hold a duplicate of the whole data set including networks covering all events. This provides coordinated and stable information that is impervious to manipulation and trading [10].

Another crucial factor that is rapidly gaining ground is artificial intelligence, which enables a device to comprehend, draw conclusions from, and modify cognitive abilities in response to data. Recent market estimates (Economics of Ai Technology) predict that by 2030, AI will be worth $13 trillion [11]. Due to extensive research and information creation via sensor systems, IoT devices, social networks, and online apps, artificial intelligence has advanced that data can be used by machine learning algorithms. A centralised training model leveraging clusters or clouds services offered by businesses like Google, Amazon, Oracle, etc. is the foundation of most of AI's algorithms for machine learning. The sensed data is error- and security-prone, so keep that in mind. AI can benefit greatly from the decentralised architecture, privacy, and authenticity of blockchains. Although they operate on shared data, intelligent algorithms are secure, dependable, and genuine. Decentralized learning is used in blockchain-based artificial intelligence approaches to assist secure the exchange of information and trust across numerous agents who can participate and contribute on decision-making [12].

According to their respective natures, blockchain with artificial intelligence exhibit various properties. A learning security behaviour that can detect and eliminate risks much like people do, but a thousand times faster, can be created using all solutions. The digital ledger platform used by Blockchain, on the other hand, is secure, fully encrypted, and only accessible by authorised peers [13].

We'll discuss cybersecurity, blockchain technology, artificial intelligence, and their interactions in more detail in the sections that follow. Beginning with the literature study in Sect. 2, we talked about the numerous aspects of cyber security in Sect. 3, along with the risks that are associated to them. Section 4 involves the concept of blockchain Technology in term of cyber security and role of AI in cyber security is covered in Sect. 5. Convergence of blockchain and AI in cyber security system is listed in Sect. 6 and Sect. 7 concludes the chapter.

2 Literature Review

The world is getting more modest in the present culture with regards to association, correspondence, and particularly interconnection. The Internet of Things is quickly extending, with everything becoming associated through the accessible organizations. To construct availability across assorted areas, various organizations are teaming up. As per an exceptional examination by Cybercrime Magazine EIC Steve Morgan, by 2021, there will have been an expected $6 trillion put resources into network protection. That's what these measurements show, to prevent assaults and get a good deal on answering a break that has proactively occurred, individuals and associations ought to do whatever it may take to ensure that the security controls set up inside their frameworks are solid [14]. Surprising cyberattacks every so often simply become visible after they have previously occurred. This is a shortcoming in online protection applications.

They are unaware of the value of cybersecurity since so many individuals use the Internet every day. Most people only do it when absolutely required, particularly when it comes to the accounts they must set up for their applications. Even yet, they don't make them as robust as they could. The exchange of information enables the establishment of a physical infrastructure in cyberspace. People who do not understand the value of cybersecurity applications do not take sufficient steps to secure their systems and devices. Since we now keep all of our data in the cloud, anyone with access to the cloud can render someone technologically helpless. Such occurrences result in losses including identity theft and financial losses [15]. It is necessary to raise awareness so that people can realise how important cybersecurity is. The benefit is that those working in IT departments are fully aware of the risks posed by cyberthreats and are implementing the necessary security measures. Even still, this is insufficient for a firm because black hackers would always search for the weak spot and take advantage of it. According to [16], a lot of people hold fast to the conviction that every aspect of the cyber-physical environment will balance itself out. They are unaware also that harm done may also have an immediate impact on them. Since ignorance is bliss, we must work to eliminate it by raising knowledge of the value of cybersecurity and employing the most effective methods for doing so.

In light of this, end-user safety is quite important. People frequently install malware to their systems without even realising it. Only those who are experts in this field can spot malware rapidly. Understanding this can also be beneficial. There may be training materials available for staff who are able to recognise malware and how to effectively increase their security in a corporate setting. Applications and end-user security protocols are available. Because cybersecurity is a haven't really problem, as soon as we stop addressing one threat, another one will inevitably start to emerge. Because of this, cybersecurity must be a top priority [17]. To have a chance against cyber dangers, we must adapt to the rapidly changing technological world. Because adversaries are constantly evolving, we must, too, in the realm of cyberspace.

3 Security for Cyberspace

As a result of the rapid development of technology, several research on cybersecurity have been carried out to further understand the issue and even provide potential solutions. If it isn't already, we are entering a time when technology and subsequently the Internet will permeate every part of our life. Look at social media platforms for a moment, for instance, where anyone devotes time these days. For this reason, among other industries, it is the one that is able to unite people and holds a large amount of data, of which some are sensitive and may be exploited for criminal purposes if it ends up in the wrong hands. We need to have a thorough grasp of the dangers that are present in order to more fully understand cybersecurity and its use. The only way we can attempt to prevent and control an issue is by first understanding how something works [18]. In order to develop the best course of action, we also need to be aware of the vulnerable areas. Knowing the factors that contribute to cyber dangers provides us an edge since we can try to stop crimes before they start by implementing security measures.

Three categories are used to categorise cyber threats, which makes it easier to decide how to respond and also what security precautions to take. The first is cyberterrorism, which is receiving more attention. It seeks to instil fear of technology and panic among the general public. It is employed by radicals who oppose technological advancement. They use fear and terror to interfere with society's normal functioning. The majority of our cyberattacks are motivated by political motives and share material that is not intended for the general public. The last type of crime is cybercrime, which targets different systems and is motivated by financial gain [19]. Cyber risks currently exist, but as technology develops and progresses, more threats will emerge.

Every technological improvement is accompanied by changes in cyber dangers. They have phishing, spyware, ransomware, as well as social engineering among the major cyber threats. Social engineering depends on human interactions in which the attacker engages or manipulates the victim to bypass security precautions designed to protect data [20]. In these situations, the victim is left to deal with the attack's consequences. Planning is necessary for such assaults because the majority of security precautions also call for physical access to the computers or buildings in which the information has been entered.

Additionally, phishing makes use of fraud, in which con artists send false emails and utilise them to obtain the information they want. The emails are intended to entice the victims, and they resemble emails from reputable sources. Because of the ransom that is demanded once the attacker has access to the data or system files, ransomware gets its name. Malware is injected into to the computer system as part of ransomware attacks. Malware is the introduction of malicious software that corrupts system files or gives others access to the data stored inside. To access the information, cybercriminals utilise a structured language query to install viruses or harmful software. SQL sends a rogue SQL statement, giving hackers access to the targeted database [21]. A man-in-the-centre assault and a refusal of administration are two different models. The potential harm of cyberattacks can be better understood by taking a look at previous incidents.

The United States Department of Justice prosecuted a commander of an organised cybercrime team in December 2019 with being responsible for the Dridex malware attack. The malware had an impact on the public, the government, and infrastructure all around the world. Because of malware and our current level of connectivity, attackers can simultaneously harm a large number of users. Zero-day threats, which are new threats that lack any recognisable digital signatures, are on the rise right now. When an attack occurs, specialists heavily rely on the electronic signatures the hacker left behind to triangulate information and track down the offender. Threats without a recognisable digital signature are challenging to defend against. They particularly triangulate and look into the weaknesses the attacker used in cases of present hazards when dealing with cyberthreats and enhancing cybersecurity. The usage of digital signatures provides information on how the network or system files were accessed by the hackers [22]. By eliminating the weaknesses and enhancing system security, they may now work on preventing such attacks. Hackers or even malicious individuals who want to harm the firm can launch cyber-attacks. Even those who we do not and will never suspect may be involved. Because they are aware of the security apps and procedures in place, this is why the problem becomes urgent. In order to prevent such attacks, that security should be very powerful and legitimate. Unfortunately, there is always a risk management override for any safety precaution.

3.1 Cyber Security Components

Cybersecurity affects so many economic areas, it has been separated into several categories to accommodate them. The different components are as follows:

3.1.1 Information Security

Because it encompasses numerous sectors and divisions within the sector, it may be conveniently covered by utilising the distinctive divisions developed in each context. One thing that is crucial in the world of technology is information. Every product we buy and every piece of technology we use requires a specific amount of information we must provide. The information can range from basic protocol at times to private information that aids in locating the product's owner at other times [1]. In such circumstances, the data needed to be securely secured to prevent it from falling into the wrong hands, which is where the need for cybersecurity comes into play.

3.1.2 Network Security

Network security is the term for safeguarding the internet network, whether it is being used by computers, phones, or other devices. Network security sometimes doesn't mean preventing unauthorised users from accessing the system, but virus can tamper with device data or completely erase it. Network security refers to the protection of the network's physical infrastructure. We have a number of cybersecurity methods that fall under network security. Knowing the multiple ways we could disregard network security offers us a procedure for managing them. One of the disastrous assaults is the forswearing of administration assault, which is exceptionally broad in network security [23]. At the point when a danger entertainer programmer—keeps the expected clients from involving an organization for some time, this is known as refusal of administration. Since more often than not it includes compromising an organization, which is a stage utilized by numerous clients consistently, it is exceptionally difficult to do.

3.1.3 Application Security

Obviously, we utilise many applications on the devices we own for a variety of purposes. Therefore, the initial step after purchasing a gadget, such a cell phone, is to download all of the different applications that one deems necessary. Games, fitness apps, and social media apps are all things we have. Each app asks you for personal information before you can use it and sign in as a user [3]. A few applications are more critical than others and call for more verifiable data. For example, in light of the fact that the data is so significant, applications connected to banks request exceptional keys. Everyone needs to make sure they have the appropriate security measures installed for the different applications. A bible app and other ordinary apps such as the workout app don't often need much security. However, there are apps that deal with things like your home's security system. Then it needs to be securely fastened.

Some hackers have historically depended on the applications on their devices to gather the data they require to exploit them. Typically, it gives these programmes security measures before they're even made available to the general public. Even then, taking extra precautions is never a bad idea. Every component of cybersecurity must be handled to the best of its ability due to its diversity. Most people have to cope with hazards like this, especially in today's rapidly expanding Internet of Things [24]. Now, if someone has access to the app that makes your device interconnected, they have access to every device you own. As a result, maintaining the applications' security measures is crucial.

3.2 Threat to Cybersecurity on a Large Scale

A cybersecurity assault occurs in the United States every 39 s, according to University of Maryland research [25]. Numerous people fall prey to the cyber menace every day and must deal with the repercussions. According to research, the majority of breaches are made possible by the weak passwords and usernames we employ. People typically choose the simplest username and password possible in order to remember them. As soon as they can recall them, those outside of the IT department don't spend much time coming up with the most secure login or password. Hackers take advantage of this excellent chance since it presents itself to them. For one, hackers are accustomed to how electronics and networks work; providing a less rigorous security measure makes it easier for them. Cyberattacks are problematic because they usually target a wide range of industries. As long as the industry has technology and networks, it is highly susceptible to cyberattacks [26]. Even the financial and healthcare sectors have embraced digitization today, and without the required security measures, they are vulnerable to attack.

As of September 2019, there have been around occurrences of information spillage because of breaks across different areas. Contrasted with the earlier year, there was a critical increment. In the United States, public establishments and medical services offices are among those that are habitually focused on the wellbeing foundations save fundamental information that is incredibly important, especially on the black business sectors. The programmers could try and get the opportunity to exchange the information they got and continue on toward other weak regions when a break is distinguished and fixed. Networks are the most current models, and innovation use in the medical services framework has expanded altogether [27].

What preferred method for doing it over to embrace the utilization of innovation? It promotes healthcare to improve its health care offerings. Unfortunately, they also assert that their sector receives the least amount of financing for cybersecurity. Cybersecurity funding from the federal government of the United States grew from prior years to $18 billion. They spent 5% of their budgets, or the least amount, on cybersecurity with healthcare organisations [28].

One change that happens concerning network safety innovation is that the makers are progressively prepared to deliver their items. Cybersecurity is an afterthought in today's devices and networks. As soon as a product is finished, that is. It is made available to the public so that consumers can use it and the business can continue its profitable trend. Later, when the equipment and technologies are available on the market, cybersecurity application possibilities are provided. Cyberattacks have already caused damage to certain persons. As previously said, cyber-attacks and threats are developing at the same rate as cyberspace. Every day, a new modern technology is developed, which creates a growing cyber danger [9]. The minds behind the black and grey hackers are the same ones who develop technological updates and advances. Consequently, whenever a new gadget or system is made available to the public. Before they are noticed, they rapidly spot the weaknesses and take advantage of them. Because of this, some cyber-attacks always include a cat and mouse game.

Since they are still out before they are found, dark programmers are continually out in front of white programmers and other digital specialists who are attempting to create and carry out network protection. Network protection safety measures are ordinarily taken to stop an assault that has previously happened [29].

Modern issues call for modern answers, which is why cybersecurity applications are so important right now. The world is transitioning to a realm where technology will rule. We need to be ready for our actions and our response to online attacks. We must set up a cybersecurity system that is effective.

As indicated by Parrend et al. [30], multi-step and zero-day takes advantage of are progressively normal in digital assaults. It is trying to recognize the two assaults or even to stop them before hurt is finished. With the guide of such an assault, the assailants or exploiters distinguish the powerless spots of the weaknesses. They then speak with different groups of people in their main subject area by means of coded messages or alarms. Because of this, the end user is unaware of a vulnerability in their system, although other organisations are acutely aware of it. They can all take advantage of the weakness and gather the data they require or desire for their objectives [31]. Because of this, they recommended the adoption of two strategies in their research and study to guarantee the implementation of cybersecurity. The two methods made use of machine learning and statistics as well as artificial intelligence. Artificial intelligence makes it feasible to find weaknesses before malicious hackers do. If not, a new addition into the system will be instantly recognised by machine and statistical learning. Artificial intelligence, which finds the weaknesses, is the ideal ideology in this situation. As systems gain new users, especially on channels like social media and website platforms, it may become difficult for authorised or new users to access the systems. This system will monitor the event sequences and identify any unusual behaviour inside the systems. It is simpler to maintain and even implement cybersecurity with such systems. In a world where technology is assuming dominance, we must be extremely prepared to tackle the clear threat that is just waiting to happen. Use of more dependable cybersecurity measures is the best method to do it.

3.3 Cybersecurity Challenges

Cyberattacks and challenges related to cybersecurity pose a significant number of problems. There is also the issue of some people not taking cybersecurity seriously or doing so seldom. They must therefore learn more about the benefits of cybersecurity applications. A system's ability to protect data on an individual or corporate level is crucial. The information kept on the discs or in the systems is both crucial and private. When it comes to businesses, their rivals can utilise it to ruin or even bankrupt the organisation [32]. It may divulge private information, such as banking information, which may put a person in financial ruin due to private concerns. In both scenarios, all non-public data needs to be adequately secured, which can only be done with cybersecurity tools.

A high level of speed is preserved with cybersecurity where the system has malware and viruses, it becomes slower, which reduces its efficiency. Because of the rapid cyber speed, it is crucial to apply cyber security measures. Particularly if a company's billing department relies on the speed of the system, slow systems might result in significant losses and damage. After assaults, a lot of resources have already been used to restore cybersecurity [33].

Cyberattacks are difficult to manage, but having cybersecurity in place beforehand can dramatically reduce costs. Cyberattacks on large corporations may result in a loss of clients and goodwill. People are a component of the cyber world, and when they interact with a business, they can share some sensitive data. They count on the confidentiality being upheld. As a result, when cyberattacks occur and their data and information are stolen, customers lose faith in the organisation, which lowers their standing in society. Businesses greatly rely on public perception of them. Companies must maintain their reputations in order to succeed, and if they do, they will inevitably lose clients. It is preferable to safeguard our systems than struggle to restore equilibrium after an individual has hacked them, which is true of other industries as well as cybersecurity. In such cases, it takes some time until the stolen data is located [34].

An excellent illustration is how much data businesses have stored in their systems. Attacks like these target the trivial and benign data while breaches and hackers look for important information. In these situations, the attacker is already on the loose, the harm has already been done, and it takes time for them to locate the actual damage.

3.4 Cyber Security Applications

Effectively developing plans and tactics to stop cyber threats based on what we already know. This will require a dependence on previous attacks and threats and also how we dealt with them. The strategies used here will try to prevent having the very same vulnerabilities as it ever was. That is referred to as protection against the recognized instead of the unknown. It ought to include all the elements that form the basis of cybersecurity that we are aware of or presume to exist. Second, cybersecurity demands ongoing warnings about fresh vulnerabilities. In other words, getting ready for the unexpected Notwithstanding settled gambles, new ones frequently create over the course of time, hence experts should be totally ready for them [35]. This could mean routinely utilizing white programmers to find framework weaknesses with the goal that they can be fixed before they fall into some unacceptable hands. It constructs new network safety frameworks and systems utilizing the laid-out ones, updating them so they can endure the new kinds of dangers that are expected to show up. The development of security techniques in the first section is based on already known information. The second half, however, calls for new approaches because it deals with problems that were anticipated to be conceivable due to system developments.

3.5 *Barriers in Cybersecurity Technologies*

Regrettably, cybercriminals have capitalised on chances created by online crime. Black markets are where professional black hackers sell the tools they use to other users, particularly for zero-day attacks. With several individuals having access to them, these technologies have the potential to disrupt many computer systems and online activities. People have started using mobile phones in nearly everything over time. These days, we have applications that we may utilise to speak with in every area of our lives, including encounters at work, within bank systems, etc. [36]. This is in addition to how widely used mobile phones have grown and how inexpensive they have gotten. Hackers can now easily gain access to mobile telephone system and change them to their advantage. To strengthen security in these situations, the systems must take drastic methods like voice recognition and facial recognition.

The Internet of Things is used effectively all over the world. The advancement of technology has advanced quite a bit. The sector that is now evolving the fastest is the Iot, where all devices are connected to better control and monitoring. While this is fantastic for people, hackers can also gain from being able to access and manage all of your devices from a single location. They can now access the one gadget that has control over all the other devices in a human's life thanks to their skill in assaulting systems. We are unable to eliminate this problem because technology and cyberspace must advance [37]. In order for end users to benefit from technological breakthroughs, we must develop cybersecurity tactics and strategies that are more robust and effective. In the next ten years, there should be about 125 billion internet users in the World of things. As a result, the challenge will continue.

As was mentioned, cybersecurity challenges are ongoing and always change. Whatever changes is how people respond to it, therefore the methods for handling attacks and other situations should likewise continue to advance to better and much more dependable iterations [38]. Another difficulty is that at the moment, hackers use third-party vendors as targets for their assaults. That refers to individuals who are not protected by the business. One who delivers to a corporation, for instance, may be treated differently within the firm. Since the employees of the organisation are the most protected locations, prevention of these kind of breaches can be challenging to handle.

The readiness to confront cybersecurity issues varies noticeably from one place to the next. Many businesses are aware of the security risks and cyberattacks they may encounter, yet when faced with a challenge, they either respond inadequately or too slowly. They take a very long time to detect and react to attacks, and they never seem ready.

4 Blockchain Technology

Bitcoin's fundamental innovation, blockchain, has grown rapidly as of late, and its purposes are currently not just restricted to advanced monetary forms. A blockchain is a conveyed information base that is available to general society and monitors past occasions of computerized exchanges. A hash capability connects each Blockchain block to its ancestor to keep up with the chain of exchanges on the conveyed record [39]. Upon joining the network, the internet backbone components will acquire a pair consisting of the public key & private key. Each element's public key serves as a distinctive identifier. The network uses the private key for encryption and decryption as well as to sign transactions. All nodes get the transactions, which are then verified. Several nodes assigned as miners collect them into a time-stamped block. A block is browsed among the many blocks delivered by the excavators and transferred to the Public blockchain utilizing a "No Central Authority" agreement process known as blockchain. Prior to rolling out any improvements to the ongoing block of information, all hubs in the organization complete methods to assess, affirm, and contrast the exchange information and Blockchain history. In the event that most of hubs endorse the exchange, another block is added to the current chain.

4.1 Blockchain Technology Process

The use of Blockchain has advantages including data security, error reduction, reliability, improved integrity, and effectiveness [40]. An illustration of how Blockchain functions is shown in Fig. 1. Currently, there are three ways to handle the purchase between the buyer (on the left) as well as the seller (on the right).

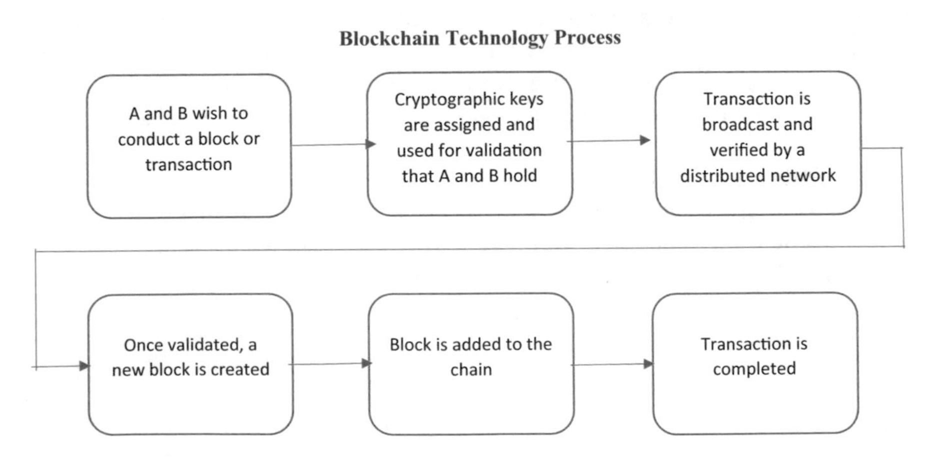

Fig. 1 How a blockchain transaction works

4.1.1. Both parties can manage it because they have each other's trust. Depending on their friendship, if they complete the transaction; if not, the buyer has the right to immediately withhold payment.

4.1.2. The purchaser and merchant sign an agreement; on the off chance that the purchaser neglected to pay, the court would become involved, and the settlement methodology would take more time.

4.1.3. Introduce a third party to oversee the procedure, but be aware that they might pocket the cash instead of paying the seller, returning us to the previous two choices.

4.2 Broker with Blockchain

We can furnish the broker with Blockchain; however it is guaranteed, secure, quick, and reasonable. The exchange is safeguarded (hashed) and kept in scattered data sets, as displayed in the figure (information sources). The potential chance to monitor the exchange's advancement is accessible to both the purchaser and the dealer consistently. The activity of blockchains is found in Fig. 1. By gathering exchanges into blocks, every one of which contains a specific number of exchanges and associations with blocks before it, the blockchain network coordinates exchanges. An ordered chain isolates the blocks [41].

Blockchain has four key attributes including permanence, disseminated data set structure with practically ongoing exchanges, and control obstruction. Solid agreement ensures exchanges with almost no extortion. The probability of having all or most hubs partake in misrepresentation is low [42]. Not the Blockchain's all's points of interest are what we desire to offer. We are regardless inquisitive about its capability in network safety and security applications and frameworks, including those for the Internet of Things, the electrical matrix, the monetary area, and more.

One of the fundamental benefits of blockchain innovation is security. We can use its distributed ledger to make secure millions of records inside its platform by making a few design adjustments. Its proof of work system is designed to need user approval for all additions and changes. The thrustless principle, which allows for anonymous transactions that are nevertheless recorded in the chain, is also supported by this system. These actions all work to preserve the accuracy of the data. One of Blockchain's drawbacks or problems is this. When a transaction is accused of being fraudulent, a method must be developed to reveal the identity (Fig. 2).

Another block containing a "hash" or finger impression from the blockchain record is produced by the actual stage each time a block on a blockchain stage is settled. Never is a similar exchange recorded two times. Subsequently, there could be as of now not a requirement for a focal mediator.

Also, blockchain stages require client proprietorship verifications and utilize evidence of stake (PoS) and confirmation of work (PoW), two kinds of agreement between all individuals, to affirm any changes made on some random block. A

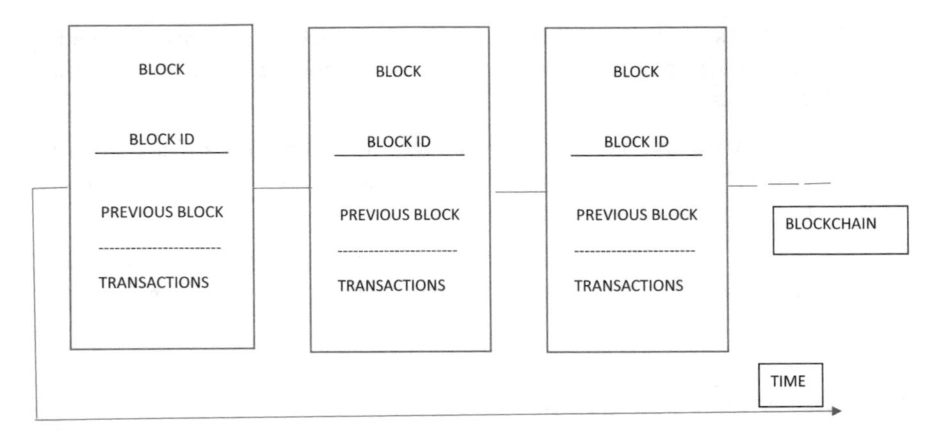

Fig. 2 The operation of blockchains

blockchain network utilizes verification of stake (POS) as a system to arrive at disseminated agreement. Also, blockchain stages require client proprietorship evidences and utilize verification of stake (PoS) and verification of work (PoW), two kinds of agreement between all individuals, to affirm any changes made on some random block.

While blockchain innovation is still in its early stages, as per Ed Powers, Deloitte's US Cyber Risk Lead, it can possibly assist organizations with tending to unchanging Cyber Risk difficulties like safeguarding information respectability and overseeing computerized characters [43]. Blockchains can possibly help increment network safety by safeguarding information, staying away from misrepresentation through agreement cycles, and identifying information altering thanks to their characteristic characteristics of irreversibility, straightforwardness, auditability, information encryption, and functional versatility (counting no weak link). Be that as it may, as Cillian Leonowicz, Senior Manager at Deloitte Ireland, brings up, "blockchain's qualities don't cause it an invulnerable panacea to all to digital ills, to accept the equivalent would be credulous, best-case scenario, and on second thought blockchain plan and execution and roll-outs should incorporate common frameworks and organizations network safety controls, an expected level of investment, practice, and strategies."

Blockchain innovation can upgrade network safety and settle issues welcomed on via imprudent or careless clients. By and by, we should be careful about the opportunity that superior handling power and unscrambling techniques will empower the hash code to be broken.

5 Artificial Intelligence

Conventions, programming, and even uncompiled code are utilized to give man-made consciousness-based security arrangements. The capacity of man-made reasoning

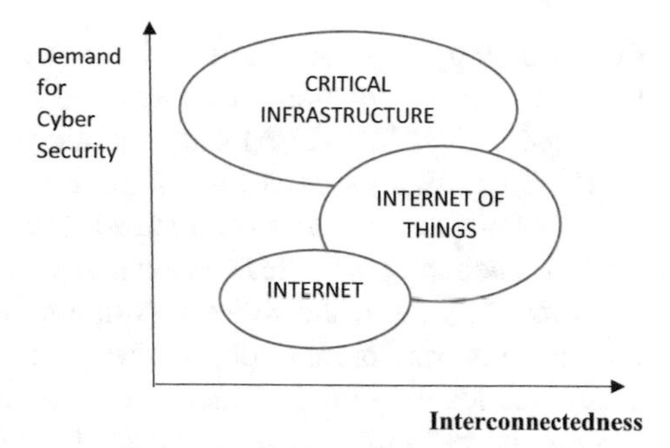

Fig. 3 Role of AI in cybersecurity

(AI) to gain from dangers, security blemishes, and different information assembled through its frameworks adds layers of safety. They can subsequently forestall security breaks in the future because of them. Considering this, the more goes after a framework is presented to, the more solid its security will become in its ability to shield itself over the long haul. The use of AI in cybersecurity across the Internet, the Internet of Things (IoT), and critical infrastructure was covered by Zeadlally et al. in their paper. Figure 3 [44] provides a list of how Al functions in these three areas. With the development of the Internet, Al's importance in cybersecurity is growing. In essential systems for human welfare and national security, all methodologies are employed. To programme computers to think and act like people, various techniques are employed to help them solve problems logically.

5.1 Role of AI in Cybersecurity

Humans experience fishing attacks on the internet domain area. Thus, automatic phishing detection is how AI resolves this. Both the network and application levels experience service denial and message semantic changes. AI manages this by recognising trends, adapting, and creating more intelligent classifiers. IoT domain is vulnerable to impersonation, unreliable data gathering, and unreliable data exchange. AI provided levels of security for cloud and dispersed environments. AI employed a logic-based policy and framework in vital infrastructure, where all assaults fall under the category of cyberattacks.

6 Convergence of Blockchain and Artificial Intelligence

The first thing we think of when we discuss Blockchain is its decentralised architecture. A decentralised artificial intelligence is created when we combine blockchain technology and artificial intelligence. Artificial intelligence (AI) with machine learning has been present since the 1950s. It is not a recent invention. Although the technology is quite old, the essential concept has never changed. We have a bunch of info information. Since we have an objective as a main priority and something we believe that the AI should achieve—for this situation, decide whether an article (suppose a pine tree) is a pine tree or not—we feed it a preparation set of photos of different sorts of trees. The idea is that the Individual would be expected to separate each trademark that characterizes a pine tree, including its shape, its leaves, the level of its stem, and so forth. Although it was a highly taxing operation, doing it by hand is what humans do, so after extracting those features, we fed the data to a machine learning model. There are many things that are out; gradually, it would acquire the mapping and receive a fresh tree image before identifying it as a pine [45]. Recent years have seen the addition of many deep layers, also known as deep learning, to neural networks, which are machine learning models. As a result, these neural networks have surpassed all other machine learning models in terms of performance. Nowadays, deep learning outperforms all other AI applications, including self-driving cars and medicine discovery.

Bitcoin was created by Satoshi Nakamoto and distributed in a report on a cryptography email bunch. A framework permits two individuals to exchange cash online without the contribution of an outsider, like a bank. The idea is that the miners must accept the transaction and give their approval when someone (let's pretend it's me) transmits value to (you). As a result, there is a group of people known as miners serving as the third party rather than a bank. Anyone can start mining; all you need is some sort of computing device. So that each miner does have a record of each and every transaction which has taken place in the network and can determine whether or not a transaction is genuine, let me examine this list of transactions. You might be worried that someone will pretend to be several distinct people in order to approve a transaction as though they were separate miners.

To stay away from this, Satoshi Nakamoto fostered the evidence of work calculation (PoW) [46], which requires every excavator to demonstrate that they have tackled an irregular numerical issue. This truly intends that as well as claiming the main part of the figuring power in the Bitcoin organization, you additionally need to have handling power more noteworthy than the quickest supercomputers on the planet. Since nobody has very as much figuring power as Bitcoin does, which has a market capitalization of more than 350 billion as of this distributing, nobody has had the option to hack it. Bitcoin is an exceptionally impressive innovation that has been around for an age.

These two advancements don't blend, similar as oil and water, in spite of the way that they significantly praise each other. Because of their interesting construction, changeless and unchangeable information design, and verification of work calculation, blockchains are deterministic.

We forever know about what's happening and things can't be changed. Models of likelihood and forecast are the underpinning of man-made consciousness. Nothing remains at this point but to blend blockchain and man-made brainpower. We could have an AI speak with the Blockchain and send information to a decentralized stockpiling framework utilizing this permanent record. On the off chance that we can empower AI to live on the Blockchain, it will have full admittance to the upsides of a decentralized, permanent engineering and an extremely safe mark. On the off chance that AI is available there and no human middle person is there, we are giving AI whole opportunity to explore, adjust, and create prior to dominating! (Startling, huh?) Knowing the PoW approach and the requirement for supercomputers to assault and hack makes it extreme to stop such a colossal monster attempting to assume command. This union of blockchain innovation with AI should be standardized and managed to safeguard control. Furthermore, a model called SecNet that offers framework wide security utilizing AI and Blockchain innovation has been proposed in [45].

This design guarantees that the different framework individuals' trades of information are kept secure and safeguarded. The plan went through a security survey by the creators and was demonstrated to be impervious to a Distributed Denial-of-Service (DDoS) assault.

Blockchain and man-made reasoning innovation can be consolidated to deliver decentralized calculations and applications that approach a similar perspective on a protected, trustworthy, and shared foundation of information, logs, data, and choices. The decentralised nature of blockchain technology can defeat centralised organisational structures, improving data security. Trust is improved because of the blockchain's deterministic property, which defeats al probabilistic (changing). Data integrity on the blockchain supports decentralised intelligence in AI [47]. Overall, Blockchain enables AI to reach a choice that is open, reliable, and comprehensible. As we all know, the architecture and functioning of a blockchain need countless considerations and trade-offs between security, effectiveness, decentralisation, and many other factors. AI can swiftly simplify those choices and increase Blockchain's effectiveness. Aside from that, AI is essential for preserving customer privacy and security since all Blockchain data are openly available.

6.1 Conceptual Model

Blockchain and AI both have advantages and disadvantages on their own. Salah et al. great's overview of the difficulties in integrating AI with blockchain is found in [48]. In some cases, artificial intelligence (AI) can help with blockchain characteristics and operations. The goal is to give faster solutions with as much transparency and dependability as is practical by creating new digital systems that harness the potential of both Blockchain and Artificial Intelligence. Taking into account the overflow of created frameworks available [49]. All of these frameworks have an emphasis on a specific application or industry, like banking, energy, medical care, or money. There will constantly be an interest for a general model that can accomplish this union while using the capability of shared informational indexes and machine knowledge models to assist with finishing clients settle on instructed choices. Ponder how these enormous organizations might team up to create shared datasets and models that would be open to the more extensive public and end clients. Enormous stowed away (private) informational collections and AI models are most likely present in every one of these critical organizations.

Training artificial intelligence systems on larger, more diverse datasets can enhance their performance. Therefore, there is a requirement for cross-organizational sets of data and models for machine learning that don't interfere with these businesses' and their clients' right to privacy and secrecy. Blockchain offers encryption, dependability, transparency, and validation, making it a reliable option for data storage. Smart contracts can be used to safely and securely accomplish actions such as uploading the data, updating data, including data analytics by participants to the dataset and AI models. Individuals have access to trained models and shared data as a blockchain end user. In order to effectively use Blockchain for data exchange, storage, and ownership transparency as well as systems for rewarding systems, we are working to develop a convergence model. A strategy to stop dishonest participants from flooding datasets with false data and degrading the performance of training models is on the forefront of that list. The strategy is based on the idea of using a collaborative dataset and the decentralised, secure storage offered by Blockchain to harness the potential of AI. This model is hypothetical and not thoroughly vetted. Future study will use Ethereum blockchain technologies to validate the model [50]. The well-known cryptocurrency Bitcoin established the blockchain concept, which is now used by Ethereum, a decentralised blockchain network.

6.1.1 Blockchain Framework

In Fig. 4, we outline our proposed situation for how AI and blockchain will unite. The three levels of the framework are as per the following:

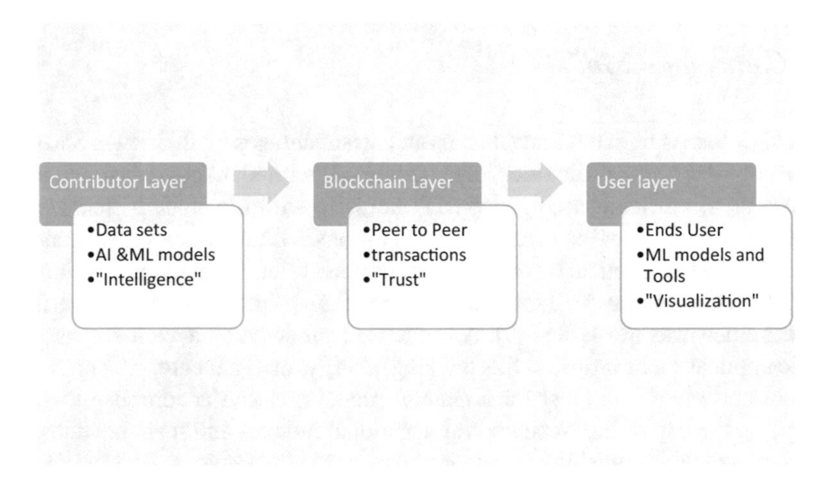

Fig. 4 Conceptual model

The Contributors Layer and Data Set for AI Models

In this tier, the company includes its training datasets and AI/ML models. It has many different features and functions, including data addition, data updating, data validation, incentive mechanisms, prediction or data development, and analytics models. Depending on its function and the application being studied, we can refer to it as the admin user layer. For instance, in healthcare systems, various hospital systems could be the actors in this tier [51]. Each design can share a highly accurate prediction model with a trustworthy and well-trained dataset. Depending on the quality of the medical record and the biological implications, each form of data as well as its combinations have important implications for particular illness situations.

Blockchain Layer

Records every transaction, communication, peer-to-peer (P2P), encryption, censorship, policy role, etc. This layer is perceived as Blockchain as A Service by Users and Contributors (BaaS). They don't have to be concerned about the underlying computing, network, or storage architecture. Execute the transaction, authenticate the transaction, display the outcome, and interact with a lesser level of system implementations are the primary variations in the Blockchain layer's operations. For instance, roles for defining and validating regulations for access, system implementation in a health system.

User Layer

It shows several users of the system interacting with it. The ranks and functions of these users can vary. For instance, users in the health care system can be categorised as physicians, patients, employees, and executives, each of whom has a specific role to play.

The model is carried out utilizing open-source structures, and many types of AI models are thought of, including directed and unaided models along with grouping, assessment of spammers, positioning, and motivation frameworks [52]. There are a couple of interesting points in our future work, as was depicted previously. Blockchain's decentralized nature can be a bottleneck because of the enormous measure of information and expected to store the entire data set on every hub. Do we need to openly distribute information, or are there security and protection worries that simply permit us to adjust the information and model? With regards to AI and blockchain assembly, we search for nonstop information updates, train and test information to make ends, stockpiling, and procedures to screen out spammers and limit or reject their belongings.

6.2 *Blockchain and AI Convergence Use Cases*

Each system has unique demands and requirements, as we discussed in the previous part, but they all fit inside the suggested model.

6.2.1 Power Grids

Renewable power sources and the rising interest in green power have been the driving factors behind numerous advancements in the energy sector, including how utility companies interact with their customers. One of those advancements is the use of smart grids, which effectively merges the traditional electricity infrastructure with the IT industry. It experiences more cyberattacks. These threats can be lessened with the aid of AI and Blockchain [53]. The communication system of the Power System is susceptible to cyberattacks since wireless connections are open and the Smart Energy Infrastructure (AMI) is distributed. As seen in Fig. 5, we can observe how AI and Blockchain function via the lens of our model.

The infrastructural components are what make up the user layer. However, the Blockchain layer is capable of performing a number of functions:

Things' Identity

Throughout a device's lifetime, the ownership can change or be revoked. Each gadget has characteristics including its maker, type, and GPS coordinates. With the help of its distributed ledger, blockchain will be capable of registering connected devices, give them identities based on their attributes, and preserve that information.

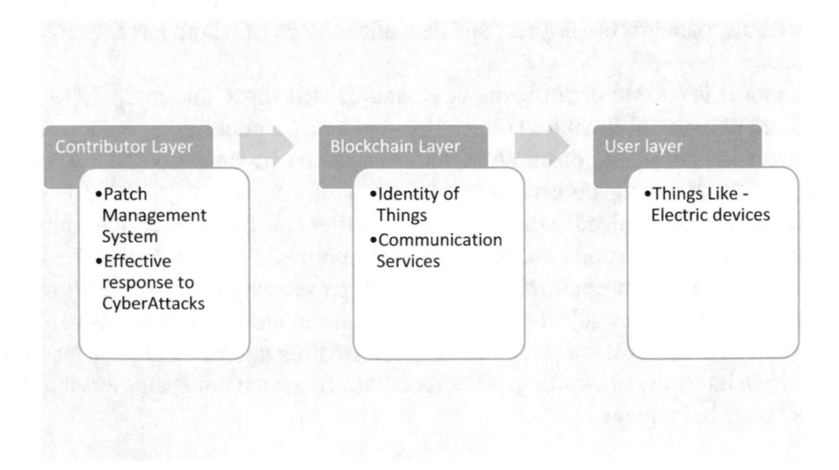

Fig. 5 AI-powered smart grid and blockchain model

Data Integrity

All transactions are encrypted; therefore, the send will cryptographically sign and check all data sent by the devices for accuracy. Each node will have a distinct private key and public key, and by timestamping each information transaction, it will assure the integrity of the data.

Secure Communication

When a node joins a network, Blockchain generates an asymmetric key pair using a unique universal identity, or UUID. When compared to PKI certificates, this will result in a quicker handshake.

As part of patch management, several suppliers can employ smart contracts in the Contributor layer to confirm that the correct patch was transported to the metres.

The model, firmware form, and gadget explicit information will be the premise on which the agreement will work. The contraption will be told to refresh by the shrewd agreement, which will likewise choose whether to do as such. The savvy agreement would bring down its position and inform the energy provider that we could have a compromised gadget assuming that the gadget dismissed the redesigns (seller).

6.3 Electoral Process

Decisions are viewed as a festival of a majority rule government since they give each citizen the potential chance to utilize their entitlement to pick the best individual to get down to business. The decisions should likewise be dealt with fairly, considering no the interests of any one party or applicant [54]. Most of decisions are held at actual democratic spots. Casting a ballot on the web or on a telephone would be more pragmatic and speed up the outcomes system. By doing this, we can lessen how much time it takes to report political race results. The speedy detailing of results is an extreme issue with the ongoing democratic frameworks.

The strategy is shown in Fig. 6. Below is a description of each layer.

6.3.1 User Layer

Electors utilize a cell phone application to project their voting forms. A personal ID or some other biometric recognizable proof must likewise have the option to be recorded by this application. Political decision authorities can utilize a dashboard to screen elector turnout and extortion.

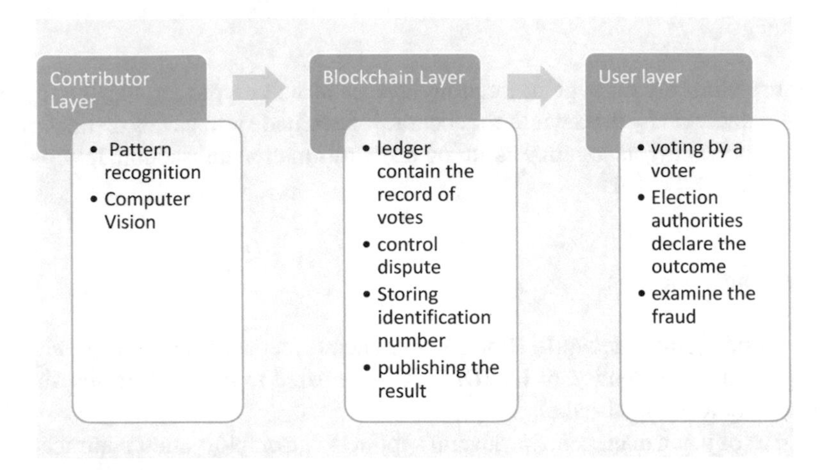

Fig. 6 AI and blockchain paradigm for the electoral system

6.3.2 Blockchain Layer

Every voter will have a wallet thanks to the blockchain network. Voting takes place within a smart contract, and each candidate's votes are counted using a mapping system. In the event that the AI detects or alerts a voter who has submitted fraudulent evidence, the Blockchain also can enable dispute management.

6.3.3 Contributor Layer

Verify the evidence provided to the authorities using the classifier and tools that are AI or machine learning capable.

6.4 Cybersecurity Recommendations

Ironically, we begin with the most fundamental necessity, which is ensuring that each firm has a multi-layered security system in place to thwart any attempts to assault or access by authorised personnel. That is the use of encryptions with more than two types of access authentication. With the mobile technologies we now incorporate into our daily activities, such a protection system can function quite effectively. Handle the outside vendors here in the same vein. We can make sure they follow the organization's security guidelines. Considering that they are routinely communicating with the firm however are not its laborers, it tends to be accomplished by conveying the assistance level arrangement so they completely get it and concur with terms [55]. It could likewise be allowed to incorporate Identity and Access Management (IAM).

Its subtleties who in an association will approach what under both the clients' and representatives' agendas. With character and access the executives, it is simpler to screen representatives' exercises, including how regularly and what data and frameworks they are getting to. In sorting out when somebody has gotten to a framework without authorisation, it very well may be useful. It is easier to find the weakness after a break and to make a move to close it as fast as could be expected.

The partnership might be certain they are totally educated regarding how things are going by leading customary reviews of the instructive innovation and mechanical areas. They should hire a white hacker who will employ any method at their disposal to try to break into the systems. In order to bypass the system, this exercise discovers any potential weaknesses. A business or an individual can prevent risks from entering as much as they possess real-time intelligence. Think of a circumstance where you are aware of impending danger. You would have plenty of time to prepare for the threat heading your way during that period. Similar to the audits, which provide real-time data that black hackers may otherwise use. With the information, they may improve their areas of weakness and put sensible solutions into action. Systems contain important information and data that, in the wrong hands, might be very harmful [56].

On an individual level, this can be supplanted by the standard security framework refreshes that are incorporated into the items we use. They can dispose of any malware or infections found inside the gadget framework with routine framework refreshes. Computerized reasoning and AI have both been lauded as predominant administration instruments for network safety applications. One of the best ways of overseeing cyberattacks and dangers is to comprehend the way in which the assailant thinks and acts inside the framework. It is an unexpected on such countless levels to overlook the collaborations between the contending assaults and the protective demonstrations. The issue with online protection is that new dangers are continuously fostering that weren't there before [57]. Therefore, the cybersecurity programs team should continually be on guard because strategies that worked a month ago might not work now against a current threat. To combat unexpected attacks that have never occurred before, defensive designs are developed that are foreseeing and effective. If they do, they can deal with them and safeguard the relevant data, information, or network. To achieve this level of understanding, it is necessary to apply several attackers with various actions and contextual enhancements. Although it is challenging to create, genetic programming makes it conceivable. An excellent understanding of how assaults can occur and the best strategies to stop them, if not completely avoid them, can be gained via the combination of genetic programming and adversarial evolutionary algorithms.

Genetic programming requires cybersecurity investigation, which will help collect the information required to determine the best line of protection to employ against attackers. In these situations, it is utilised to identify impending weaknesses or attacks in order to develop a defence mechanism that will assist in doing so [58]. Examples of the kind of programs it associates with are provided below. By removing places that aren't likely to be attacked, it can eventually find weak areas. Investigation into network defences focuses on cybersecurity topics including distributed denial of service and isolation defences. To ascertain how the systems interact, the physical infrastructure must be protected. The identification of anomalies and vulnerability testing are the next steps, which will help pinpoint potential access points for hackers. Risk management is also a key component of cybersecurity applications. The dangers are novel to the systems, as was previously mentioned. Before experiencing any attack from black hackers, organisations must assume the risk by implementing the security procedures they think are most appropriate.

Prior to that, they must implement strong identity and access management, understand the vulnerability's characteristics, and identify the data that may be accessible if the vulnerability were to be exploited. It is not easy to implement risk management strategies, and they must ensure that they have the right data on an informational component in order to act on the known variable. One difficulty primarily experienced by cybersecurity applications is lack of knowledge. For this reason, businesses should hold training sessions for their staff. So long as they are aware of cybersecurity best practises, it will be simpler for them to spot or stop an incursion. When dealing with cyberattacks, restoration and discovery times can be sped up with training and awareness. The most crucial thing is getting the most people. Everybody around here regularly uses technology and the Internet. Because of this, cybersecurity must be maintained by everyone. To begin, we must make sure that the information on our devices is securely safeguarded and that the passwords on our gadgets are complicated. Try not to share an excessive amount of data on open discussions. As the globe changes to Internet of Things organizing, it will be basic that the overall population regards their right to information security. An association organization can become powerless on the off chance that a weak link causes it, opening the entryway for programmers to target different pieces of the framework. Today, particularly web-based entertainment networks are helpless against digital dangers, consequently security should constantly be kept up with. Little contemplations like using public Wi-Fi can immensely affect how network safety is utilized.

7 Conclusion

Everyone who uses technology or the Internet is impacted by cybersecurity. It is crucial that we play our parts because corporations must develop strategies to thwart cyberattacks. Because cyberattacks are unpredictable, they are always in motion and available for investigation as technology develops. Youth of today should be motivated to pursue careers in cyber-related fields in order to better manage the problem. There is insignificant information on the mechanics of data frameworks, including network safety, as per research. Since they depend on human ways of behaving that have generally exhibited achievement, like expectation, responsiveness, and approval, blockchain and computerized reasoning (AI) are two of the most encouraging security advances for the not-so-distant future. Blockchains may be used in information security applications to maintain digital identities, secure edge devices, and safeguard vast volumes of data. Blockchain is among the most secure providers of cyber defence thanks to a few important components. It has the ability to offer security for financial transactions like those using cryptocurrencies. Decentralization, immutability, and accountability are a few features that make Blockchain incredibly secure. Numerous cryptographic protocols are used by blockchain, the most of them are challenging to crack. AI can be strengthened by the combination of Blockchain & AI technologies by creating a safe and reliable environment.

References

1. Lee J, Azamfar M, Singh J (2019) A blockchain enabled cyber-physical system architecture for industry 4.0 manufacturing systems. Manuf Lett 20:34–39
2. Zheng Z, Xie S, Dai H, Chen X, Wang H (2017) An overview of blockchain technology: architecture consensus and future trends. In: 2017 IEEE international congress on big data (bigdata congress), pp 557–564
3. Renuka KM, Kumari S, Zhao D, Li L (2019) Design of a secure password-based authentication scheme for M2M networks in IoT enabled cyber-physical systems. IEEE Access 7:51014–51027
4. Genge B, Haller P, Duka A-V (2019) Engineering security-aware control applications for data authentication in smart industrial cyber-physical systems. Future Gener Comput Syst 91:206–222
5. Nkenyereye L, Adhi Tama B, Shahzad MK, Choi Y-H (2019) Secure and blockchain-based emergency driven message protocol for 5G enabled vehicular edge computing. Sensors (Basel Switzerland) 20(1). https://doi.org/10.3390/s20010154
6. Vangala A, Bera B, Saha S, Das AK, Kumar N, Park YH (2021) Blockchain-enabled certificate-based authentication for vehicle accident detection and notification in intelligent transportation systems. IEEE Sensors J 21(14):15824–15838
7. Tawalbeh LA, Tawalbeh H (2017) Lightweight crypto and security. In: Security and privacy in cyber-physical systems: foundations principles and applications, pp 243–261
8. Bahalul Haque AKM, Bhushan B, Nawar A, Talha KR, Ayesha SJ (2022) Attacks and countermeasures in IoT based smart healthcare applications. In: Balas VE, Solanki VK, Kumar R (eds) Recent advances in internet of things and machine learning. Intelligent systems reference library, vol 215. Springer, Cham. https://doi.org/10.1007/978-3-030-90119-6_6
9. Srinivas J, Das AK, Kumar N (2019) Government regulations in cyber security: Framework standards and recommendations. Future Gener Comput Syst 92:178–188
10. Chen H, Pendleton M, Njilla L, Xu S (2019) A survey on Ethereum systems security: vulnerabilities attacks and defences. CoRR abs/1908.04507, pp 1–29
11. Su Z, Dai M, Qi Q, Wang Y, Xu Q, Yang Q (2018) Task allocation scheme for cyber physical social systems. IEEE Trans Netw Sci Eng 7(2):832–842
12. Feng Q, He D, Zeadally S, Khan MK, Kumar N (2019) A survey on privacy protection in blockchain system. J Netw Comput Appl 126:45–58
13. Xu Q, Su Z, Lu R (2020) Game theory and reinforcement learning based secure edge caching in mobile social networks. IEEE Trans Inf Forensics Secur 15:3415–3429
14. Hathaliya JJ, Tanwar S, Tyagi S, Kumar N (2019) Securing electronics healthcare records in healthcare 4.0: a biometric-based approach. Comput Elect Eng 76:398–410
15. Kim H, Ben-Othman J (2020) Toward integrated virtual emotion system with ai applicability for secure cps-enabled smart cities: Ai-based research challenges and security issues. IEEE Netw 34(3):30–36
16. Sarker IH, Furhad MH, Nowrozy R (2021) AI-driven cybersecurity: an overview security intelligence modelling and research directions. SCI 2:173. https://doi.org/10.1007/s42979-021-00557-0

17. Sravanthi K, Kumar SM, Amit T (2019) Cyber physical systems: the role of machine learning and cyber security in present and future. Comput Rev J 4
18. Wajid MA, Zafar A (2021) Pestel analysis to identify key barriers to smart cities development in India. Neutrosophic Sets Syst 42:39–48
19. Sarker IH, Kayes ASM, Badsha S, Alqahtani H, Watters P, Ng A (2020) Cybersecurity data science: an overview from machine learning perspective. J Big data 7(1):1–29
20. Basso T, Silva H, Moraes R (2019) Extending a re-identification risk-based anonymisation framework and evaluating its impact on data mining classifiers. IJCBCS 9(4):348–378
21. Saha R, Misra S, Deb PK (2021) Fog FL: fog-assisted federated learning for resource-constrained IoT devices. IEEE Internet Things J 8(10):8456–8463
22. Reham MF (2018) Security vulnerabilities of cyberphysical unmanned aircraft systems. IEEE Aerosp Electron Syst Mag 33(9):4–17
23. Pollard T, Clark J (2019) Connected aircraft: cyber-safety risks insider threat and management approaches. In: Proceedings of the 52nd Hawaii international conference on system sciences
24. Guo R, Tian J, Wang B, Shang F (2020) Cyber-physical attack threats analysis for UAVs from CPS perspective. In: 2020 international conference on computer engineering and application (ICCEA)
25. Kumar A, Bhushan B, Nand P (2022) Preventing and detecting intrusion of cyberattacks in smart grid by integrating blockchain. In: Sharma DK, Peng SL, Sharma R, Zaitsev DA (eds) Micro-electronics and telecommunication engineering. Lecture notes in networks and systems, vol 373. Springer, Singapore. https://doi.org/10.1007/978-981-16-8721-1_12
26. Onyema EM, Dalal S, Romero CAT, Seth B, Young P, Wajid MA (2022) Design of intrusion detection system based on cyborg intelligence for security of cloud network traffic of smart cities. J Cloud Comput 11(1):1–20
27. Mistry I, Tanwar S, Tyagi S, Kumar N (2020) Blockchain for 5G-enabled IoT for industrial automation: a systematic review solutions and challenges. Mech Syst Signal Process 135
28. Bagga P, Sutrala AK, Das AK, Vijayakumar P (2021) Blockchain-based batch authentication protocol for Internet of Vehicles. J Syst Architect 113
29. Anusha R, Yousuff M, Bhushan B, Deepa J, Vijayashree J, Jayashree J (2022) Connecting blockchain with IoT—a review. In: Sharma DK, Peng SL, Sharma R, Zaitsev DA (eds) Micro-electronics and telecommunication engineering. Lecture notes in networks and systems, vol 373. Springer, Singapore. https://doi.org/10.1007/978-981-16-8721-1_14
30. Parrend P, Navarro J, Guigou F, Deruyver A, Collet P (2018) Foundations and applications of artificial intelligence for zero-day and multi-step attack detection. EURASIP J Inf Secur 2018(1):4
31. De Bruijn H, Janssen M (2017) Building cybersecurity awareness: the need for evidence-based framing strategies. Gov Inf Q 34(1):1–7
32. Maalem Lahcen RA, Caulkins B, Mohapatra R, Kumar M (2020) Review and insight on the behavioral aspects of cybersecurity. Cybersecurity 3:1–18
33. Misra S, Mukherjee A, Roy A, Saurabh N, Rahulamathavan Y, Rajarajan M (2021) Blockchain at the edge: performance of resource-constrained IoT networks. IEEE Trans Parallel Distrib Syst 32(1):174–183
34. Budhiraja I, Tyagi S, Tanwar S, Kumar N, Rodrigues JJPC (2019) Tactile Internet for smart communities in 5G: an insight for NOMA-based solutions. IEEE Trans Ind Inform 15(5):3104–3112
35. Wist K, Helsem M, Gligoroski D (2021) Vulnerability analysis of 2500 docker hub images. In: Advances in security networks and internet of things. Springer, pp 307–327
36. Raj RK, Anand V, Gibson D, Kaza S, Phillips A (2019) Cybersecurity program accreditation: benefits and challenges. In: Proceedings of the 50th ACM technical symposium on computer science education, pp 173–174
37. Tanwar S, Parekh K, Evans R (2020) Blockchain-based electronic healthcare record system for healthcare 4.0 applications. J Inf Secur Appl 50
38. Zheng Z, Xie S, Dai H-N, Chen W, Chen X, Weng J et al (2020) An overview on smart contracts: challenges advances and platforms. Future Gener Comput Syst 105:475–491

39. Alladi T, Chamola V, Naren (2021) HARCI: a two-way authentication protocol for three entity healthcare IoT networks. IEEE J Sel Areas Commun 39(2):361–369
40. Merabet F, Cherif A, Belkadi M, Blazy O, Conchon E, Sauveron D (2020) New efficient M2C and M2M mutual authentication protocols for IoT-based healthcare applications. Peer-to-Peer Netw Appl 13(2):439–474
41. Masud M, Gaba GS, Choudhary K, Hossain MS, Alhamid MF, Muhammad G (2021) Lightweight and anonymity-preserving user authentication scheme for IoT-based healthcare. IEEE Internet Things J 1–1
42. Deebak D, Al-Turjman F (2021) Smart mutual authentication protocol for cloud based medical healthcare systems using internet of medical things. IEEE J Sel Areas Commun 39(2):346–360
43. Gao Y, Li X, Peng H, Fang B, Yu P (2020) HinCTI: a cyber threat intelligence modeling and identification system based on heterogeneous information network. IEEE Trans Knowl Data Eng
44. Zeadally S, Adi E, Baig Z, Khan IA (2020) Harnessing artificial intelligence capabilities to improve cybersecurity. IEEE Access 8:23817–23837. https://doi.org/10.1109/ACCESS.2020.2968045
45. Salah K, Rehman MHU, Nizamuddin N, Al-Fuqaha A (2019) Blockchain for AI: review and open research challenges. IEEE Access 7:10127–10149. https://doi.org/10.1109/ACCESS.2018.2890507
46. Srivastava T, Bhushan B, Bhatt S, Haque AKMB (2022) Integration of quantum computing and blockchain technology: a cryptographic perspective. In: Kumar R, Sharma R, Pattnaik PK (eds) Multimedia technologies in the internet of things environment, vol 3. Studies in big data, vol 108. Springer, Singapore. https://doi.org/10.1007/978-981-19-0924-5_12
47. BlockchianInsight: Top Blockchain Technology Companies 2021 (2021). https://www.leewayhertz.com/blockchain-technology-companies-2021/. Accessed Feb 2021
48. Hassani H, Huang X, Silva E (2018) Big-crypto: big data, blockchain, and cryptocurrency. Big Data Cogn Comput 2(4):34
49. Sun CC, Joseph-Duran B, Maruejouls T, Cembrano G, Meseguer J, Puig V et al (2017) Real-time control-oriented quality modelling in combined urban drainage networks. IFAC 2017 world congress, pp 4002–4007
50. Rathi R, Sharma N, Manchanda C, Bhushan B, Grover M (2020) Security challenges & controls in cyber physical system. In: 2020 IEEE 9th international conference on communication systems and network technologies (CSNT). https://doi.org/10.1109/csnt48778.2020.9115778
51. Zafar A, Wajid MA (2020) A mathematical model to analyze the role of uncertain and indeterminate factors in the spread of pandemics like COVID-19 using neutrosophy: a case study of India, vol 38. Infinite Study
52. Hosseinzadeh M, Sinopoli B, Kolmanovsky I, Baruah S (2022) ROTEC: robust to early termination command governor for systems with limited computing capacity. Syst Control Lett 161
53. Roy D, Hobbs C, Anderson JH, Caccamo M, Chakraborty S (2021) Timing debugging for cyber-physical systems. In: Proceedings of the 2021 design automation and test in Europe conference and exhibition, pp 1893–1898
54. Lavaei A, Soudjani S, Abate A, Zamani M (2021) Automated verification and synthesis of stochastic hybrid systems: a survey. arXiv preprint
55. Benveniste A, Caillaud B, Nickovic D, Passerone R, Raclet J-B, Reinkemeier P et al (2018) Contracts for system design. In: Foundations and trends® in electronic design automation, vol 12, no 2–3. Now Publishers, Inc, pp 124–400
56. Madaan G, Bhushan B, Kumar R (2020) Blockchain-based cyberthreat mitigation systems for smart vehicles and industrial automation. Stud Big Data Multimedia Technol Internet Things Environ. https://doi.org/10.1007/978-981-15-7965-3_2

57. Cao R, Hao L, Gao Q, Deng J, Chen J (2020) Modeling and decision-making methods for a class of cyber-physical systems based on modified hybrid stochastic timed petri net. IEEE Syst J 14(4):4684–4693
58. Hsieh FS (2021) A dynamic context-aware workflow management scheme for cyber-physical systems based on multi-agent system architecture. Appl Sci 11(2030)

Scalable Blockchain Architecture of Internet of Medical Things (IoMT) for Indian Smart Healthcare System

Ravinder Kumar, Ritu Rana, and Sunil Kumar Jha

Abstract The state-administered Indian Healthcare system involves both public and private participation. The system is transitioning to a cost-effective, quality-based care system that is becoming overburdened due to population growth, resulting in low-quality healthcare for the poor. Additional issues include drug counterfeiting, clinical trials, and healthcare data management. Blockchain technology promotes healthcare disintermediation, patient data transfer, and system automation by standardising data collection, preventing data tampering, and standardising data collection. The next version of blockchain technology, blockchain 5.0, allows for one million transactions per second or more, meaning that it will be able to support use cases like the Internet of Medical Things (IoMT). As healthcare becomes increasingly digitised, more affordable, and more efficient, blockchain technology is being employed more frequently. This chapter proposes a blockchain-based, scalable encrypted health record management system as a solution to the aforementioned issues plaguing the healthcare sector. This chapter further investigates how well it communicates with other health record systems to fulfil our needs for openness, safety, and scalability. Consensus techniques for optimising blockchain performance will also be discussed.

Keywords Indian healthcare system · Blockchain 5.0 · IoMT · Decentralization · Scalability · Privacy traceability · Smart contract

---

R. Kumar (✉) · R. Rana
Skill Department of CSE, Shri Vishwakarma Skill University, Palwal, India
e-mail: ravinder.kumar@svsu.ac.in

R. Rana
e-mail: ritu.rana@svsu.ac.in

S. K. Jha
School of Computer and Software, Nanjing University of Information Science and Technology, Nanjing, Jiangsu, China

B. Bhushan et al. (eds.), *AI Models for Blockchain-Based Intelligent Networks in IoT Systems*, Engineering Cyber-Physical Systems and Critical Infrastructures 6,
https://doi.org/10.1007/978-3-031-31952-5_11

1 Introduction

The term "healthcare system" is used to describe the collective effort of individuals, institutions and resources to address the medical needs of a population. There are as numerous healthcare systems and histories as there are countries. Despite the prevalence of primary healthcare and public health programmes, nations must build their own health systems based on their needs and resources. In some countries, market participants design healthcare systems. Others provide organised health care services to their communities through governments, unions, charities, religious groups, or other coordinated institutions [1].

Physical and mental illnesses are identified, treated, and prevented by medical professionals, support staff, hospital systems, etc. [1]. Accidents, illnesses, and other medical crises need reliable care. Diagnosis, therapy, and aftercare have all benefited from recent medical advances. Access to quality healthcare is crucial to a nation's development. Access to quality healthcare is guaranteed as part of the WHO Constitution [2]. Studies have shown that developing countries often lack adequate healthcare infrastructure. Lack of access to healthcare is a leading cause of poverty. The availability of medical treatment differs from one nation to the other. The economy of India is growing but its healthcare system still needs work.

After three decades of real per capita GDP growth of more than 5%, India joined the lower-middle income category in 2009. A demographic dividend might emerge from a large and growing number of individuals of working age, as well as an ageing population. Non-communicable illnesses and reproductive health difficulties are on the rise. The IMR has decreased from 88 in 1990 to 32 in 2020, a considerable decrease. The MMR declined from 556 in 1990 to 113 in 2018 per 100,000 live births over a three-year period from 2016 to 2018. Despite increases in life expectancy and child and maternal mortality, India's healthcare system has to be improved. Economic, demographic, and epidemiological changes have an impact on the country's health-care system development. Although overall rates have risen, people in economically challenged areas have benefited the most [3].

There has been an increase in medical costs due to underfunding and the expansion of the private provider sector. Out-of-pocket (OOP) costs account for a staggering 66% of all family health care expenses. Catastrophic medical bills are a reality for 17% of Indian households, putting 55 million people below the poverty line every year. The health care workforce has grown significantly over the past two decades due to an expansion in medical, nursing, and technical schools. In 2019, India had 9.28 medical physicians, 23.89 registered nurses and midwives, 2.04 dentists, and 8.89 pharmacists for every 10,000 people. Both the quality and the dispersal of such workers across different cities and regions is not uniform [3].

Conditions that persist over time can be treated at either private clinics or public hospitals. In India, private healthcare providers account for over 70% of all medical visits, 58% of all hospitalizations, and 90% of all pharmaceutical and diagnostic services. Competing vendors provide varying degrees of affordability, quality, and effectiveness. Medical schools, health departments, 30% of outpatient care, and 42%

of hospital stays receive support from the government. Evidence of low importance is shown in chronic underfunding of the healthcare system and inadequate regulatory frameworks. Since 2004, participation of the public sector has increased, in part because of NRHM/NHM investment.

In India, there have been a number of proposed measures to improve healthcare. Initially, the focus of NRHM/ NHM was on maternal and newborn health and infectious disease control, but the results were mixed. Healthcare providers, the pharmaceutical industry, and interconnected systems all need to be managed for the interest of patients and the healthcare system as a whole. Private sector players in India are regulated in ways that are both vague and inconsistent. The majority of Indian states have not implemented the Clinical Establishments Act, 2010. The Act mandates the registration of all healthcare institutions and establishes standards for the provision of medical care. In India's subnational drug regulating system, regulation compliance is low because of the weak infrastructure, lack of competent staff, confusing law, and several administrations. Existing prescription price caps are fair to both pharmaceutical companies and patients. Limitations on private sector medication coverage and cost savings have persisted since its inception in 2013. Medical education and practice can both benefit from revamped professional councils. In 2020, NMC took over for MCI. This new organisation would streamline regulations for medical education, improve efforts to rank medical schools, establish guidelines for tuition, fees and prioritise medical research of private medical colleges. There should be greater funding, better management, and higher quality services in the public health care system.

Public health expenditure at the federal and state levels should treble over the next five years, rising to 8% of total government spending. In addition, improvements are required to the public financial management system to ensure a fair and effective allocation of funds. The newly allocated monies for health sector investment must be utilised to recruit, train, and hire health workers, especially nurses and allied healthcare professionals, to offer world-class primary care given the current shortages. Consolidating medications and simplifying supply chains may help improve efficiency and reduce costs. The illness load in India has been shown to be twice that of other countries. Although there has been progress in reducing infectious illnesses and maternal and child mortality, the benefits differ from state to state [4]. Mother and infant mortality are exacerbated by malnutrition, food poverty and nutritional deficits such as anaemia. Cardiovascular disease, cancer, chronic obstructive pulmonary disease, diabetes, mental illness and accidents are all increasing as a public health problem in India. Only China's neighbouring country, India, has a larger population. India's total population will hit 1.41 billion in the year 2022. The percentage of Indians aged 65 and older is projected to rise from 3.8% in 1990 to 6.6% in 2022, while the percentage of Indians aged 0–14 is projected to reach 25.5%. A staggering 79.1% of Indians relied on people 65 and older in 1970, but by 2022, that number had dropped to only 38.1%. From 5.9% in 1970, the proportion of the elderly who rely on others is expected to increase to 10.1% in 2020 [5]. If India prioritises public spending in areas like healthcare, education and infrastructure, the country's burgeoning working age population might be a demographic blessing.

The rest of the chapter is organised as follows. Section 2 presents the status and use of Information technology in the Indian Healthcare System. Key characteristics, types, generations and applications of Blockchain are presented in Sect. 3. Internet of Medical Things is elaborated in Sect. 4. Section 5 describes the horizontal and vertical scalability of the blockchain. Section 6 presents the related work on the topic of the chapter. Section 7 describes the proposed Blockchain based IoMT Architecture. The conclusion is presented in the last Sect. 8 of this chapter.

2 Indian Healthcare System

Technology and infrastructure both improved as India rose to prominence. India's healthcare environment is improved by market liberalisation and private sector involvement. Following the independence of India in 1947, public healthcare dominated. A large population, inefficient resource distribution, and a lack of coordination between the organised and unorganised sectors limit the effectiveness of India's healthcare system. The Indian healthcare system is hampered by ineffective resource allocation, unskilled labor, financial constraints, inflation, and low quality. Poor healthcare is a result of imbalanced medical therapy in India. Rich people choose private healthcare because it offers reliable medical services, excellent healthcare, professionals, superior infrastructure, diagnostics, a quick supply chain for medications, etc.

2.1 Organisation and Governance

Under India's federal system, both the central government and the individual states may make policy decisions. Decisions, plans, and services in the healthcare industry are impacted by both the central and state governments. The Indian Constitution gives authority over healthcare policy making to the individual states. The State governments manage and finance the healthcare initiatives. Healthcare systems, administration, funding, and oversight are all coordinated at the state level. A combination of governmental and private sectors provide healthcare. State-level public health service providers are under the jurisdiction of local administrative authorities. Health care facilities and private doctors are regulated in different ways.

After ratifying Alma-Ata in 1978, India had its first National Health Policy (NHP) in 1983. The NHP 1983 emphasised treatment aimed at avoiding sickness, enhancing health, and rehabilitating the sick. By forming alliances with the business sector and non-profit organisations, the NHP 1983 was able to reduce public spending on primary care without sacrificing quality or accessibility. Inadequate public funding, pro-private sector policies, and increased individual wealth all contributed to the private health industry's explosive growth throughout the 1980s. The growing incidence of inequality makes it harder for people to pay for necessary medical care. The rural–urban gap develops as more highly skilled public and commercial providers leave rural regions for urban centres. In 2002, the National Health Program advocated for a modernization of the healthcare sector. Differences in health care access between rural and urban regions are the primary focus of NHP 2002.

Increases in public health spending were called for by the National Health Plan (NHP) in 2002, with a goal of 2% of GDP by 2010. Since the original NRHM did not appear until 2003, it was published a full year after the NHP in 2002. India has the lowest healthcare spending as a share of GDP among developing countries in 2005. India spent 1.1% of its GDP on health care in 2014–2015, despite having one of the world's fastest-growing economies. In 2018, the Indian government implemented Pradhan Mantri Jan Arogya Yojana (PM-JAY), the successor to RSBY. More than 107 million low-income families would receive $7,000 apiece (INR 0.5 million). Primary healthcare (PHCs and SHCs), secondary healthcare (CHCs, taluka and district hospitals), and tertiary care (other institutions) (medical colleges and teaching hospitals) make up India's public healthcare system [5].

2.2 *Health Status*

India's life expectancy is estimated by experts at 70.19 by 2022. Despite its progress, India's average lifespan of 69 years falls short of that of other middle-income countries like Sri Lanka (74), Brazil (74), China (75), and Costa Rica (79) [5]. The average life expectancy of an Indian woman has increased by 24 years between 1970 and 2016, while that of an Indian man has increased by 20 years. There is a 15-year gap in the life expectancy of men at age 60 and that of women of the same age. Reduced rates of baby, child and maternal mortality have contributed to an increase in life expectancy.

The Health Performance Index, created by NITI, allows for comparisons between states. Between 1970 and 2020, life expectancy rose fastest in the Indian states of Uttar Pradesh, Tamil Nadu, Odisha, Himachal Pradesh, Gujarat, Bihar, Assam and Andhra Pradesh [5]. An increase of 22 years in life expectancy over 50 years was recorded in the Indian state of Uttar Pradesh. By 2020, there will be 32 infant deaths for every 1,000 births, down from 142 in 1970. India's maternal mortality ratio (MMR) decreased to 113 in 2016–18 from 556 in 1990, representing a 15% reduction from the global MMR total (ORGI & CCI).

The prevalence of chronic diseases is rising in India, as it is everywhere in Asia. Stroke, chronic obstructive pulmonary disease, and ischemic heart disease are all leading causes of death worldwide. In India, fewer children are falling ill with infectious diseases and other paediatric conditions. The leading causes of death in children and young adults include complications after birth, coronary artery disease, lower respiratory tract infections, diarrhoea, and chronic obstructive pulmonary disease. Stroke, tuberculosis, self-harm, traffic accidents, and congenital disorders are major causes of death in India.

The use of tobacco products and the abuse of alcoholic beverages are major contributors to the prevalence of chronic diseases in India and the rest of the world [5]. Many Indians partake in tobacco use via smoked cigarettes, bidis, or chewing tobacco. The average annual alcohol consumption per Indian is 2.30 L. Self-reported consumer surveys frequently underestimate alcohol usage compared to estimates based on sales, tax collections, manufacture, or commerce. The use of processed meals and sugary beverages leads to poor tooth health. Injuries and illnesses including gum disease and atrophy can play a role in the mortality rate of the elderly.

Rapid urbanisation has increased the need for water, sanitation, and hygiene (WASH) services, which are essential for promoting health and well-being. WASH and access to bathrooms have all increased in India. Open defecation fell from 56 to 46% between 2006 and 2016. In 2021, 70.2% of homes with toilets were also equipped with modern sanitation facilities, according to a WHO/UNICEF survey. NFHS-5 households have better drinking water, according to reports. Having access to safe drinking water and sanitary facilities can help reduce the prevalence of disease and hunger.

There are two health concerns that might arise. Human health in both cities and the countryside is threatened by air and water pollution. Air pollution is the second highest risk factor for disease burden, according to the 2017 Global Burden of Disease (GBD) India Report. It is a major cause of cardiovascular illness, chronic respiratory disease, and lower respiratory tract infections. Surface water in India, particularly in major cities, is highly polluted. The water pollution problem in India is the cause of 4.6% of the country's sickness burden due to contaminated water, lack of sanitation, and improper handwashing.

Because of national insurance reform initiatives, the cost of providing healthcare to the poor has been reduced. People in the city's higher strata may be able to afford private health insurance. Lower levels of end-user compassion can be attributed to a number of factors, including the presence of hidden clauses, a lack of understanding, a convoluted legal system for settlements, and indifference on the part of intermediaries and big insurers. Ineffective resource allocation due to a lack of trust and transparency wastes money, causes unnecessary patient suffering, and costs lives. Cybercriminals want to steal patient information from the healthcare industry. Information about patients from different groups is not pooled together in Indian healthcare. A lack of efficient methods for storing medical records, as well as a lack of tools for reducing duplication of tests and reports. Healthcare has been impeded by a lack of interoperability. In India, doctors and hospitals mostly use paper and visual records. It's slow, sloppy, and rife with mistakes, leading to misdiagnosis and subpar care for patients.

The diagnostic precision gained by data interchange is improved by peer review. In other words, it prevents medical errors. The counterfeit drug problem is also a worry in the Indian healthcare system. Other difficulties that pharmaceuticals in Indian healthcare face include setting reasonable prices for medications, providing complete and accurate information about potential drug side effects, and conducting sufficient clinical trials.

Everyone in India, regardless of where they reside or who they identify as, should have equal access to high-quality healthcare that doesn't break the bank. The government places a premium on health care quality. The federal government has launched a variety of healthcare programmes. India's efforts to achieve the Millennium Development Goals (MDGs) are bolstered by the country's national health policy, national population policy (NPP), and national AIDS prevention and control strategy. Healthcare professionals and hospitals that offer many specialties are free from taxes for five years in all but eight jurisdictions. There is a serious lack of progress in healthcare in India.

India has the world's lowest public spending on healthcare, despite the fact that a quarter of its population is poor. Despite recent advancements, India still has a poor life expectancy. As a result of budget shortfalls and resource mismanagement, many states have subpar health management systems. Healthcare is a major issue for many individuals. Only a limited number of beds are currently available. India has a lower ratio of hospital beds to citizens than the other BRICS countries. Trained professionals are necessary in the healthcare industry. According to the WHO, private insurance covers 60% of healthcare costs.

No of a person's location, gender, or identity, the healthcare system in India must prioritise access, efficiency, and quality. The provision of healthcare is one of the federal government's highest priorities. The general population has access to several healthcare options. The National Population Policy (NPP) of 2000, the National Health Policy of 2002, and the National Aids Prevention and Control Policy of 2004 all support the MDGs. There is a five-year tax break for private medical clinics and hospitals that treat many specialties. Improvements are needed in Indian healthcare on a global scale.

India has lower healthcare expenditures compared to other countries in the medium income range. India has the lowest public healthcare spending in the world, while having the world's largest poor population (World Bank 2018). There have been some positive changes, but overall, India still has a low life expectancy. Inadequate health management is the result of poor fiscal and resource management in many jurisdictions. Unfortunately, many people lack access to adequate medical treatment. India is just one of a few BRICS countries that does not have access to a large number of hospital beds. Humans trained in the medical field are required.

It is possible that recent technical developments might help India's healthcare system. The application of new technologies has the potential to improve the healthcare system in a number of ways. This research examines the possible effects of blockchain technology on India's healthcare system, and proposes a plan for adopting it, including input from a wide range of stakeholders. The information systems perspective enhances the approach for addressing issues in the Indian healthcare ecosystem by putting an emphasis on stakeholder centricity and the most suitable demand elicitation for key stakeholders.

2.3 Information Technology

India's health care system has been bolstered and institutionalised thanks to the rapid growth of Internet connectivity and IT use. There are now 692 million Internet users in India. The CUBE 2020 research by IAMAI and Kantar projects that by 2025, there will be 900 million internet users in India. There were 26 Internet users for every 100 people in India in 2015. Telemedicine, teleradiology, and other forms of cutting-edge medical technology hold great promise for improving patient care and reducing healthcare costs.

Primary health care centres (PHCs) are strongly encouraged to implement the usage of Internet-connected computers by the International Primary Health Care Research Group (IPHS). The Mother and Child Tracking System and online portal (consolidated facility-based reporting) and the District Health Information System are two methods by which districts transmit data to regional and state headquarters. NACO developed the Strategic Information Management System in order to collect and centralise information about the HIV/AIDS epidemic. Since 2004, the IDSP has been collecting illness data from healthcare institutions every week [6]. These facilities include both public and private hospitals and medical schools.

Public health clinics, community health centres, district hospitals, block/taluka hospitals, private hospitals, academic medical centres, and specialised hospitals all provide data into the HMIS to create a comprehensive picture of the country's health. A comprehensive health database for any of the state's regions is available on NRHM website [7]. Despite its prominence in the IT industry, India lacks a standardised network. Some of the HMIS available lacked necessary documentation, user capacity, frequent training, and process standards. Patient information recorded in HMIS does not match what is reported on paper from the SC to the district. In order to collect data, district-level employees must use the paper-based HMIS system [8].

For the purpose of cutting down on data redundancy and fragmentation, NHA created NDHM. NDHM is permitted to maintain, access, and exchange patient medical records. Consumers, healthcare providers, and insurance companies may all benefit from the use of IT services. There are electronic health records available through NDHM. The National Directory of Health Providers (NDHP) streamlines the identification of healthcare providers through the use of a standardised health ID, electronic medical records, and patient-centric health information. Health programmes receive funding from MoHFW. Programs and policies for health are developed by DoHFW. NHM provides support to DoHFW. Research, treatment, and prevention are all coordinated by DHR. Regional and state offices of DGHS manage the nation's health programmes.

There have been three major periods of Indian government health care policy. In the wake of India's independence, the government adopted five-year plans and a mandate for couples to have just one child. As population control became more of a priority, efforts to improve maternal and child health and reduce the spread of illness were put on the back burner. Financial resources were reduced between 1983 and 2001 as a result of economic shifts and macroeconomic stabilisation efforts that favoured private curative therapy at the expense of public health care. Between 2002 and 2020, there were six major changes to health policy. Health care and its finance have recently undergone significant changes. Pharmaceutical reforms (drug and medical device pricing control, transitioning from process to product patent system, etc.), the Clinical Establishments Act, 2010 and Rules, 2012 [9], the National Health Protection Act, 2017 [10], and Ayushman Bharat, 2018 [11] are just a few examples. To adequately serve the population, 1.5 million hospitals are required.

Health, risk reduction and a prompt response are all global health priorities. Health care systems are designed to prioritise equity, efficiency, efficacy, accessibility and cost. By 2025, healthcare in India is projected to grow by 31%. Healthcare demand will increase due to factors including wealth, ageing populations, health literacy, and shifting attitudes toward preventative treatment. When it comes to research and development, India is in the forefront because of the low cost of its clinical studies. The committed budget for 2022–23 is Rs. 86,200.65 cr [12]. Health care providers in the United Kingdom are regulated by the National Commission for Allied and Healthcare Professions Bill, 2021 [13].

Incorporating more telemedicine into India's healthcare system may enhance distant healthcare management, education, and training. Telepathology, teleradiology, and e-pharmacy are prioritised in India. PPPs exist at Apollo, AIIMS and Narayana Hrudayalaya. By 2025, according to EY and the Indian Pharmaceutical Alliance, telemedicine will be worth $5.5 billion. 12 million patient-doctor consultations have been conducted via eSanjeevani by September 21, 2021 [14].

Patients may directly communicate with doctors and receive treatment alternatives by using AI applications. TrakItNow joined the IHF in April 2021. TrakItNow Technologies is developing a mosquito-control solution using IoT and AI (SDIL). COVID-19 vaccines had been provided in 2020 using the digital vaccine platform CoWIN. This smartphone app records immunisation information for all states and UTs' "Healthcare Workers" database. CoWIN is open source in 2021, and 76 countries have filed CoWIN requests for COVID-19 vaccines [15].

Popular technologies include Electronic Medical Records (EMR), Mobile Healthcare, Electronic Health Records (EHR), telemedicine, Hospital Management Information System (HMIS) and softwares by PRACTO. Uttar Pradesh is working on having automated medicine distribution. "Health ATMs" will be available in all 75 districts of Uttar Pradesh covering 4600 health centres by 2022. AstraZeneca India and Docon Technologies of Bengaluru signed an agreement to digitise 1,000 clinics in India. Eka Care can print Health ID cards, download certifications, and schedule immunizations. The medical tourism industry in India is growing. NHM assists STN under PIP in developing a dependable, comprehensive, high-speed network backbone.

As of November 18, 2021, as part of "Digital India" initiative, e-Hospitals are established. On August 15, 2020, Prime Minister Narendra Modi inaugurated the National Digital Health Mission (NDHM). By May 2021, there are 3,106 doctors, 1,490 facilities, and 11.9 million Health IDs [16]. India's health-tech startup received $4.4 billion in venture finance between 2016 and 2021, including $1.9 billion in

2021. In January 2022, HealthifyMe gained 500,000 monthly customers and had a $40 million ARR. Tata Digital Limited purchased 1 mg in June 2021.

Because of the government's emphasis on Ayushman Bharat, digitising health data, and Mission Indradhanush, technology advancements and digitally enabled services will dominate the country's healthcare business. To improve patient care, digitise medical records, automate manual procedures, employ data analytics, and integrate ML/AI/IOMT. RPA automates cognitive functions based on rules. Digital healthcare is efficient, engaging, relevant, and long-term. To establish tech-savvy teams, healthcare providers must invest in digital technologies and digital literacy. Personnel who are technologically knowledgeable increase the productivity of healthcare practitioners.

Healthcare organisations that are digitally equipped require a digital strategy and a CIO. Healthcare has been transformed by the Internet, mobile devices, social media, and technology. Patients get more interested as they improve their health. Patients make decisions about their health management, treatment, and results. Intelligent patients require innovative services and business tactics. Simple episodic treatment will become obsolete; instead, clinicians must focus on comprehensive care in order to retain and engage patients. As healthcare technology becomes more widely available, it will aid R&D and the development of drugs for diseases such as Covid and lifestyle issues like diabetes. India's human development index is 0.647%. According to the United Nations, India ranks 131st in terms of healthcare growth. India has a shorter life expectancy than the United States, the United Kingdom, Japan, Australia, and China [3].

According to the OECD, India spends 3.6% of its GDP on healthcare. Healthcare spending will fall by 5% by 2025 [17]. India is looking for a dependable healthcare system for its citizens, healthcare providers and the government. The healthcare system in India is disorganised as a result of squandered resources, questionable compliance, and unknown costs. Inequitable access to medical facilities in rural and urban areas has an influence on healthcare delivery. Inadequate healthcare is one of the government's top objectives.

The 2004 National Strategy for HIV/AIDS Prevention and Control was developed as part of the right to health and the MDGs [18]. In 2015, India ceased enforcing the second National Health Policy. The New Health Paradigm, 2015 posits a link between health and progress. Healthcare may improve as a result, and health goals may be accomplished.

Patients have challenges in a variety of areas, including choosing a healthcare provider, obtaining appropriate care, and maintaining their own records. Issues with data traceability, availability, and integrity beset India's healthcare system. Data missing due to weak integration and inaccurate health records may save time by reducing the number of people you need to contact. Conflicts may be lethal and result in deaths. There is frequently a dearth of infrastructure to facilitate the transfer of medical records.

More study and knowledge are required for India's healthcare system. Blockchain features include immutability and data immutability, in addition to non-repudiation, disintermediation, and auditability. Researchers are investigating the possibilities of blockchain technology in the healthcare business. Remove any intermediates, ensure the data is reliable, and look for any problems. Blockchain technology might be beneficial to the healthcare industry. The protocol-level possibilities provided by the Blockchain ecosystem, as well as the application-level foundations for stakeholder or human agent interaction, behaviour, and economics, are investigated by information theory. The usage of blockchain technology in the medical business in India is quite restricted.

3 Blockchain

Blockchain maintains an unchangeable distributed digital ledger that records peer-to-peer transactions. As middlemen are eliminated, transparency and traceability are increased. It is disseminated to network nodes after being duplicated. Consensus is used by users to approve transactions in this shared ledger. A hash is produced by encrypting transactions in a data block. Each block is tamper-proof because it keeps the hash from its predecessor. A single data modification might alter the hash value and prevent the transaction.

3.1 Key Characteristics

Blockchain is decentralised, which means that no single entity controls the data that is submitted to it. Instead, the entries posted to the blockchain are agreed upon by a peer-to-peer network using various consensus processes. Another crucial aspect of blockchain technology is persistence. Because of the distributed ledger's numerous node storage, removing entries after they have been accepted into the blockchain is extremely difficult. Furthermore, some blockchains consider anonymity (or pseudonymity) to be a desirable attribute. Blockchains offer auditability and traceability by identifying each new block in a chain of blocks with the hash of the previous block.

3.2 Types of Blockchain

The three most prevalent types are public (permissionless), private, and consortium (public permissioned) blockchains [19]. The two differ in terms of who may read, write, and access data on the blockchain. On a public chain, all users may access and use the same data, and anybody can participate in consensus decisions and protocol modifications. Bitcoin and Ethereum are two famous instances of the public blockchain, which is widely used in the cryptocurrency business (the main chain).

They are part of a network type that does not require specific authorisation to access the public internet.

While the consensus process in a consortium blockchain is theoretically decentralised, it is frequently limited to a small number of carefully selected groups of organisations in practise. A private blockchain network is frequently centralised while remaining decentralised. Only a few nodes are permitted to join in these networks, which are frequently managed by a single entity. There is continuing debate regarding how to appropriately characterise and identify the many types of blockchains covered here. There is continuous debate over what properties and consensus processes are required for a system to be classified as a blockchain.

Decentralized apps (dApps) may be created using current blockchain platforms and frameworks. The two most prominent solutions at the present are Ethereum (a decentralised platform) and Hyperledger (a framework), which allow developers to connect their own blockchain apps into existing networks and construct their own test networks using the protocols already developed by those networks.

Consensus Mechanisms

A critical component of blockchains is the consensus mechanism for confirming data entries before they are added to the distributed ledger. While numerous approaches to obtaining an agreement have been proposed and implemented, the three most common are Proof-of-Work (PoW), Proof-of-Stake (PoS), and Practical Byzantine Fault Tolerance (PBFT) [20].

Because of its implementation in Bitcoin, Proof-of-Work (PoW) has emerged as the consensus mechanism most closely tied to blockchain technology [21]. During the PoW protocol's operational phase, miners compete to be the first to complete a difficult computational task. The mining procedure entails employing brute force to find a hash of the proposed block that is less than a particular threshold. A hash value must be calculated in order to validate the transactions (or other items) in the block, and the miner who calculates this hash value first gets rewarded. When applied to a large blockchain, PoW's wasteful usage of energy becomes obvious. The current power requirements for Bitcoin mining, for example, are comparable to those of a small country.

The node with the largest stake in the blockchain is chosen to operate as the validator using a consensus mechanism known as Proof of Stake (PoS). One's account balance in any particular cryptocurrency represents that person's cryptocurrency holdings. However, the "wealthiest" node may unjustly gain from this. This problem has prompted the creation of a number of hybrid PoS systems, all of which incorporate a combination of the stake and randomness in picking the approving node. In September 2022, Ethereum, the second-largest cryptocurrency, finally made the long-awaited move from Proof-of-Work to Proof-of-Stake.

Practical Byzantine Fault Tolerance (PBFT) is based on a protocol for Byzantine agreements [19]. Because it demands that every node be made publicly aware, PBFT can only be used on a public blockchain. The PBFT consensus method is divided into three stages: pre-prepared, prepared, and committed. To move to the next step, a node must receive permission from two-thirds of the other nodes. In its current incarnations, Hyperledger Fabric employs PBFT.

Smart contracts

Ethereum is only one example of a blockchain that may facilitate smart contracts. These are legally binding contracts written in code that automatically carry out the terms that have been agreed upon. Smart contracts eliminate the need for a mediator or arbitrator by enforcing its terms automatically. This smart contract functionality may be activated by a blockchain transaction, and its potential use in the healthcare industry is exciting.

3.3 Emergence of Blockchain

Hash trees, also known as Merkle trees, were created by computer scientist Ralph Merkle in the late 1970s, which is when blockchain technology first emerged. These trees are a kind of computer science structure used to store data by cryptographically connecting blocks. Stuart Haber and W. Scott Stornetta utilised Merkle trees to build a system that prevented tampering with document timestamps in the late 1990s. Based on prior work by Stuart Haber, W. Scott Stornetta, and Dave Bayer, a blockchain was constructed in 2008 by a person (or group of individuals) using the name (or pseudonym) Satoshi Nakamoto to serve as the public distributed record for bitcoin cryptocurrency transactions [22]. The Merkle tree used to represent the transactions within the blocks may be confirmed from the known root to each leaf value (transaction) [21]. The tree's root might be kept on the distributed ledger to confirm the validity of data (blockchain).

BlockChain 1.0 (Cryptocurrency)

Hall Finley introduced BlockChain Version 1.0, the first cryptocurrency-based programme based on Distributed Ledger Technology (DLT), in 2005. This enables bitcoin to be used for DTL, or financial transactions via blockchain technology. Any participant may carry out a legal Bitcoin transaction in this permissionless form. The majority of these applications deal with money and payments.

BlockChain 2.0 (Smart Contracts)

Because Bitcoin mining was inefficient in BlockChain version 1.0 and the network could not be expanded, the current version was made accessible. Version 2.0 has therefore improved the problem resolution. This blockchain will support smart contracts in addition to currency.

As a result, tiny contracts are composed of chains of blocks comprising small computers. These small computers run free software that lowers transaction costs while automatically validating previously stated needs like facilitation, verification, or enforcement. Ethereum has taken the position of Bitcoin in BlockChain 2.0. As a result, BlockChain 2.0 handled a large number of transactions on the open network quickly.

BlockChain 3.0 (DApps)

DApps, or Decentralized Applications, were introduced in a later version than 2.0. A DApp is comparable to a traditional app in that it may have a frontend written in any language that connects with its backend, which operates on a distributed peer-to-peer network. Several distributed data transit and storage technologies are employed, most notably Ethereum's Swarm.

There are various decentralised programmes, such as BitTorrent, Popcorn, Tor, and BitMessage. The capacity of DApps to perform transactions without the need of a third-party mediator ensures data privacy. Blockchains employ cryptography to safeguard network data. By avoiding duplication, blockchain speeds up transactions. DApps have certain downsides, including the possibility of human error. Bitcoin transactions are expensive. Because blockchain technology is immutable, data cannot be changed.

BlockChain 4.0 (Industry)

Blockchain 4.0 is predicted to be the superior blockchain platform, after Blockchain 3.0. Businesses use Blockchain 3.0 as a component of Blockchain 4.0. It intends to completely mainstream the technology by making blockchain relevant in business environments for the development and usage of apps. Automation, corporate resource planning, and the integration of multiple execution systems are the three primary features of industries. Blockchain technology might aid in achieving the highest levels of privacy and trust required for the Fourth Industrial Revolution. Financial transactions, supply chain management, condition-based payments, Internet of Things (IoT) data collection, health monitoring, and asset management are just a few of the numerous applications for blockchain.

Security, automatic record-keeping, immutability, and the ability to pay bills, wages, and invoices in a totally secure environment are just a few of the advantages that previous incarnations of blockchain technology have demonstrated for businesses. There is room for improvement in both the speed and limited simplicity with which blockchain innovations may presently be created. Blockchain 4.0 is a project aimed at better serving the demands of users.

Blockchain 5.0

Relictum Pro is a blockchain platform for public, private, and commercial companies. Relictum Pro, a distributed registry with a smart contract architecture, formalises every event in a person's life, such as sales and purchases, logistical accounting, copyright monitoring, and connections with legal authorities. It features smart contracts in every sector (self-executing transactions). Smart contracts might regulate any event-based activity.

A smart contract is used to independently verify contract criteria. Relictum Pro protects against code tampering. As a result, a smart contract that has been confirmed by two nodes cannot be altered by an attacker. Various smart contracts may be employed. Create additional forms of smart contract property. Ten parties can sign a smart contract at the same time. Relictum Pro blocks are 120 bytes in size, or 8000 times smaller than Bitcoin blocks. By combining AI, data analytics, and Industry 4.0, Blockchain 5.0 becomes intelligent. Blockchain 5.0 employs blockchain technology to enable intelligent and self-contained activities.

Working of blockchain

The working of blockchain consists of several steps as following:

1. Keep a record of the transaction. A blockchain transaction documents the transfer of real or digital assets from one party to another inside the blockchain network. It is saved as a data block and may contain information such as who was involved in the contract, what occurred throughout the transaction, when did the transaction take place, where did the transaction take place, why did the transaction take place, what proportion of the asset was exchanged, how many of the transaction's preconditions were met etc.
2. Obtain consensus. The majority of distributed blockchain network stakeholders must agree that the recorded transaction is valid. The terms of agreement might vary depending on the kind of network, but they are normally defined at the start of the network.
3. Connect the blocks. Transactions on the blockchain are recorded into blocks analogous to the pages of a ledger book after the stakeholders have reached a consensus. A cryptographic hash is applied to the new block along with the transactions. The hash serves as a connection between the blocks. If the contents of the block are deliberately or accidentally updated, the hash value changes, allowing data tampering to be detected. As a result, the blocks and chains are securely linked and cannot be edited. Each successive block reinforces the prior block's verification and hence the whole blockchain. This is similar to stacking wooden blocks to build a tower. You can only stack blocks on top, and removing a block from the centre of the tower causes the whole structure to collapse.
4. Distribute the ledger. The system distributes a copy of the most recent central ledger to all stakeholders.

3.4 Blockchain in the Healthcare Ecosystem

Blockchain technology naturally incorporates elements like distributed ledgers, decentralised storage, authentication, security, and immutability. In recent years, it has graduated from the realm of marketing hype to find actual use in important industries like healthcare. Health Insurance Portability and Accountability Act, 1996 (HIPAA) imposes stronger authentication, interoperability, and information sharing standards for blockchain applications used in the healthcare business.

The key areas of attention for the implementation of blockchain technology are digital health records and individual health records. Among the problems that blockchain technology intends to address in this area are those of data integrity, provenance, access control, and interoperability. Many data-driven industries, including healthcare, stand to be upended by this innovation.

IBM found that 70% of healthcare executives believe that decentralised frameworks for exchanging electronic health information and improvements in clinical trial administration are on the horizon. In addition, it is estimated that by 2022, the global market for blockchain technology in healthcare would have grown to over $500 million. It is possible that this technology might help healthcare organisations analyse patient data and gain a deeper knowledge of their patients.

There is a lot of potential for blockchain technology to increase data efficiency in the healthcare industry and reduce the prevalence of fraudulent clinical research. Potentially easing worries about data tampering in the healthcare industry, it offers a one-of-a-kind data storage pattern with the highest level of security. It enables adaptable, standardised, interoperable, responsible, and authenticated data access. There are several reasons why patients' medical records should be kept private. Blockchain's decentralised data security and ability to help mitigate some hazards make it a promising technology for the healthcare industry.

Modern healthcare has been eroded by a lack of trust and transparency caused by a provider-centric approach to care delivery made possible by technology. The necessity to safeguard patients' personal information and the threat of losing a competitive edge due to data sharing are only two examples of how this development has worsened existing difficulties with data. Decentralizing this paradigm through the use of trust-enabling technology is vital as the Indian healthcare industry transitions to a patient-driven care delivery model by including preventive healthcare and building a patient-payer-provider ecosystem.

Because blockchain puts the patient at the centre of all data collection and exchange, it has the potential to contribute to the development of a trust-based healthcare ecosystem. It has the potential to make healthcare safer, more transparent, and more efficient, all of which might lead to a reevaluation of fundamental practices. The safe and efficient capture and exchange of data is one way in which blockchain-enabled technology platforms may improve the patient experience, health outcomes, and healthcare insights.

The healthcare industry is becoming interested in blockchain technology as a method of increasing trust, accountability, and transparency in complex procedures. Provider credentialing and certification, clinical trial administration, pharmaceutical supply chain monitoring, health data management, health information exchange, and management of clinical trials are all areas where some healthcare organisations have adopted considerable application use cases. It's possible that data management and sharing for pharmaceuticals might be improved with the use of blockchain technology.

Data access, manipulation, and trustworthiness are critical capabilities in the people- and data-intensive healthcare business. A healthcare system that achieves its goals in terms of patient health requires a multidisciplinary group to work together, utilising the most advanced information and diagnostic tools at their disposal. Triage, health problem resolution, clinical decision making, implementing and assessing therapy are all examples of knowledge-based healthcare operations. In order to develop their skills, students majoring in the healthcare field require clinical experience and exposure to real patients. In return, they supply an army of capable workers.

Scientists and engineers working in the medical field should have simple access to clinicians, patients, data, and other tools. Participating healthcare institutions should have a say in study design, methodology, and data collection and analysis. Due to the efforts of research and engineering centres, the healthcare business now has access to state-of-the-art knowledge, techniques, and technology. Biomedical research, engineering, and the training of healthcare personnel rely heavily on the smooth functioning of healthcare facilities. Consents, patient data and proof, and payment processes must be easily shared between institutions for any work to be done. The information contained in a patient's medical file is very personal and must be kept secure at all times.

Access control, provenance, data integrity, and interoperability must be implemented to preserve patient privacy and allow collaboration across healthcare organisations. The fundamental assumption of traditional access control approaches is mutual confidence between data owners and storage organisations. The majority of these establishments are servers, and they control who may access what. Individual and community health may be enhanced through the interoperability of information systems, devices, and applications, which allows for the sharing and use of data from a variety of sources. The provenance of a dataset is the history of how it came to be. Provenance has the potential to improve the auditability, openness, and user confidence of electronic health records (EHRs). An individual's data quality should match the expectations placed upon it. How closely the quality of the actual data meets or surpasses expectations is what is meant by "data integrity level".

Healthcare organisations might benefit from having access to more commercial and research data. The public's trust in medical institutions has been eroded as a result of the leakage, infiltration, and theft of private data. Last but not least, medical fraud takes advantage of helpless people (e.g. the problems with counterfeit drugs, procedures, skills and patients). This issue requires a novel approach and different instruments to tackle. Blockchain technology offers promise in facilitating interoperability, information sharing, access control, provenance, and the integrity of data, all of which might be used to establish and sustain trust between participants. Some examples include the ability to store data without a central server, distribute it to several users, and guarantee its accuracy. Healthcare is the world's most data-intensive industry. Unfortunately, this information is underutilised because of concerns about privacy and exploitation. In the healthcare industry, blockchain technology might usher in a new era of efficient data use, leading to better patient outcomes and lower costs.

Global use cases

Medical records may be stored and shared securely using blockchain technology. Electronic health records based on the blockchain were implemented in Estonia. In the years after its independence in 1991, Estonia made extensive use of technology to improve public services, most notably healthcare, and to give its citizens a digital society. Estonia's National Health Information System uses blockchain-enabled EHRs and an exchange platform. As a result, data may be shared between the public and commercial sectors with confidence, and internal data dangers can be mitigated. The e-patient portal allows the e-health record to collect information from a wide variety of healthcare providers. Prescriptions, health records, and other patient information are all digital in Estonia thanks to blockchain-enabled programmes.

India use cases

Businesses in India's healthcare sector are creating, implementing, and adopting a blockchain-based solution to address a number of issues. Blockchain-based solutions are being used by both private and public healthcare organisations to associate patients' identities with their own health information. Prototypes of a decentralised pharmaceutical supply chain are now being developed by technology and pharmaceutical companies. Notably, a group has been established to explore blockchain's possible use in the insurance industry. Blockchain technology has many potential uses in the healthcare industry, including data protection, personal health record management, point-of-care genomics management, data management in electronic medical records, interoperable electronic health records, tracking of disease and outbreaks, genomics security, and many others [23].

4 Internet of Medical Things (IoMT)

Internet of Things gadgets use efficient low-power wireless media. Biometric data may be located, modified, collected, analysed, and protected with the use of these tools. This set of little sensors keeps tabs on the patient's blood pressure, heart rate, oxygen levels, temperature, exhaustion level, and sleep length. The development of a network of interconnected, intelligent gadgets and products that can automate mundane tasks is being fueled by recent technological and cultural advances. We refer to it as the Internet of Things (IoT). Improvements in computing power, wireless networking, and miniaturisation have allowed IoT technologies to be of use in the medical field.

Everything from information to hardware to software to networking tools to services is considered part of the Internet of Medical Things (IoMT). Amazon Web Services may be used by IoMT gadgets for data storage and analysis. Wearable mHealth devices that communicate with caregivers are a part of IoMT, as are the tracking of prescription orders and the mobility of hospitalised patients.

Hospital beds fitted with sensors for monitoring patients' vital signs, and infusion pumps fitted with analytics dashboards, might benefit from IoMT. There are now more use cases for IoMT than ever before, much like IoT. NFC RFID tags are increasingly commonplace in consumer mobile devices, facilitating information exchange with computer networks. Medical goods and equipment might be RFID-tagged for inventory management purposes. Patients can be monitored remotely using IoMT as part of telemedicine. Patients may now handle their own health care without having to visit the hospital or doctor for every little thing. Concerns concerning the privacy and confidentiality of patient information and other HIPAA-protected data that may be sent through IoMT are widespread among healthcare professionals.

IoMT's impact on healthcare will persist for quite some time. Deloitte projects the IoMT market to grow from $41 billion in 2017 to $158 billion in 2022. Health problems can be better diagnosed, monitored, and treated thanks to medical technology (medtech). There are 21 distinct types of medical devices, all of which have been designated by the Global Medical Devices Nomenclature (GMDN) Agency.

Bandages, syringes, surgical instruments, monitoring technology, and imaging devices are all examples. Software for managing patients, test kits for diagnosing illness, and other forms of medical technology are all part of this category. There is support for 21 different types of medical gadgets. The advancement of medical technology has led to a sea change in the medical industry.

5 Scalable Blockchain

Scalability is the degree to which a blockchain network can accommodate a growing number of nodes and a higher throughput of transactions. Adding additional computers to an existing network (horizontal scalability) refers to node and client scalability, whereas adding more resources (of any sort) to an already working system (vertical scalability) relates to throughput, block creation rate, latency, and storage scalability [24]. Despite blockchain's rising popularity, one of the biggest obstacles to the technology's disruptive potential remains the difficulty of scaling blockchain-based solutions. In this part, we take a look at the key characteristics that may be used to characterise a blockchain system's scalability, and define them in detail.

5.1 Horizontal Scalability

The advancement of the blockchain as a distributed ledger depends on its horizontal scalability, or its capacity to add more nodes and clients. Expanding the number of nodes or clients a blockchain can support without compromising performance or efficiency is known as "horizontal scaling". In contrast to proof-of-work (PoW) based consensus systems, which may maintain performance stability even as the number of nodes rises, Byzantine fault tolerance (BFT) is a distributed consensus methodology in which system performance degrades as more nodes are added to the network.

Clients are the application programmes that submit transactions on behalf of a user, and client scalability is the ability of a blockchain system to accommodate a growing number of clients without decreasing the performance of the system as a whole. A blockchain that can safely add nodes without compromising network security is said to be growing.

5.2 *Vertical Scalability*

We need vertical scalability, or the ability to expand the capacity of individual nodes, to guarantee timely and accurate transaction processing. Vertical scalability may be improved by adjusting several factors like block size, parallel mining, lightening, sharding, etc. Improvements in throughput, block creation speed, latency and storage capacity are all examples of vertical scaling.

Throughput

One of the most important aspects of scalability is transaction throughput, which is measured in transactions per second (TPS). To put it another way, it's the pace at which valid transactions are committed and appended to a block once consensus has been reached among the blockchain's stakeholders (miners) [24]. The exponential growth in transaction volume has rendered current implementations like Bitcoin inadequate to the needs of a transaction-intensive environment. Seven transactions per second is the maximum for Bitcoin's original blockchain. Examples of high-performance cryptocurrencies are Ethereum and Bitcoin Cash, which can process up to 20 and 60 TPS, respectively. The processing speed is still significantly lower than that of current digital payment systems such as VISA, which can handle up to 24,000 TPS.

Block Generation Rate (BGR)

The block size determines how many individual transactions are combined into a single block in a typical blockchain setting. The BGR details the pace at which blocks are mined, created, and added to the blockchain. The quantity of computational resources required to generate blocks is a direct result of the transaction mining process, which in turn is determined by the size of blocks and the effectiveness of the consensus mechanism. In Bitcoin's case, the Nakamoto technique is used, which mandates a new hashing solution for each new block in the chain and a constant block size of 1 MB. Building a single block will take at least 10 min. Ethereum's BGR is far quicker than those of competing networks, requiring only 10–20 s.

Latency

Latency, in the context of computer science, is the time elapsed between an input and an output. A blockchain's ability to function depends heavily on the network latency being as low as possible. Both the network and individual transactions on a blockchain might be slow. Network latency is the delay in a transaction being requested and confirmed. The processing and execution of large numbers of transactions relies heavily on the latency at which they are confirmed.

Storage scalability

A blockchain's storage scalability is measured by its ability to store an enormous amount of data without jeopardising its integrity as that data grows. Having control over block size and chain size are also key features of scalable storage.

The maximum number of transactions and associated data that may be contained in a single block is referred to as the block size. It is possible that more than 500 separate Bitcoin transactions might be stored in a single megabyte-sized block on a typical Bitcoin network. Although increasing TPS requires increasing the size of individual blocks, this may slow down the rate at which new blocks are created. However, if the block size is reduced too much, forks may arise. This trade-off between throughput per second and block creation rate impacts the block size selection.

The size of a blockchain, or distributed ledger, is proportional to the number of transactions that have been recorded on it. In order to become part of the network, a node must have enough storage capacity to save the chain's download. Since the blockchain for the Bitcoin network is around 280 GB in size, miners have a lot of information to download. This has important repercussions for systems with limited resources, such as the Internet of Things, and must be considered to promote its wide adoption.

6 Related Work

Blockchain's traditional Proof of Work (PoW) approach relies on individual mining nodes, which results in poor transaction processing rates and restricted scalability. As a solution to the problem of low throughput, Hazari et al. [25] devised a parallel PoW strategy that takes advantage of the benefits of parallel mining to speed up the operation of the classic PoW method. Each epoch, a management node is selected to oversee the miners inside a single block, ensuring that no two miners within that block are assigned the same work. The method's incentives system and ability to prevent forks make it a good choice. Because of this, not even the most powerful miner can get their reward by solving all of the blocks, but just the blocks that follow the one they created.

However, the method is still unjust to miners who use less robust gear, as those with quicker hardware will solve more blocks and have a larger chance of becoming a manager in subsequent blocks, and hence will receive more rewards. Another issue with the strategy is that it puts too much faith on managers to solve complex challenges. Hashes and nonces for transactions can still be generated by miners even if a management node fails; however, they will have to be generated manually, one at a time. Since the manager of the present block was chosen in a previous block, the parallel approach will be used again in the blocks that follow. If there aren't many miners working at once, a parallel mining approach performs no differently than a standard PoW method.

Gao et al. [26] increased output by facilitating the simultaneous execution of smart contracts. The sequential execution of smart contracts in most systems limits a distributed ledger system's capacity to handle smart contracts. To prevent others from using up all the system's processing power, smart contracts should be sufficiently sophisticated. Acceleration is achieved by parallelization of smart contracts. SCom was created to facilitate the division of smart contracts. If the smart contract is approved for execution by SCom, it will be carried out. It's possible for subcommittees to coordinate their efforts. To prevent hostile takeovers of the SCom, we employ majority voting and public keys that haven't been used in a while. The short-term fairness of a system is ensured by smart contract load balancing. Integer programming problem solutions provide long-term equity.

It was the combination of blockchain and IoT that Anusha et al. [27] investigated. Typical blockchains like Bitcoin can only handle 7 TPS. Due to the fact that IoTs connect billions of unique, low-power devices, their transaction rate requirements are far higher than Bitcoin's. In order to solve this issue, the authors designed an architecture that uses a local registration system to associate each IoT device with a certain business.

Distributed storage with authentication and auditing was first proposed by Li et al. [28] for Internet of Things-based applications. Edge computing allows IoT devices to execute cryptographic computations close to the network's periphery. It is useful for DHT data preservation and data collection. The blockchain serves as a trusted third party to validate read/write requests to the DHT. A certificateless cryptosystem safeguards credentials by generating a partial private key based on the user's identification and fusing it with a secret value. Disabling key escrow for identity-based encryption is the same as turning off that encryption method altogether. The proposed technique has potential applications in healthcare, smart grids, and smart metres. Data can be freely exchanged by the owner. A user might potentially utilise blockchain to exchange ECG data with a researcher. A scalable design is essential for effective usage of distributed storage. The authors make no mention of the security of IoT devices or edge servers. Due to the lack of a reliable defence mechanism, the system is vulnerable to majority takeover attacks.

Integration of IoT devices necessitates enhancements to the blockchain's architectural design to enable greater scalability. A gateway, external storage, or delivery network may alleviate some of the burden placed on Internet of Things devices. In any case, this causes costs associated with centralization. In the literature, many efforts to construct scalable, decentralised applications utilising blockchains are centred on adopting tactics unique to that particular app's use case. These apps use blockchain capabilities like immutable data storage and decentralised organisation, but there are steps that must be taken in order to increase their capacity.

In order to prove that the decentralised Access Management System (AMS) is scalable, Oscar Novo [29] compared it to the conventional AMS. While centralised access control solutions are limited in scope due to their reliance on a single server, decentralised systems may scale to include an unlimited number of Management Hubs and nodes in the blockchain. The simulation results favour horizontal scalability, which offers significant scaling benefits over typical blockchain situations by distributing the load over many management centres. Because of the proliferation of Internet of Things devices, there is a growing need to optimise transaction throughput.

A Service Oriented Architecture (SOA) for effective resource discovery in Hyperledger Iroha for IoT-based CPS was given by Ruta et al. [30]. Using blockchain technology and smart contracts, this SOA further improves upon its predecessor.

Decentralized blockchain technology based on Hyperledger Fabric, which allows for fast ledger updates and high levels of stability, was used by Grabatin et al. [31] to address this issue. For better data management, the authors developed a Hyperledger-based system that limits access to transaction data. A federation first creates a micro federation, then uses chaincode for provider registration. After signing up with an SP, patrons will be able to provide input on how best to operate a micro federation. If all SPs agree to this event, their data will be added to the peer ledger of the network as a byte array visible only to other SPs in the same micro federation. Even though federations may employ a variety of connection points, SP is able to analyse data with little storage and processing requirements. While the technique is more effective than the standard setup, the system's efficiency is still dependent on the SP's response time and the block time (s) configured in Hyperledger.

Delay was cut from 10 min to 3 s once Proof of Authentication (PoAh) was introduced by Puthal et al. [32]. With ELGamal, both public and private keys may be generated. In the context of cryptocurrency mining, validators are reliable miners. When a block is successfully mined, the trust value increases, allowing more miners to participate in the network. In order to validate a block's signature, a reliable mining node will employ asymmetric cryptography. Using PoAh ID for broadcasting requires verifying the MAC address first. Raspberry Pi and Python simulations are used to test the framework. The proposed fix can minimise latency in some use cases, but it has to be extensible before it can be implemented in practical contexts.

Tan et al. [33], proposed a BHMV platform that combines the benefits of encryption and blockchain technology to provide a secure and useful platform for the storing and exchange of medical data. This paper highlights blockchain technology incorporation into healthcare management systems and searchable encryption. The BHMV supports the confidentiality of the outsourced EHR and its index, keyword search capabilities, user verifiability, immutable storage, and dynamic EHR updates.

In order to construct a distributed solution using Hyperledger Fabric, Westphal and Erik [34] proposed a system that integrates the current Healthcare management systems with consortium-based blockchain. To protect patient privacy, the data is used in conjunction with a proxy re-encryption method. To handle business logic that has been approved by network members, use various chain codes. Several traits include: Problems with DDOS attacks have been fixed. This technique allows for the incorporation of biofeedback data into the patient's medical record and gives them ownership of the data back.

For an efficient healthcare management process, Panwar et al. [35] presented a new architecture for managing personal health records (PHRs) utilising the IBM cloud data lake and blockchain platform. They suggested a method that aims to reduce latency and increase throughput. The suggested system bases its calculations on a number of matrices, including the F1 Score, Recall, and Confusion matrices. As a consequence, the suggested work was very accurate and produced superior outcomes than those of earlier methods.

In order to establish resource access management in a reliable, auditable, and scalable way by the owner, Hao et al. [36] provided a framework for managing access based on smart contracts. To manage resource access rules based on characteristics, flexibility and credibility in access decisions for clients, a contract for access control is created on Blockchain. Owner signs a collection of qualities and provides them as off-chain signatures to the clients. On the Ethereum network, a test prototype is created, and theoretical and experimental evaluations are done to evaluate its effectiveness and scalability.

In the context of cloud-based IoMT, Hao et al. [37] suggested a secure strategy for information exchange that grants a client privileged access. The ability to use re-encryption techniques using ABE as the primary structural block gives Cloud Server the ability to re-encrypt the ciphertext. Patients can effectively share their information and control their rights. To demonstrate the productivity and security of the suggested task, the author examined the performance.

Due to inherent security vulnerabilities in the current IoT systems, Bhushan et al. [23] presented a comprehensive survey on improving security using blockchain based IoT systems. Although blockchain provides huge advantages over current IoT systems, several obstacles are there as the blockchain technology is still under development and IoT devices are limited in resources.

As the Blockchain technology provides distributed security, privacy, confidentiality and other security features, Kumar et al. [38] proposed a blockchain-based framework to share electronic health records securely using Ethereum based smart contracts, the InterPlanetary File System (IPFS) as an off-chain storage system and key management showing promising results over existing systems. It depicted methodology to prevent unauthorised users from accessing the data.

Kumar [39] presented a survey on how blockchain can resolve numerous security and privacy issues. Due to the sensitive nature of health information, sharing information using centralised Electronic Health Record systems (EHR) raises several concerns. He proposed a blockchain-based framework to resolve these concerns and the limitations of the use of blockchain were also provided.

Bhushan et al. [40] performed a survey on blockchain technology and considered several concepts such as distributed ledger, cryptography, fork and consensus to analyse security threats and the mitigation strategies. Through these, they provided directions to implement various blockchain systems for the real-time large-scale applications.

Several promising security features of blockchain are leading to adapt this technology for IoT and Software-Defined Networking (SDN). During the survey Chaganti et al. [41] highlighted that the Distributed Denial of Service (DDoS) attack is attacking vulnerable IoT devices and southbound channel saturation in the SDN architecture. Several DDoS mitigation solutions are reviewed and strategies to implement blockchain to mitigate DDos attacks are provided.

7 Blockchain Based IoMT Architecture

Our idea is based on the expansion of healthcare monitoring outside hospital walls. Thus, the patient's temperature, heart rate, blood pressure, and oxygen saturation are all measured. For the purpose of tracking a patient's movements and detecting any falls, more sensors might be put in their house. Our home and wearable sensors upload their data to a central database. When anomalies are detected, a real-time monitoring system notifies doctors, who may then determine whether or not to take any corrective action remotely. Such information is recorded so that doctors can keep eyes on their patients' health and quickly respond to any changes. Whenever we interact with one another, we share bits of knowledge about ourselves that may be used to harm us. A safe and secure method is needed to ensure the privacy of these health records so that no one has to go through anything unpleasant.

All of these objectives are possible with blockchain-based remote patient monitoring. As depicted in Fig. 1, blockchain-based architecture is composed of three distinct layers: the data acquisition layer, the network layer, and the IoMT application layer.

Information is gathered from the source and conclusions are drawn in the data acquisition layer. As it is, the data access sublayer is a part of the data gathering layer itself. The primary function of the data collection layer is to extract information of value from the data gathered via the use of different medical perception devices and signals acquisition gear. The most common means of gathering signals are graphic codes, radio frequency identification (RFID), general packet radio service (GPRS), and related technologies. With the help of Bluetooth, Wi-Fi, and ZigBee, as well as

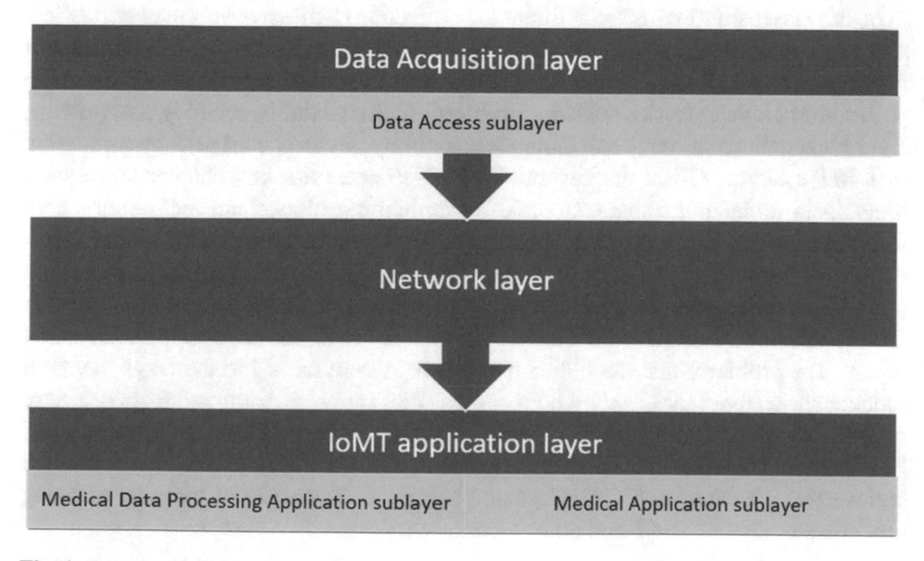

Fig. 1 Layers of blockchain architecture

other low-power wireless technologies, the data access sublayer transfers information between the data-gathering layer and the network layer.

Services related to platforms and interfaces, as well as other types of data transmission, are within the purview of the second layer, also known as the network layer. The data acquired at the data collection layer must be transmitted to the network layer in a timely, consistent, and barrier-free manner via various means such as mobile communication networks, wireless sensor networks, the internet, etc. This layer also facilitates the integration of previously separate data sources and storage facilities. In order to facilitate these kinds of connections, the platform offers not just open interface services but also a slew of other services inherent to the platform itself.

The IoMT application layer is in charge of the patient's medical record and uses data collected from the network layer to do so. Both the medical data processing application and the medical application sublayers make up this level. The medical data processing application layer is concerned with the analysis of information pertaining to patients, diseases, medications, diagnoses, treatments, etc., while the medical application sublayer stores medical equipment and other materials related to information for maintaining patient information, such as inpatient, outpatient, medical treatment, etc. records.

8 Conclusion

The COVID-19 pandemic is unprecedented, yet new advances in information and communication technology provide solutions. In particular, the broad use of IoMT in healthcare institutions may facilitate the collecting of massive volumes of medical and healthcare data that may be used by medical practitioners for the purpose of making accurate diagnoses and therapeutic recommendations. So IoMT can help with some healthcare tasks, but it has problems with scalability, security, and privacy. But blockchain technology can enhance scalability, security, and privacy protection in IoMT systems. Given the current COVID-19 scenario, this chapter suggests a blockchain architecture that can scale by combining blockchain technology with Internet of Medical Things (IoMT) gadgets. This chapter begins with a brief introduction to the history and current state of India's healthcare system, before moving on to discuss the potential applications of blockchain technology and IoMT in this sector. Then, we introduced a blockchain-based, IoMT-centric system with many layers. The problems that the Indian healthcare system faces and the ways in which blockchain-enabled IoMT might help address them are also examined from a variety of perspectives, including the provenance of medical data and the use of telemedicine. Finally, the potential for blockchain-enabled IoMT in the future is laid forth. We believe that we can give affordable medical care to the people of India by utilising blockchain-enabled IoMT and other technologies.

References

1. Kharya S, Onyema EM, Zafar A, Wajid MA, Afriyie RK, Swarnkar T, Soni S (2022) Weighted Bayesian belief network: a computational intelligence approach for predictive modeling in clinical datasets. Comput Intell Neurosci
2. Constitution of the WHO. https://apps.who.int/gb/bd/PDF/bd47/EN/constitution-en.pdf
3. Zafar A, Wajid MA (2020) A mathematical model to analyze the role of uncertain and indeterminate factors in the spread of pandemics like COVID-19 using neutrosophy: a case study of India, vol 38. Infinite study
4. Global Burden of Disease (GBD) Statistics, Institute for Health Metrics and Evaluation (IHME). https://www.healthdata.org/gbd/2019
5. Ghouali S, Onyema EM, Guellil MS, Wajid MA, Clare O, Cherifi W, Feham M (2022) Artificial intelligence-based teleopthalmology application for diagnosis of diabetics retinopathy. IEEE Open J Eng Med Biol 3:124–133
6. IPHS Guidelines, National Health Systems Resource Centre (NHSRC). https://nhsrcindia.org/IPHS2022
7. World Health Research Report 2013, National Health Systems Resource Centre (NHSRC). https://nhsrcindia.orgsites/default/files/NATIONAL%20HEALTH%20ACCOUNTS-%20Estimates%20for%20India-2013-14(2)_0.pdf
8. Study of Public Health IT Systems in India, National Health Systems Resource Centre and Taurus Glocal Consulting, New Delhi. https://hispindia.org/docs/indian-public-health-it-system-study.pdf
9. The Clinical Establishment (Registration and Regulation) Act, 2010, Ministry of Health and Family Welfare (MoHFW). https://www.indiacode.nic.in/bitstream/123456789/7798/1/201023_clinical_establishments_%28registration_and_regulation%29_act%2C_2010.pdf
10. Clinical Establishment (Central Government) Rules, 2012, Ministry of Health and Family Welfare (MoHFW). http://clinicalestablishments.gov.in/WriteReadData/386.pdf
11. Ayushman Bharat—National Health Protection Mission. https://www.india.gov.in/spotlight/ayushman-bharat-national-health-protection-mission
12. Budget 2022–23, Ministry of Finance (MoF). https://www.indiabudget.gov.in/doc/bh1.pdf
13. National Commission for Allied and Healthcare Professions Act 2021, Ministry of Health and Family Welfare (MoHFW). https://egazette.nic.in/WriteReadData/2021/226213.pdf
14. Healthcare goes mobile: evolution of teleconsultation and e-pharmacy in new normal, Sept 2020, EY Times. https://assets.ey.com/content/dam/ey-sites/ey-com/en_in/topics/health/2020/09/healthcare-goes-mobile-evolution-of-teleconsultation-and-e-pharmacy-in-new-normal.pdf?download
15. Gupta PK, Siddiqui MK, Huang X, Morales-Menendez R, Pawar H, Terashima-Marin H, Wajid MS (2022) COVID-WideNet—a capsule network for COVID-19 detection. Appl Soft Comput 122:108780
16. National Digital Health Mission 2020, Ministry of Health and Family Welfare (MoHFW), https://www.niti.gov.in/sites/default/files/2021-09/ndhm_strategy_overview.pdf
17. OECD Health Statistics 2022. https://www.oecd.org/els/health-systems/health-data.htm
18. The HIV AND AIDS (Prevention and Control) Act, 2017, Ministry of Health and Family Welfare (MoHFW). http://naco.gov.in/sites/default/files/HIV%20and%20AIDS%20Act-%20English.pdf
19. Bhushan B, Sahoo C, Sinha P, Khamparia A (2020) Unification of blockchain and internet of things (BIoT): requirements, working model, challenges and future directions. Wirel Netw. https://doi.org/10.1007/s11276-020-02445-6
20. Queralta JP, Westerlund T (2021) Blockchain for mobile edge computing: consensus mechanisms and scalability. In: Mukherjee A, De D, Ghosh SK, Buyya R (eds) Mobile edge computing. Springer, Cham. https://doi.org/10.1007/978-3-030-69893-5_14
21. Tang S, Wang Z, Jiang J et al (2022) Improved PBFT algorithm for high-frequency trading scenarios of alliance blockchain. Sci Rep 12:4426. https://doi.org/10.1038/s41598-022-08587-1

22. Nakamoto S (2008) Bitcoin: a peer-to-peer electronic cash system (PDF). Accessed 25 Dec 2022
23. Saxena S, Bhushan B, Ahad MA (2021) Blockchain based solutions to Secure IoT: background, integration trends and a way forward. J Netw Comput Appl 103050. https://doi.org/10.1016/j.jnca.2021.103050
24. Nasir MH, Arshad J, Khan MM, Fatima M, Salah K, Jayaraman R (2022) Scalable blockchains—a systematic review. Future Gen Comput Syst 126:136–162. ISSN 0167-739X. https://doi.org/10.1016/j.future.2021.07.035
25. Hazari SS, Mahmoud QH (2019) A parallel proof of work to improve transaction speed and scalability in blockchain systems. In: 2019 IEEE 9th annual computing and communication workshop and conference (CCWC). IEEE, pp 0916–0921
26. Gao Z, Xu L, Chen L, Shah N, Lu Y, Shi W (2017) Scalable blockchain based smart contract execution, pp 352–359. https://doi.org/10.1109/ICPADS.2017.00054
27. Anusha R, Yousuff M, Bhushan B, Deepa J, Vijayashree J, Jayashree J (2022) Connecting blockchain with IoT—a review. In: Sharma DK, Peng SL, Sharma R, Zaitsev DA (eds) Microelectronics and telecommunication engineering. Lecture notes in networks and systems, vol 373. Springer, Singapore. https://doi.org/10.1007/978-981-16-8721-1_14
28. Li R, Song T, Mei B, Li H, Cheng X, Sun L (2019) Blockchain for large-scale internet of things data storage and protection. IEEE Trans Serv Comput 12(5):762–771
29. Novo O (2018) Blockchain meets IoT: an architecture for scalable access management in IoT. IEEE Internet Things J 5(2):1184–1195
30. Giao J, Nazarenko AA, Luis-Ferreira F, Gonçalves D, Sarraipa J (2022) A framework for service-oriented architecture (SOA)-based IoT application development. Processes 10:1782. https://doi.org/10.3390/pr10091782
31. Grabatin M, Hommel W (2018) Reliability and scalability improvements to identity federations by managing SAML metadata with distributed ledger technology. In: NOMS 2018—2018 IEEE/IFIP network operations and management symposium, 2018, pp 1–6. https://doi.org/10.1109/NOMS.2018.8406310
32. Puthal D, Mohanty SP, Nanda P, Kougianos E, Das G (2019) Proof-of authentication for scalable blockchain in resource-constrained distributed systems. In: 2019 IEEE International conference on consumer electronics (ICCE), 2019, pp 1–5
33. Tan TL, Singh M (2022) Blockchain based healthcare management system with two-side verifiability. PLoS ONE 17(4):e0266916
34. Westphal E (2021) Digital and decentralized management of patient data in the healthcare system using blockchain implementations. Front Blockchain 36
35. Panwar A, Bhatnagar V, Khari M, Salehi AW, Gupta G (2022) A blockchain framework to secure personal health record (PHR) in IBM cloud-based data lake. Comput Intell Neurosci 2022:3045107. https://doi.org/10.1155/2022/3045107. PMID: 35463293; PMCID: PMC9019420
36. Hao J, Huang C, Tang W, Zhang Y, Yuan S (2022) Smart contract-based access control through off-chain signature and on-chain evaluation. IEEE Trans Circuits Syst II Express Briefs 69(4):2221–2225
37. Hao J et al (2022) Secure data sharing with flexible user access privilege update in cloud-assisted IoMT. IEEE Trans Emerg Topics Comput 10(2):933–947
38. Kumar A, Kumar R, Sodhi SS (2022) A novel privacy preserving blockchain based secure storage framework for electronic health records. J Inf Optim Sci (Taylor & Francis) 43(3):549–570. https://doi.org/10.1080/02522667.2022.2042092
39. Kumar R (2021) Blockchain in EHR: a comprehensive review and implementation using hyperledger fabrics. Blockchain technology for data privacy management. CRC Press, pp 275–293. ISBN 978100313339

40. Bhushan B, Sinha P, Sagayam KM, Andrew J (2021) Untangling blockchain technology: a survey on state of the art, security threats, privacy services, applications and future research directions. Comput Electr Eng 90:106897. https://doi.org/10.1016/j.compeleceng.2020.106897
41. Chaganti R, Bhushan B, Ravi V (2023) A survey on blockchain solutions in DDoS attacks mitigation: techniques, open challenges and future directions. Comput Commun 197:96–112. https://doi.org/10.1016/j.comcom.2022.10.026

Higher Education: AI Applications for Blockchain-Based IoT Technology and Networks

R. S. S. Nehru, Ton Quang Cuong, Achanta Ravi Prakash, and Bui Thi Thanh Huong

Abstract Building a blockchain-based architecture on the Artificial Intelligence and IoT platform for higher education institutions could increase 5G/6G network communication efficacy. 5G/6G wireless networking allows users to communicate reliably at higher speeds. Today's world has high-speed, intelligent, durable networks with modern technology like low power utilization. IoT cloud can store and handle Blockchain and IoT data. Designing a blockchain-based virtualization framework for IoT architecture in mobile communications is challenging. This article links a blockchain-based educational system to cryptographically secure Internet devices and the Indian New Education Policy (NEP-2020) with the current research. Some colleges utilize machine learning to predict student interests. Complex analytic systems calculate "demonstrated interest" by tracking website, social media, and email interactions. Schools analyze how quickly email recipients open and click links; This research improves blockchain and IoT to create an online interactive system for students, faculty, educators, developers, facilitators, recruiters, and accreditors. This complex framework would be valuable to the education sector and higher education.

Keywords Educational system · Internet of Things · Wireless communication · Blockchain · Cloud computing · Higher education

---

R. S. S. Nehru
Department of Education, Sikkim University, Gangtok, Sikkim, India

Sambalpur University, Sambalpur, Odisha, India

T. Q. Cuong (✉)
Faculty of Educational Technology, University of Education, Vietnam National University, Hanoi (VNU-UEd), Hanoi, Vietnam
e-mail: tonquangcuong@gmail.com

A. R. Prakash
Department of Management, Sikkim University, Gangtok, Sikkim, India
e-mail: arprakash@cus.ac.in

B. T. T. Huong
VNU School of Interdisciplinary Studies, Vietnam National University, Hanoi, Vietnam

B. Bhushan et al. (eds.), *AI Models for Blockchain-Based Intelligent Networks in IoT Systems*, Engineering Cyber-Physical Systems and Critical Infrastructures 6,
https://doi.org/10.1007/978-3-031-31952-5_12

1 Introduction

Blockchain, IoT, and A.I. can improve corporate processes, create new company models, and disrupt industries. Blockchain's digital ledger boosts a company's trust, transparency, security, and privacy. Funds and personal data can be stored in a blockchain [3]. The Internet of Things automates and simplifies German and European manufacturing. Pattern-spotting AI improves operations [19]. In actual use, blockchain, IoT, and A.I. are disregarded. The technologies of the future need to integrate. The Internet of Things gathers information, blockchain creates a decentralized ledger and interaction rules, and artificial intelligence enhances procedures and legislation [19, 23]. Three innovations function in tandem with one another. Merging disparate technologies into a single system can boost BPA and data management.

This paper signifies that blockchain technology and A.I. benefit the 21st-century campus. The blockchain divides data into encrypted blocks, ensuring anonymity and security. Higher education security specialists are concerned about protecting student profiles and certifications. As part of Education 4.0's digital transformation toward the university of the future, a few higher education institutions are experimenting with and using blockchain technology. Academic credentials like degrees, certifications, and transcripts were checked as one of the first steps to stop credential fraud.

Blockchain is a shared, unchangeable record that lets many parties exchange encrypted data quickly, openly, and in a way that everyone can see. Blockchain tracks orders, payments, accounts, production, and more. When members can trade with other businesses, they have more confidence and trust, as well as more opportunities and better ways to do things. A.I. uses computers, data, and machines to simulate human problem-solving and decision-making. Machine learning and deep learning use data to teach A.I. algorithms to make predictions or groups that improve over time. A.I. automates tedious activities, improves decision-making, and improves the customer/student experience.

In this paper, all parameters are systematically arranged and all the dimensions are discussed in depth: higher education and digitization; intelligent contract automation; industry and education (4.0); a model of the 4th Industrial Revolution versus higher education (4.0); Digitization initiatives in NEP-2020; NETF's functions include: artificial intelligence, blockchain-based IoT applications, university reform, What about blockchain technology? What is IoT?; Creating trust with blockchain technology Identity, Blockchain, and Academic Transparency: A Case Study; Blockchain and the New Pedagogy; Costs and blockchain technology (student debt); Blockchain and the Meta-University; Incentives to Change; Technology integration in higher education in Indian education Below are some of the policy's vital technological features: Higher education; administration; A.I. adaptation; digital India; priorities; technological innovations in education; a technology integration model; are properly synthesized in the paper with a conclusion.

2 Higher Education and Digitalization

Regulatory and funding changes in higher education and a greater focus on quality and outcome create a more scrutinizing environment. Digital technology would improve communication, interconnection, information flow, support systems, innovation and cooperation, governance, and 360-degree performance evaluation. Short-term and long-term goals must be set for university outputs. Open and honest performance management and reporting would boost trust and camaraderie.

There is a wealth of free data available on the internet. There have been millions of people throughout the world who have benefited from online education given by for-profit and nonprofit organizations since the early 2000s. New businesses and economic structures have emerged as a result of distance learning. Students can benefit from a quality education through flipped classrooms, V.R., labs, digital simulations and models, electronic documents, textbooks, online assignments, and free resources. Students in both virtual and brick-and-mortar classrooms can benefit from this kind of instruction thanks to its low marginal cost and the platform's inherent malleability, adaptability, and personalization [1].

Growing student populations help universities promote new, adaptable offerings. The digitalization of education enables new business models such as e-universities, virtual universities, digital universities, creative universities, agile universities, and universities. 4.0. University products must be revised and distributed more efficiently. Technology integration will be needed to facilitate product upgrades, research, and modular certification.

3 Smart Contract Automation

Applying blockchain, IoT, and A.I. together can be very promising for automating corporate processes. Smart contracts connect these three advances. Smart contracts are digital agreements that automatically execute their provisions. Smart contracts are like "if–then" functions that define actions if a particular event happens. If products are successfully delivered ("if"), payment is automatically made ("then"). Smart contracts integrate IoT, A.I., and blockchain. Despite their potential, industrial enterprises do not use smart contracts. Classical intelligent contracts require crypto assets like Ether or EOS and transfer quantities. Regulatory and economic concerns make organizations hesitant to adopt crypto assets.

The recipient of an Ether smart contract faces considerable exchange rate risk. Crypto asset prices can rise or fall by 10% in a day. Even though stablecoins can solve the excessive volatility of "traditional" crypto assets, industrial enterprises and B2B situations will not employ them. Stablecoins are unregulated. Risk-averse corporations avoid unregulated instruments. Companies' accounting and I.T. systems use fiat currency like the Euro. Converting stablecoins to "system-based" currencies is a headache for corporations. This conversion requires time and money (transaction fees, price hedging).

Only blockchain-based fiat money that "fellows through" intelligent contracts can fully leverage their promise. Only a blockchain-based digital Euro would enable Euro-denominated smart contracts, so machines, cars, or sensors could offer pay-per-use, leasing, and factoring. Due to a digital blockchain-based Euro, new business models could emerge: fully automized gadgets making decisions on their own using A.I. and "economically surviving" by using blockchain for financial transactions while implementing a profit centre logic on the device level Technologies [13].

3.1 Industry and Education (4.0)

The fourth industrial revolution's guiding principles (Industry 4.0) Understanding the fourth industrial revolution requires studying past industrial processes. Oil and steam engines powered modest manufacturing during the first industrial revolution (Industry 1.0). "Industry 2.0" emphasized organized work and how electricity facilitated mass manufacturing. The third industrial revolution was based on integrating electronics and I.T. to automate production (Industry 3.0). As a result of Industry 3.0, intelligent industrial systems are emerging due to the integration of cutting-edge technologies (A.I., cloud computing, IoT, big data, robots) (Industry 4.0). Industry 4.0 merges the physical, biological, and digital worlds to create intelligent cyber-physical systems. The third industrial revolution was based on integrating electronics and I.T. to automate production (Industry 3.0). As a result of Industry 3.0, intelligent industrial systems are emerging due to the integration of cutting-edge technologies (A.I., cloud computing, IoT, big data, robots) (Industry 4.0). Industry 4.0 merges the physical, biological, and digital worlds to create intelligent cyber-physical systems. Industry 4.0 must better link the physical and digital worlds to manage products, business processes, and services (Fig. 1).

Industry 4.0 supports customized goods and services based on consumer needs. The authors say Industry 4.0 is the industry's use of cutting-edge technologies like 3D printing, robotics, big data, cloud computing, the Internet of Things, and A.I. to automate and digitalize manufacturing processes. All facets of human existence have undergone tremendous change due to the numerous industrial revolutions, particularly in schooling [16].

Higher education institutions may lead this quickly changing world by taking the initiative. It is still being determined how and when higher education will change, even if some of the above pressures are present. Universities must incorporate University 4.0's emphasis on student employability and experience. Educational institutions must be adaptive, responsive, and robust in challenging circumstances. Neglecting higher education is destructive to society and must change.

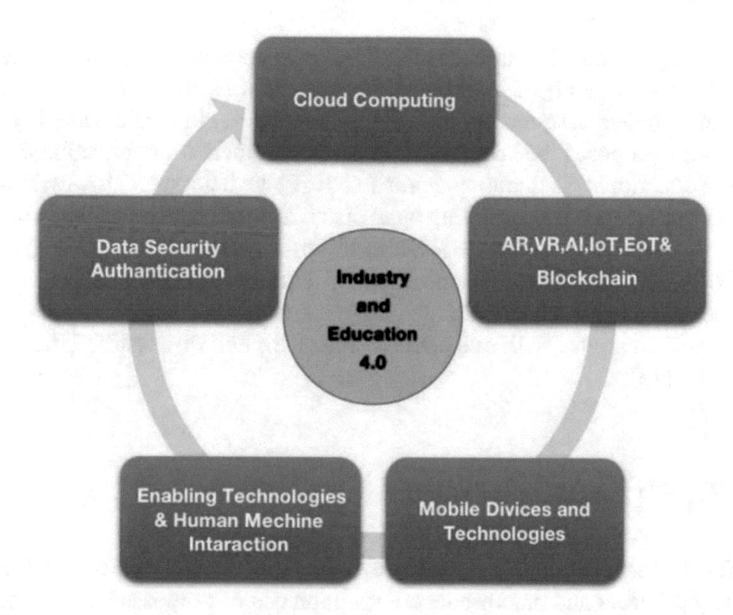

Fig. 1 4th Industrial Revolution versus higher education (4.0) (*Source* Conceived and Proposed by Authors)

4 Digitisation Initiatives in NEP-2020

A group led by Dr. Kasturirangan, the former head of the Indian Space Research Organization, has proposed a new method of teaching (ISRO). The policy paper has four parts: elementary and secondary education, higher education; other significant areas; and execution [13].

Section III of the Policy, headed "Use and Integration of Technology," elaborates on the policy's perspective on these issues. To make India more competitive in the global economy, the government has launched the Digital India Campaign. Consequently, technological advancements only serve to enhance the quality of education. The internet's consequences in today's digital age are vast. Online platforms have made it easier for teachers and students to connect. Taking courses via the internet is becoming increasingly common.

The malleability of the technology supporting online education has contributed to its meteoric rise in popularity in recent years. More individuals may have access to a high-quality education, regardless of location or time, thanks to the advent of online learning. These changes have made possible more advanced learning environments.

The New Education Policy 2020 acknowledges that technology will only impact education in various ways, some of which can be envisioned now due to the rapid rate of technological progress and the inventiveness of tech-savvy instructors and entrepreneurs. Especially for student entrepreneurs, as long as they are thoroughly assessed and transparently adopted before being used on a larger scale, the policy backs and encourages the use and integration of technology to improve various elements of education. Educational planning, management, and administration will be simplified as a result of technological interventions that boost the quality of teaching and student assessment (including admissions, attendance, and reviews).

The policy proposes the creation of a non-governmental organization called the National Educational Technology Forum (NETF) to help steer this technological revolution in education by promoting open discussion and debate about the best ways to implement technical tools in pre-K through 12th-grade classrooms and universities.

The NETF hopes that people in positions of authority in higher education, as well as their counterparts in state and federal governments and other interested parties, would be able to confer with one another and share and discuss the latest findings and research [13].

4.1 NETF's Functions Include

The NETF will rely on various researchers to examine factual data from educational technology pioneers and practitioners. Paragraph (23.6) proposes building software for all ages and skills pupils.

For DIKSHA's e-content-based teacher professional development, all States, NCERT, CIET, CBSE, NIOS, and other bodies/institutions shall produce E-content in all regional languages. The DIKSHA app proudly announces "The National Initiative for School Head's and Instructors' Holistic Advancement," the first-ever online training program for school administrators and teachers (NISHTHA). Three of the planned training modules still need to be finished. The following [13]

1. To advise Central and State Government agencies on technology-based interventions;
2. To strengthen intellectual and institutional capacities in education technology;
3. To envision strategic thrust areas in this domain;
4. To articulate new research and innovation initiatives.
5. Set standards for online/digital content, technology, and pedagogy. These standards will help States, Boards, Schools, HEIs, etc., develop e-learning norms.
6. Maintain the regular flow of legitimate data from many sources, including educational technology innovators, and engage diverse scholars to analyze the data.
7. Hold regional and national conferences, workshops, etc., to solicit input from researchers, entrepreneurs, and practitioners.
8. Identify technological interventions to improve teaching–learning, evaluation, teacher preparation and professional development, educational access, and educational planning, management, and administration (admissions, attendance, assessments, etc.).
9. Categories emerging technologies based on their potential for disruption and offer this analysis to MoE.

The policy suggests giving teachers the tools to incorporate digital material into classroom education to improve the efficacy of DIKSHA/SWAYAM. K-12 and higher education will use user ratings and reviews to create accessible, engaging content.

Have faith in our capacity to meet the challenges of the following educational period. Educators should familiarise educators with online teaching tools such as video conferencing, online laboratories, mentoring, and testing proctoring. The success of NEP-2020, a radical departure from the current educational system, hinges on how well it is implemented. The strategy calls for concerted and systematic work by numerous institutions to attain its goals; The Policy is implemented with a timeline and evaluation plan by the Ministry of Education (MoE), the Council of Academic Boards of Education (CABE), the Union Government, the State Governments, the education-related Ministries, the State Departments of Education (SDEs), the Boards, the National Teachers' Association (NTA), the regulating organizations of schools and universities (SCERTs and HEIs) [13].

5 A.I., Blockchain-Based IoT Applications University Reform

Which emerging technology do you predict will have the most significant influence on universities in the following years? Not massive amounts of data, social media, online courses, virtual reality, or A.I. These are revolutionary new capabilities made attainable by blockchain technology. It sounds more like a strategy for college football than it does a game-changing piece of technology. The blockchain, which is effectively the Internet 2.0, has the potential to disrupt several sectors, including the financial sector, the corporate sector, the government, and even higher education [22].

There are four options open to those who seek to innovate in the field of higher education.

(1) The identification of pupils, the protection of their privacy, the measurement, recording, and certification of their achievements, and maintaining confidentiality records.
(2) The New Pedagogy: Tailored instruction and the development of innovative learning models.
(3) How do we place value on education, how do we finance it, and how we reward students for their efforts.
(4) The Meta-University plans to redesign higher education to make former MIT president Chuck Vest's vision a reality. The blockchain may help us with this AI application (Fig. 2).

Fig. 2 A.I., blockchain-based IoT applications

5.1 What About Blockchain Technology?

With the advent of the internet, billions worldwide now have access to a platform ideal for online interaction and cooperation. However, technology has remained the same in how we conduct business because it is designed to transfer and store information rather than value. Emails, lecture notes, PowerPoints, and audio recordings of lectures are all copies, not the originals that professors distribute to their students. Having a hard copy of your PowerPoint presentation is helpful. It is still not acceptable to print money or academic credentials. The Internet of Information necessitates using authoritative third parties to facilitate the transfer of value. Governments, banks, digital platforms (such as Amazon, eBay, and Airbnb), and educational institutions all play a role in establishing our identities, attesting to our reliability, and providing us with the documents we need to acquire and transfer assets and settle transactions [2].

In general, they are practical; however, they could be better. The reliance on a small number of centralized servers makes them vulnerable to attack. They demand a 10% percentage for an international money transfer as compensation for their services. They collect our information without our knowledge or consent, compromising our privacy in the process. Such go-betweens are notoriously slow and sluggish. They leave out two billion people who cannot afford necessities like food and shelter, much less high education. The most severe issue is that they need to share equally. What if there were a global, distributed, highly secure platform, ledger, or database where we could store and trade valuable items on trust with one another without the need for trusted third parties?

By embedding the community's self-interest in this new native digital medium for value, the blockchain may ensure our digital transactions' integrity, security, and privacy. Blockchain is the Trust Protocol because trust is fundamental to the system. Why make such a huge deal out of it? As music professors, we believe creative people should be able to make a living from their craft. Are you an immigrant who has grown weary of paying exorbitant transfer fees so that children might go to school in the land of your ancestors? Maybe you are a parent fed up with the lack of transparency and accountability from state lawmakers and their appointees in charge of higher education in your state. You may use social media but worry about who sees your data, and developers are hard at work on blockchain-based apps to address these concerns. The samples given here are just the tip of the proverbial app iceberg [3].

Everyone from corporations and universities to governments and citizens stands to gain significantly. Consider the corporation, a bedrock of the capitalist system we live in today. With the emergence of a global peer-to-peer network for identity, trust, reputation, and transactions, we can re-engineer the foundations of businesses to encourage innovation and equitable value generation [4]. We are discussing how to build companies in the information age that are less hierarchical and more networked than those of the industrial era. The blockchain will bring revolutionary change to many industries, not only the financial services industry. Consider how well students at today's colleges and universities are prepared for that future.

Emergence and influence of blockchain in Industry 4.0

Blockchain will revolutionize the management of distributed digital assets. Centralized systems are vulnerable to scams, hacking, data loss, and poor transparency; Blockchain provides a decentralized peer-to-peer network and stores data in blocks on multiple nodes. Immutable and publicly available blockchain records keep user identities private. Smart contracts have made it possible for healthcare, insurance, supply chain management, public donations, asset registration, verifying IoT devices, online gambling, and more. Smart contracts work on the blockchain to save time, eliminate intermediaries, prevent paperwork, avoid human errors, maintain transparency, and prevent fraud. Blockchain has problems with how much power it uses, how slow transactions are, how few people use it, and how professionals and the government regulate it. Few companies use their strengths. We need a lighter consensus algorithm to solve power and transaction speed issues, and it is recommended that businesses use computer-animated models based on fundamental design ideas to implement this technology. Due to its many advantages, blockchain technology is the future.

5.2 *What Is IoT?*

Shortly, billions of intelligent things in the physical world will be able to feel, respond, communicate, share essential data, generate, purchase, and sell their electricity, and carry out any task, from regulating our health to protecting the environment. It looks like this Internet of Everything will require some centralized [9].

A significant benefit of blockchain technology's potential to solve the perplexing prosperity problem is a substantial benefit. Fewer and fewer people are sharing in the fruits of economic expansion; arts and sciences and the agricultural sector could use this technology to increase their share of the profits from their endeavours [6]. This approach is more long-term and effective than simply transferring wealth to combat growing social inequality.

5.3 Creating Trust with Blockchain Technology

Digital assets—Financial assets, securities, intellectual property, creative works, loyalty points, and academic records are disseminated throughout a global ledger using the most robust encryption technologies. A purchase is broadcast to millions of computers worldwide. "Miners" have access to 10–100 times Google's server capacity. Miners consolidate the preceding 10 min' transactions into a single block every 10 min, like a heartbeat. Miners compete to solve a challenging challenge; the winner verifies the block and gets a bitcoin. Winners get Bitcoin from the blockchain.

A chain is formed when the new block is connected to the ones before it. Like a digital wax seal, time is embedded in each block. Attackers would need to break into millions of computers simultaneously, all employing the most robust encryption, in broad daylight to hack a block and distribute the same Bitcoin to numerous people. Trying. Modern computer security is inadequate. Ethereum was created by Canadian teen Vitalik Buterin, who was inspired by the success of Bitcoin. Ethereum is one of a kind because of its capabilities and features. It facilitates the development of "smart contracts," legally binding agreements written as computer code to manage enforcement, execution, and payment. There are efforts on the Ethereum blockchain to make government more transparent and accountable to the people [11].

5.4 Identity, Blockchain, and Academic Transparency

In today's world, you need an organization with endowed rights to provide you with an identity, WISeKey's Carlos Moreira stated. The first step in establishing one's identity is acquiring a birth certificate from a physician who is registered with the state. The newborn's data, which eventually will include academic accomplishments, begins to accrue on that day. Protecting the confidentiality of student information while it is housed digitally at universities is the first obstacle to overcome. Institutions have gotten quite good at collecting data about students and alums to use in fundraising efforts, as detailed in a 2013 report published by the Education Advisory Board (EAB). However, when protecting this information, universities are just as susceptible to data breaches as any other large corporation. Kirkwood Community College, the University of Wisconsin–Milwaukee, the University of California–Berkeley, and the Ohio State University have all been the targets of hacking attempts in recent months. Yale University exposed sensitive data online accidentally, and Indiana University stored similar information on an unsecured server. Laptops and cassettes with sensitive information were stolen from the University of Miami, Stanford University, and the University of Utah Hospitals and Clinics [5].

The blockchain can record anything stated in code, starting with birth certificates and moving on through transcripts, I.D.s, loans, etc. The blockchain depends on reliable public critical infrastructure (PKI) to maintain a secure transaction environment. Public-key cryptography (PKI) is a sophisticated form of asymmetric cryptography that uses two keys that do not share a standard function to encrypt and decrypt data. That is why it is fair to call them unequal. The Bitcoin blockchain is the largest civilian deployment of PKI to date, surpassing even the common access system used by the United States Department of Defense. Sony Global Education has adapted this technology into an open data exchange protocol to allow the secure and dependable transfer of transcripts, diplomas, and other official academic papers between institutions and individuals worldwide. However, the information is safe if hackers do not have the two matching keys.

The issue of validity is a further obstacle. Verifying a job applicant's claims is becoming increasingly critical at a time when information is plentiful, transitory, and subject to change. CareerBuilder reports that 33% of candidates lie about their education level, and 57% exaggerate their experience when applying for jobs. As may be expected, prospective employers inspect diplomas and degree certificates. Universities, however, frequently impose transaction fees while handling requests. The "base cost for a transcript" at MIT is $8.00, and there is a $2.00 processing fee for each transcript ordered online. Sony's technology has the potential to make such data transfer quick and inexpensive. Imagine all the refugees trying to start over in a new country and would profit from this system.

6 A Case Study

Only 25% of college students in the United States live on campus. Others successfully juggle their families and careers; only around one-quarter of students who attend school part-time graduate in four years. The Open Badges, Blockchain Certificates, and Learning Are Earning 2026 projects investigate several approaches to award students with credentials based on their courses' content. Like changing the oil in your car, parenting requires regular maintenance (the parent gets teaching credit). The workplace is great for learning new skills, working with others, and gaining management experience for a college application. The MIT Media Lab hashes cryptographic certificates onto the blockchain to verify community members and credit them for their contributions. Students are awarded a grade and a credential that may be useful to them professionally.

6.1 Blockchain and the New Pedagogy

So long as humankind—Let us assume that today's society values the credentials that students must pay to acquire at accredited universities since employers and governments love them. After that, the college or institution will act as a filter for prospective students. However, a university's reputation and prestige are tied to how well its students study. Let us pretend that universities and colleges start providing inferior education or hiding important information from their students. The cost of attending college or university prevents it from serving as an extended summer camp [15].

The efficiency and appeal of educational institutions have increased thanks to innovative teaching methods. Learning on a computer can free up faculty and students' mental resources to engage in higher-level cognitive activities and intellectual challenges while physically present on campus. There is a need for new approaches to teaching in higher education. Even today, many prestigious educational institutions employ the broadcast learning strategy, in which the professor gives a presentation, and the student ostensibly takes notes. When the teacher's notes bypass the students' brains and make it straight to their mailboxes, we call it a lecture. As a result of technological advancements and the emergence of a new generation of students, this approach is now obsolete. Conversation is the primary mode of education for young people. Sharers. They are so engrossed in technology that they are eager to test novel approaches. They seek classes that are interesting and fun. They need to learn to value introspection.

Some examples of how universities modernize the broadcast paradigm are essays, hands-on labs, and seminar debates. As a result, many educators are attempting to alter this paradigm. In general, it is the strongest. Instructors in the appropriate fields must abandon dry lectures to encourage more active classroom participation. Students could use self-paced, interactive computer learning tools to master knowledge (anything with a right or wrong response) outside of class, freeing up more time in class for debate, discussion, and group projects. The purpose of universities needs to be specified as well. Nothing to do with smarts or skills. Learning never stops being a necessity, as do research, analysis, synthesis, context, critical assessment of competence, application of research to problem-solving, teamwork, and communication [10].

"Bitcoin was the holy grail of combining my many passions. Mathematical. Astute with modern technology. Everything is secret. Just the numbers. There is a social and political element to this. With promises of empowerment, this group convinced me to join." Having looked around many discussion boards for advice on where to get bitcoin, he eventually stumbled upon a blogger that could help him out. "The rate for writing for Bitcoin Weekly was 5 BTC. That cost you $4 back in the day," Buterin. "Pieces I have written to the tune of twenty bitcoins. Half of it went toward the purchase of a new T-shirt for me. It was like working with the fundamentals of civilization." In what percentage of students is that a norm? An excellent case study on group collaboration in the classroom is the early Ethereum software development company ConsenSys. Holacracy, a method of management that emphasizes teamwork over hierarchy in setting work goals and allocating resources, is credited with advancing the field of management. To paraphrase one explanation of how blockchain technology works, "visible rules rather than office politics; distributed, not delegated authority; dynamic roles rather than traditional job descriptions" all apply. 12 In terms of administration, value creation, and overall structure, ConsenSys is very different from either brick-and-mortar institutions or online courses.

Those that participate in ConsenSys do not take orders from on high. Only a few people are in charge. Indirectly or not, we all have a vested interest in the success of any given undertaking. The Ethereum platform strikes a balance between complete independence and complete dependence by allowing users to issue tokens that can be exchanged for Ether and converted into any currency. Students must think quickly, be truthful, and reach a consensus in class.

Get educated on the prerequisites.

(1) Find willing students and divide the work among them.
(2) Talk about who is doing what and what they will be getting.
(3) These protections ought to be written into smart contracts.
(4) Teachers and students alike would require training before using this strategy.

6.2 Costs and Blockchain (Student Debt)

Many teachers are against the idea of schools being profit-driven enterprises. Companies like Pearson and McGraw-Hill make billions off of supplying educational materials, teacher professional development, school management software, testing materials and platforms, all of which contribute to attaining credentials such as high school diplomas and college acceptance letters, and professional licenses and certifications. These corporations devote considerable resources to influencing federal and state policymakers [20].

Melanie Swan is considering using the blockchain to deal with her student loans. Launched a think tank to study blockchain systems. Recently, she has been devoting her time to "pay for success" and MOOC validation initiatives built on the blockchain [13, 14]. The blockchain provides three tools that can aid in this endeavor:

(1) A reliable means of verifying that the students who enrolled in Coursera courses finished those courses, attempted the associated assessments and completed their course.
(2) Secondly, a means of exchange for money.
(3) Education programs that might be implemented through smart contracts.

Smart contracts are an option that programmers could consider. It seems counterintuitive that government aid would not be better served by being geared toward people's personal development. Swan made the observation. 15 Kiva-style microfinancing for coding courses rather than small businesses; complete transparency and student accountability. Sponsoring students financially in exchange for observable academic advancement is an option for companies and other organizations with skill shortages. In this scenario, you have decided you want to help a female student in Nigeria who is enrolled in Google's Training for Android developers. Students are expected to demonstrate mastery of a new instruction unit every week. This could be done mechanically using an online test, with the blockchain verifying the student's identity and keeping track of their academic progress. Then the funds for the following week are transferred into the student's "smart wallet for higher education" so they can keep paying for their classes without any additional effort. There is no need for a governmental or nonprofit organization, regardless of whether their funding is stable. Swan contended that it was inappropriate to use money aside for a sister's schooling to support a brother [19].

Jane McGonigal, the Institute for the Future, and the ACT Foundation are just a few of the forward-thinking organizations behind the Learning Is Earning initiative, which encourages students to help pay off their student loans by teaching others the skills they have recently learned or putting those skills to use in the workplace. A college diploma is not necessary for them to enter the workforce. People with a precise set of talents and knowledge instantly in demand in the workplace or the classroom will be easily discoverable via blockchain queries made by employers, fellow students, or teachers. What this means for businesses is that the blockchain will facilitate the matching of projects to students who have demonstrated skills and availability for project work. In this way, individuals may determine the exact value of each component of their education and growth, such as a particular lecture or ability.

Human resource experts can do the same for training and development costs. Some businesses go so far as to pay for a student's education in exchange for a wage split after the graduation. Because academic publishers cannot access this monitoring data on the blockchain, they may be willing to pay for some of it to enhance their learning modules for all types of students.

6.3 Blockchain and the Meta-University

The ramifications of living in a secluded academic bubble are unfavourable. It has been a slur against theory-focused academics since the nineteenth century. Sceptics see it as a sign of superior intelligence, contempt for the commoner, and obscure, irrelevant research. Try to avoid engaging with that individual. The "ivory tower" image in higher education is best exemplified by the failure of schools and universities to use the internet to foster communication and collaboration among students, faculty, and parents. Despite unprecedented connectivity, most universities continue to operate independently, especially among the young [8].

Colleges evolve into ecosystems and networks due to blockchain's decentralized data storage. When the best educational materials in the world become available online, pioneers will allow kids to chart their academic courses with the help of teachers from all over the world. Better student service requires that institutions adopt new organizational models and form new collaborations. When Vest first mentioned the "meta-university" in 2006, it was to define its structure. What we mean by "open access" is a "transcendental, accessible, liberating, dynamic, communally produced framework of open resources and platforms." The information and education system could benefit from open access to course materials. Dr. Vest theorized that this international initiative would boost high-quality education and provide instructors with substantial new resources. Students can easily access various journals, books, and periodicals on campus [9].

This online-only meta-university would only fall short of traditional education with the ability to bind individual identities to accomplishments, record and credential them over time, reward positive and cooperative community behaviour, and hold players accountable for deliverables. Only 15% of MOOC students finish them. These courses supplement more expensive online university offerings [14].

Blockchain technology can help construct a secure, transparent global higher education network. Three phases overall. Content exchange is critical. Lesson plans and classroom discussions are shared online. Second, educators use wikis and other collaborative technologies to create new lesson plans and student materials. Third, the college or university joins a global network of professors, students, and educational institutions. Name, institution, and reputation survives. International higher education exists; Brilliant minds agree that universities cannot forever change the lecture [15].

Stage 1: Content Exchange

By making their instructional materials available online, universities and colleges are giving up a competitive advantage in the global student market by making what was once a privately held product freely available to the public. Over 200 institutions have started using MIT's OpenCourseWare. OpenCourseWare is an antidote to separation by eliminating barriers to accessing educational materials.

Materials like textbooks, lecture notes, homework, quizzes, videos, podcasts, etc., are all available in digital format. Teachers and learners will look for improved asset evaluation methods and evidence of ongoing work. To encourage participation in the blockchain, smart contracts provide incentives for those who make contributions. In addition to the usual "like," "upvote," and "share" actions, users can now contribute to the community by donating tokens for use in funding research or authoring grants. The worldwide academic community will be incentivized to share intellectual property, expertise, and ideas to advance higher education quality, strengthen professional standing, and reap material or monetary rewards. This way, newcomers may see what the community loves and whom they should follow. Literary journals and educational textbook producers are among those taking part.

Stage 2: Content Co-innovation

Co-creating content is integral to collaborative knowledge development that goes beyond simple communication. Professors could co-create new instructional content, publish it, and spread the credit and rewards, just like Wikipedia's decentralized editors do.

An initiative of the Wikimedia Foundation, Wikiversity is a free, open-source, and global educational resource. Wikiversity students set their learning objectives, and the platform's multilingual community works to create related learning activities and projects. Picture a token system to recognize and encourage contributions to Wikiversity. Blockchain makes it possible. It facilitates the community's selection of high-quality ideas, forming cohesive teams, and funding each stage of development while rewarding participants.

Psychology professors in this hypothetical scenario collaborate to design a "perfect course" for their discipline. Because there are so many different worldviews, schools of thinking, and approaches to teaching, participants are likely to have differing views on what should be covered in the course. Like Wikipedia, professors could create the foundational modules, while smaller groups of similarly-minded educators could create the supplementary materials. Teachers would benefit significantly from course software that promotes student participation, facilitates class discussions in small groups, and offers assessment and grading tools.

A large group of people can create teaching aids for psychology if they can create Linux, the world's most advanced operating system. Numerous academic institutions have active open-source software development efforts underway; It is no secret that Sakai has a large fan base. Educators developed the Sakai platform for educators to facilitate cross-institutional, cross-disciplinary, and cross-classroom collaboration. Shared content innovation results in new computer programs. This helpful program can create user-generated content for in-class instruction [18]. When used correctly, blockchain platforms may improve student collaboration. Students would not need to steal the professor's work; instead, they may collaborate on its creation under little oversight and receive full credit.

Stage 3: Global Network

Fragmentation of the university is a real possibility. If a student is proficient in math in ninth grade, why not allow them to take college-level courses while still in high school? The internet has prepared young people's minds for inquiry and collaboration, undermining the lecture-driven classroom and the concept of an exclusive, walled-in institution. What are the benefits of dividing school years into grade levels? Why not suggest that an international student who plans to major in mathematics instead take English classes? There is a correlation between a degree and a single university. Students no longer need to "enrol" (or pledge loyalty) to a specific educational institution in the same way they may have in the Middle [9].

In this vision of a global network for higher education, a student receives personalized instruction from a consortium of colleges, with their progress recorded in a blockchain. The learning process, path, and outcomes for each student enrolled in primary college are uniquely tailored by a knowledge facilitator. Students can attend college in Oregon and complete their coursework at Stanford or Cambridge. These college kids have the world as their textbook. This option, however, extends far beyond simply combining classes. Future educators will provide a platform where students from all over the globe may collaborate in online forums, wikis, and blogs to share and gain information and work together to solve some of the world's most pressing problems. By collecting and harmonizing data from several schools, the blockchain may provide a comprehensive record of students' accomplishments, including their courses and skills.

Many people, not just college students, could benefit from open educational platforms. Knowledge workers must constantly retrain to launch or advance their careers in today's modern, dynamic, and technology-focused workplace. Lifelong learning initiatives may benefit from the novel; collaborative education models' increased productivity, creativity, and legitimacy. If this worldwide network is so great, why not invite corporations and governments to join it? User fees charged by businesses might finance the platform's growth.

6.4 Incentives to Change

Why should we alter our ways if all this new information benefits us? In what ways will a change in pedagogy benefit professors? The tenure system maintains the lecture format. The educational publishing industry in the USA is a major supplier of materials used in the classroom and other resources such as assessment tools and administration and student engagement software. Say this if you are a professor or administrator: "Let publishers rethink the student experience." However, that is so what! I cannot talk now; I have a lot on my plate. However, students will gain from the shift to a more modern approach to higher education, which is the only real incentive to change. College leaders should look at the fates of previous cultural groups that resisted change. The similarities between encyclopedias, newspapers, record labels, and universities are striking. Everybody makes stuff. People are hired, managed, and compensated by all of them. All of them are peddling proprietary goods and filing lawsuits against copycats. Their clients are willing to pay for their services since they provide something unique. There is a need for these businesses since there is a worldwide shortage of culture, intelligence, education, and creativity.

There are better times for encyclopedias, newspapers, or record companies. They used to have a monopoly on material, but the internet has broken that up. The advent of digital-enabled increased availability, audience engagement, and novel modes of distribution and monetization. Faster than you can buy a pizza with bitcoin, the internet has worn away its purportedly bulletproof defense's. Only two or three global behemoths are left in most industries.

The credibility and reputation of a student are still primarily determined by their college or university. Attempts to reimagine higher education run into a brick wall (i.e., the old paradigm). Sooner or later, one of the blockchain-based innovators will show that its approach to learning pays off more quickly, that employers value its credentials as much as, if not more than, traditional credentials, and that it can deliver real value to many students who cannot afford college tuition or whose cognitive or social abilities do not "fit in" with the traditional college experience.

There is no reason not to challenge prevailing ideas. If we want to build a global network for higher education that is both secure and transparent, we can do so with the help of blockchain technology. This innovative platform is an inclusion engine that makes higher education effective for all pedagogical and educational approaches. Let us use the impending internet of value and blockchain revolution to re-create our identities and incorporate our educational backgrounds. Only then can we think pedagogy and create new methods for higher education for those who want to keep learning for the rest of their lives.

6.5 Lives Technology Integration in Higher Education in Indian Education

The National Education Policy, 2020 [12], published by the Ministry of Education (MoE), is ground-breaking. An underlying theme of the policy is the relationship between education and technology, with particular emphasis on caring for young children, promoting diversity in the classroom, and revising existing curricula. The last decade has seen India transform into an "information-intensive society," increasing the demand for technological resources in the country's classrooms. The policy states that "extensive use of technology in teaching and learning, removal of linguistic barriers, improvement of access, and education planning and management" will be a central tenet of the educational system [13]. Students and educators alike will need to rethink long-held beliefs and practices as they adjust to the widespread use of online education in place of traditional classroom settings. The publication of the policy at this time is crucial because it represents the vision of education for future generations and will aid in the construction of a "self-reliant" India.

6.6 Below Are Some of the Policy's Vital Technological Features

The policy's vital technological features are intelligence and autonomy. This idea is becoming a reality because of technological improvements, but there are still hurdles, such as data reliability. Since the IoT is expected to grow in the coming years, building trust in this vast data stream is essential. Blockchain will transform information sharing. Building trust in distributed systems without authority could alter numerous industries, including IoT. Since its beginning, the IoT has used disruptive technologies like big data and cloud computing to get around its limits. Blockchain may be next. This article looks at problems with blockchain applications for the Internet of Things (IoT) and reviews relevant research to figure out how blockchain could help IoT.

Pre-school Education

The policy emphasizes the use of technology in the classroom by highlighting its role in facilitating language acquisition, reducing the language barrier between teachers and students, and facilitating the development of digital libraries (specifically for differently-abled children). Students should also learn how to code. The policy supports online courses and other forms of remote education for instructors [17].

6.6.1 Higher Education

Sharma et al. [21], Complete illiteracy elimination and the incorporation of technology into professional training (for example, in the disciplines of law and medicine) are further goals that have been proposed (through the introduction of high-quality technology-based learning opportunities for adults). Given technology's pivotal role in solving societal issues, the policy promotes interdisciplinary research and new approaches to old problems. Higher education institutions are strongly encouraged to create laboratories to incubate and develop cutting-edge innovations. A group like the National Research Foundation would significantly boost the academic research sector [13]. The policy suggests establishing the National Educational Technology Forum ("NETF") as a place for open dialogue about using technology to improve K-12 and higher education classrooms, grading systems, and administration [11].

6.6.2 Administration

The policy also creates the Academic Bank of Credit, a digital repository for accumulated credit hours obtained at different HEIs, enabling degree conferrals based on these hours. Examples of regulatory agencies that the policy aims to make more open and efficient through the use of technology are the State School Standards Authority and the Higher Education Commission of India and its four verticals (National Higher Education Regulatory Council, National Accreditation Council, Higher Education Grants Council, and the General Education Council).

6.6.3 AI-Adaptation

Limitations of A.I. are recognized by the policy ("A.I."). The importance of adopting AI-driven innovations across sectors is emphasized. The NETF has identified and classified emerging technologies based on their "potential" and "estimated timeframe for impact." After that, the MHRD will explicitly list the technologies that need to be addressed by the educational system. This policy is ground-breaking because it acknowledges the importance of educating the public and researching the next generation of disruptive technologies, which will affect many industries.

7 Digital India

The policy calls for investments in digital infrastructure, online teaching platforms and tools, virtual labs and digital repositories, training teachers to become high-quality online content creators, developing and deploying online assessments, and setting standards for online content technology and pedagogy. Establishing a department to develop digital infrastructure, digital content, and capacity building is necessary for managing the e-education needs of K-12 and higher education institutions [13].

7.1 Priorities

While the policy has done an excellent job of incorporating technological advancements into "education" in India, this does present some novel difficulties. Only 4.4% of Indian households in rural areas have computers. In comparison, 23.4% of urban households, 14.9% of urban households have internet connectivity, and 42.0% of urban households have televisions, according to official data collected between July 2017 and June 2018 and published in November 2019. Only 13.0% of those living in rural areas are proficient computer users, compared to 32.4% of those living in urban areas. Data demonstrates that rural areas are seeing a considerably higher pace of growth in internet penetration than urban ones. People in urban and rural locations could even use their mobile phones to access the internet back then.

No matter where they live, each student needs access to a computer, laptop, or mobile device for educational purposes. Most of today's low-income students have limited or no access to computers, the internet, and reliable electricity.

The policy recognizes these constraints and the need to remove them through joint initiatives like the Digital India campaign and the accessibility of low-cost computer equipment. However, workable answers remain to be discovered. Access to power, basic infrastructure, and a general understanding of the significance and usage of technology are all necessary supplements to the steps.

The "human factor" of teaching and learning can be improved with the help of technological advancements. Analyze the processes involved in creating, using, transporting, and storing technology and any safeguards to prevent data breaches.

7.2 Technological Innovations in Education

While the policy has done an excellent job of incorporating technological advancements into "education" in India, this does present some novel difficulties. Only 4.4% of Indian households in rural areas have computers. In comparison, 23.4% of urban households, 14.9% of urban households have internet connectivity, and 42.0% of urban households have televisions, according to official data collected between July 2017 and June 2018 and published in November 2019. Only 13.0% of those living in rural areas are proficient computer users, compared to 32.4% of those living in urban areas. Data demonstrates that rural areas are seeing a considerably higher pace of growth in internet penetration than urban ones. People in urban and rural locations could even use their mobile phones to access the internet back then. No matter where they live, each student needs access to a computer, laptop, or mobile device for educational purposes. Most of today's low-income students have limited or no access to computers, the internet, and reliable electricity [7].

The policy recognizes these constraints and the need to remove them through joint initiatives like the Digital India campaign and the accessibility of low-cost computer equipment. However, workable answers remain to be discovered. Access to power, basic infrastructure, and a general understanding of the significance and usage of technology are all necessary supplements to the steps.

The "human factor" of teaching and learning can be improved with the help of technological advancements. Analyze the processes involved in creating, using, transporting, and storing technology and any safeguards to prevent data breaches.

7.3 Technology Integration Model

See Fig. 3.

8 Conclusion

Blockchain, A.I., and IoT have yet to converge. The holy trinity scope and architectural designs of any of these technologies—let alone all three—are genuinely enormous and constantly evolving. Each has drawbacks, including a lack of interoperability, slow processing performance for high-volume or mission-critical applications, and incompatible standards in Higher Education. Each presents novel design paradigms in the domains of privacy and security. Each has substantial legal, political, cultural, and ethical obstacles compared to existing arrangements. There is no silver bullet for capitalizing on data or data-driven economies, but all three individuals are the subject of enormous hype and promise. Nonetheless, significant exceptions clarify the situation for future higher education.

Technology Intigration Model : AI,Block Chain and IoT onvergence

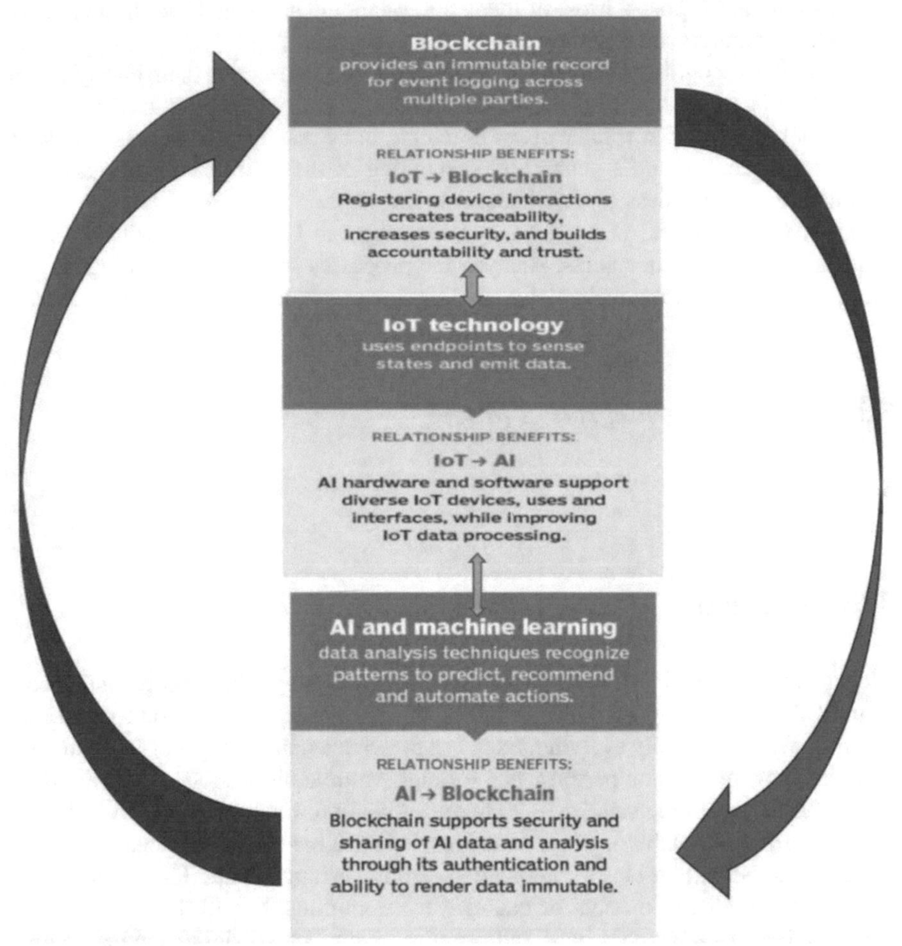

Fig. 3 Technology integration model

References

1. Barontini C, Holden H (2019) Proceeding with caution—a survey on Central Bank Digital Currency. BIS, Manek Bhawan
2. CashOnLedger (2020) CashOnLedger Website—the payment engine for the machine economy. https://cash-on-ledger.com/. Accessed 5 Aug 2022
3. Diedrich H (2016) Ethereum—blockchains, digital assets, smart contracts, decentralized autonomous organizations. Wildfire Publishing, Washington, DC
4. Dorri A, Kanhere S, Jurdak R (2017) Towards an optimized blockchain for IoT. In: Proceedings of the IEEE/ACM second international conference on internet-of-things design and implementation. IEEE, Piscataway, NJ, pp 173–178
5. Es-Samaali H, Outchakoucht A, Leroy JP (2017) Blockchain-based access control for big data. Int J ComputNetw Commun Secur 7:137–147

6. Huh S, Cho S, Kim S (2017) Managing IoT devices using blockchain platform. In: Proceedings of the 19th international conference on advanced communication technology. IEEE, Piscataway, NJ, pp 464–467
7. Indicators of household social consumption on education in India Ministry of Statistics and Programme implementation. http://www.mospi.gov.in/sites/default/files/NSS75252E/KI_Education_75th_Final.pdf
8. International Institute for Sustainable Development (2019) Tokenization of infrastructure—a blockchain-based solution to financing sustainable infrastructure (n.d.). https://www.iisd.org/system/files/publications/tokenization-infrastructure-blockchain-solution.pdf
9. IoT Analytics (2020) Industrial AI market report 2020–2025. https://iot-analytics.com/product/industrial-ai-market-report-2020-2025/. Accessed 30 July 2022
10. Karafiloski E, Mishev A (2017) Blockchain solutions for big data challenges: a literature review. In: Proceedings of the IEEE EUROCON 2017—17th international conference on smart technologies. IEEE, Piscataway, NJ, pp 763–768
11. Kumar Singh S, Rathore S, Park JH (2020) BlockIoTIntelligence: a blockchain-enabled intelligent IoT architecture with artificial intelligence. Future Gener Comput Syst 110:721–743. https://doi.org/10.1016/j.future.2019.09.002
12. National Education Policy 2020 Ministry of Human Resource Development Government of India (2020). https://www.education.gov.in/sites/upload_files/mhrd/files/NEP_Final_English_0.pdf
13. Narayanan A, Bonneau J, Felten E, Miller A, Goldfeder S (2016) Bitcoin and cryptocurrency technologies—a comprehensive introduction. Princeton University Press, Princeton
14. Nehru RSS (2022) MOOCs: the paradigm shift from traditional to new learning spaces in Indian higher education [Review of MOOCs: the paradigm shift from traditional to new learning spaces in Indian higher education]. In: Pal S, Ton QC, Nehru RSS (eds) Digital education for the 21st century: technologies and protocols. Apple Academic Press; Boca Raton, Fl, USA; Abingdon, Oxon, UK, pp 377–398. ISBN: 9781774630075
15. Nehru RSS, Chakraborty S (2020) The education of things (IoT) for smart learning through IoT intervention: a case study based analysis. In: Gunjan V, Garcia Diaz V, Cardona M, Solanki V, Sunitha K (eds) ICICCT 2019—system reliability, quality control, safety, maintenance and management. ICICCT 2019. Springer, Singapore. https://doi.org/10.1007/978-981-13-8461-5_60. ISBN: 978-981-13-8460-8
16. Pal S, Cuong TQ, Nehru RSS (eds) (2020) Digital education pedagogy: principles and paradigms, 1st edn. Apple Academic Press. https://doi.org/10.1201/9781003009214. ISBN: 9781771888875
17. Pal S, Ton QC, Nehru RSS (2022) Digital education for the 21st century: technologies and protocols. Apple Academic Press, Boca Raton, Fl, USA; Abingdon, Oxon, UK. ISBN: 9781774630075
18. Roeck D, Schöneseiffen F, Greger M, Hofmann E (2020) Analyzing the potential of DLT-based applications in smart factories. In: Treiblmaier H, Clohessy T (eds) Blockchain and distributed ledger technology use cases—applications and lessons learned. Springer, Cham, pp 245–266
19. Salah K, Rehman MH, Nizamuddin N, Al-Fuqaha A (2019) Blockchain for A.I.: review and open research challenges. IEEE Access 7:10127–10149. https://doi.org/10.1109/ACCESS.2018.2890507
20. Sandner P, Klein M, Gross J (2020) How will blockchain technology transform the current monetary system? Medium, Online. https://medium.com/the-capital/how-will-blockchain-technology-transform-the-current-monetary-system-c729dfe8a82a. Accessed 5 Aug 2022
21. Sharma A, Prakash AR, Nehru RSS (2022) Reforms in higher education in India: National Education Policy—2020. Int J Health Sci 6(S5):7220–7230. https://doi.org/10.53730/ijhs.v6nS5.11610
22. Suliman A, Husain Z, Abououf M, Alblooshi M, Salah K (2018) Monetization of IoT data using smart contracts. IET Netw 8:32–37. https://doi.org/10.1049/iet-net.2018.5026
23. Zyskind G, Nathan O, Pentland A (2015) Decentralizing privacy: using blockchain to protect personal data. https://ieeexplore.ieee.org/abstract/document/7163223. Accessed 10 Aug 2022

Data Protection and Security Enhancement in Cyber-Physical Systems Using AI and Blockchain

K. Vignesh Saravanan, P. Jothi Thilaga, S. Kavipriya, and K. Vijayalakshmi

Abstract Cyber-Physical systems termed as CPS is a system which is an incorporation of both computational algorithms and physical components. Several applications such as smart vehicles, smart cities, smart medicine, and defense systems use CPS as the basis for the development. Artificial Intelligence enables integration of information, analysis of data and uses the resulted insights for securing critical Cyber Physical Systems. In the current scenario, it is considered as a most difficult part to prevent CPS from the adversarial attacks in a world of increasing challengers and hence it is necessary to concentrate on creating resiliency in CPS. This chapter comprises the CPS characteristics, analyzes of CPSs current scenario, security threats faced by CPS, and the solutions for CPS security threats. Also, the discussions about CPS security applications to enhance the security mechanism based on block chain techniques were done. The combination of blockchain and cyber physical system will intercept the existing processes across various fields in the industries. Blockchain mainly enables the CPS to send data or information to private blockchain ledgers for inclusion in shared transactions, which could be used to improve the generalization proficiency-based security applications. This chapter proposed a novel approach for the security enhancement of the SPI incremental model with an efficient secured algorithm. This chapter finishes with a discussion of current research trends and possible future research topics, as well as an overview of recent advancements in the field of AI and blockchain established security for CPS and countermeasures.

K. Vignesh Saravanan · P. Jothi Thilaga · K. Vijayalakshmi
Department of CSE, Ramco Institute of Technology, Rajapalayam, India
e-mail: vigneshk@ritrjpm.ac.in

P. Jothi Thilaga
e-mail: jothithilaga@ritrjpm.ac.in

K. Vijayalakshmi
e-mail: vijayalakshmik@ritrjpm.ac.in

S. Kavipriya (✉)
Department of CSE, Mepco Schlenk Engineering College, Sivakasi, India
e-mail: urskavi@mepcoeng.ac.in

B. Bhushan et al. (eds.), *AI Models for Blockchain-Based Intelligent Networks in IoT Systems*, Engineering Cyber-Physical Systems and Critical Infrastructures 6,
https://doi.org/10.1007/978-3-031-31952-5_13

Keywords Resiliency in CPS · Adversarial attacks · Data privacy · Blockchain · Artificial intelligence · Regularization

1 Introduction

A computing system integrated into a physical system is known as an embedded system. An embedded system can be found in any CPS. The fundamental difference is that "embedded system" refers to a system that is primarily focused on its computer elements (that is embedded in a superior, physical system). Frequently, exploration on Embedded Systems has concentrated on issues like the recognized verification of discrete classifications (automata), hardware strategy, production costs reduction and consumption of energy, as well as development of embedded software. The CPS perspective places a strong emphasis on how crucial it is to consider the physical environment of the computational system while designing, testing, and verifying the functionality of the created things [1]. The current systems necessarily react to exterior alterations within predetermined timeframes. Numerous, but not the entire current systems are Embedded Systems.

A computerized tradeoff agent, for instance, wouldn't often be considered an embedded system despite the fact that it must adhere to rigorous timing requirements in order to be useful in reaction to quickly shifting market conditions. Real-time task scheduling in systems with periodic or aperiodic request patterns, various (interchangeable) computational resources, jobs with different priorities, and real-time communication has traditionally been the focus of real-time systems research. Naturally, worst case run time requirements are frequently the focus of this research. A real-time system can be a CPS: Concurrent restrictions apply to the control system of a car, but not always to the sound system. The capacity of a system to continue operating even when some of its components fail is known as reliability. Construction constituents from sturdier supplies, increasing idleness, and incorporating error checking to find and attempt to make up faults and letdowns are just a few approaches to increase reliability. Probabilistic approaches have been successfully applied to boost reliability while controlling costs in numerous sectors, including computational systems. But there are other approaches for creating trustworthy systems in addition to probabilistic ones. We can employ probabilistic methods more effectively if we have a firm grasp of other more fundamental ideas in systems design.

In this article, we present a thorough overview of machine learning-based cyber-physical assault detection systems. We concentrate particularly on physical system and process manipulation and damage attacks. Additionally, we focus on threat handling strategies that use machine learning to detect cyber-physical attacks. Network intrusion detection systems (IDSs) and physical model-based anomaly detection methods have been the subject of numerous surveys [2]. As a result, methodologies for IDS-based anomaly detection and traditional deviation-based attack detection in control theory are not explored.

The remainder of the chapter is structured as follows. The Problems while building an IoT network and Challenges of Cyber-physical systems are introduced in Sect. 2. The taxonomy of cyber-physical threats for each layer is provided in Sect. 3. AI Based Security in CPS are presented in Sect. 4. The potential research directions for Mechanisms Based on AI and Blockchain for Reliable CPS are discussed in Sect. 5. ML-Based Real-Time Attack Detection with Design of a Resilient CPS is explored in Sect. 6. Section 7 provides the experimentation and analysis for Secure Encryption Algorithm. Finally, Sect. 8 brings the Future Research directions and closes this chapter.

Our contribution

We have categorized the current IoT key generation, distribution, and management frameworks as well as authentication protocol-based techniques in this post [3]. It gives many researchers in the subject of Internet of things security a systematic, thorough review. This article is set up to give a basic overview of the technologies (blockchain and AI) involved in the integration of IoT, before delving into security flaws, assaults, and issues while addressing some of the potential fixes for the issues found in the research [4]. This review article's main contribution can be summed up as follows:

- It offers a succinct introduction that gives readers a general overview of the study field, explains the function of authentication in IoT security, and briefly goes over the difficulties associated with IoT adoption.
- It offers a comprehensive taxonomy of IoT attacks and groups these assaults into big categories. explains these assaults, which continue to exist during the IoT networks' initialization, deployment, maintenance, and termination stages.
- By offering a clearer grasp of the concepts, this article differentiates between authentication, authorization, and trust metric to dispel any ethical misconceptions surrounding the terminology used by new scholars.
- It lists the main session and authentication key-oriented assaults and goes into great depth on how the adversary conducts these attacks to take advantage of the network's many vulnerabilities.
- The advantages and disadvantages of conventional key generation and distribution systems with regard to IoT attacks are covered in this paper which compares several traditional authentication frameworks available. For new readers, it gradually introduces the fundamentals of blockchain and artificial intelligence technologies before describing how those technologies are integrated into the security of the Internet of Things.
- Based on cutting-edge blockchain and artificial intelligence technologies for IoT security, the work identifies the flaws that still remain in the current authentication processes.
- The article offers potential future possibilities for blockchain- and AI-enhanced IoT network security.

2 Background

Dependability requirements vary between CPSs; hence it is not necessary for every CPS to be developed with reliability in mind. In general, however, there is a growing requirement to take into account the module wide dependability consequences for every type of individual Cyber Physical Subsystem as communication between various CPSs rises. Integrity, reliability, maintainability, safety, durability, security, and availability are only a few instances of related properties that can be included under the more comprehensive concept of dependability. It is regarded as a gauge of these combined qualities in the background of the multidisciplinary area of system engineering. Because of its unique perspective between what is typically engineering and conventionally management, system engineering is frequently seen as a field of both engineering and management of engineering.

It is believed that dependability and systems engineering are of utmost importance for innovators and creators, and that the material presented in this book chapter will provide the reader with a strong basis for pursuing future research in these fields. An interactive object-based mathematical model is called a multi-agent system. It frequently relates to the mathematical field of game theory. Techniques for modeling and debating ideas like belief, knowledge, intent, competition, and collaboration are provided by multivalent systems and game theory. Multivalent systems frequently raise the question of intelligence, and there are precise representations for both hybrid and discrete (continuous/discrete) games. Game theory and multivalent systems, in divergence to the study of hybrid systems, concentrate on the behavior of groups of agents, rather than individual agents [1].

Applications for CPSs be able to found in a varied range of fields, such as highly reliable life support and medical systems and devices, advanced motorized systems, traffic management and safety, progression control, energy management, ecological control, instrumentation, aeronautics, defense systems, distributed robotics, industrial manufacturing, smart constructions, and management of crucial infrastructures [2].

2.1 *Problems While Building an IoT Network*

2.1.1 Heterogeneous Components

It might be difficult to manage and exchange different information formats among heterogeneous devices. These IoT edge devices run on many platforms and communicate with one another using various protocols for data sharing, data authentication, and data authorization. When there are many nodes, it is difficult to handle heterogeneous equipment.

2.1.2 Network Scalability

There are challenges with data management, device relocation, and integrating newer edge devices into an established IoT network while preserving forward and backward secrecy.

2.1.3 Problems with Data Sharing

These include issues with availability, reciprocal authentication, network traffic congestion, malware attack vulnerabilities, and passive attacks.

2.1.4 Durable (Efficient) IoT Algorithms

Because hardware varies, some devices function at their best while others struggle as a result of load congestion on edge devices with low computing power. The majority of the energy is lost managing network traffic and choosing better converters to transform data formats. It is difficult to develop session key generation and distribution methods with low computational overhead that perform better with greater security levels. In order to waste as little battery power as possible, the issue is to reduce the utilization required for communication among the many edge devices in an IoT network.

2.1.5 Network Components

The network component in a sensor node placed in a group of edge devices are always susceptible to physical theft by the attacker in order to access data gathered and stored in the device. When sensors are placed in wireless sensor networks with little to no security, this typically happens. In order to reduce the cost of the IoT network's adoption, these devices include less secure hardware. The edge device must be recognized using the IDs given to them during the registration step, and tracking such edge devices while preserving connectivity range is difficult. When Ad-Hoc networks are detected, it is necessary for security reasons to discard the device if it wanders out of range.

2.2 *Challenges of Cyber-Physical Systems (CPS)*

The new technological advancements have a profound impact on people's lives in various spheres, including business, agriculture, transportation, healthcare, and other comparable areas. They also introduce fresh challenges to the conventional physical gadgets at the same time. This is simple to see that outmoded information system typically rely on embedded devices through sealed features when using them as examples. The application needs for the interface, control, and expansion of physical devices cannot be met by these systems. As a result, as information technology has advanced, Cyber Physical Systems (CPSs) have gradually supplanted traditional information systems as the primary method of development [3].

CPSs are intellectual systems that incorporate integrated actuators, sensors, and controls to sustenance communication between the physical and digital worlds. Three layers of CPSs are the perception layer, the transport layer and the application layer [4]. The perception layer is utilized to gather perceptual data and carry out feedback judgments. Information and choices are transferred between various system components using the transport layer, which also includes the network layer, medium-access control layer, and physical layer. Decisions are mostly made using the outcomes of the perceptive information's analysis at the application layer, which is also identified as the control layer. In many various types of CPSs, this tiered view design is included. These CPSs be able to investigate in a variety of solicitation areas, comprising manufacturing, transportation, electric vehicles, smart grids, and more.

Researchers are beginning to highlight the combination of the network and the human society as applications of network in usual embedded information systems endure to advance, recognizing the interplay amongst the network society and the physical society [5]. Cyber Physical-Social Systems (CPSSs), also known as association systems, have arisen in this setting. They integrate physics, computing, and human resources and allow for synchronization between the cyber, physical, and social worlds [6, 7]. In the physical, informational, cognitive, and social realms, CPSSs support self-organization, comparable execution, and managerial control [8, 9], utilized, in order to create and build a strong environment with command and control organizations, it must be able to supply the ideal paradigm.

3 Security and Privacy in CPS

Figure 1 depicts the model of a significant CPS, which consists of a Physical System Cyber System, and Communication System. The state estimator and central controller make up the cyber system. There are many PLCs and sensors in the actual system. Information transmission between the cyber ad physical systems is handled by that communication system. Sensors track the physical process cycle by cycle and transmit sensory information to the cyber system [10]. The state estimator receives sensor data first, and then the state estimator assesses the system status. The estimated system state is communicated to the central controller if there are no exceptions regarding the sensory data. The central controller then sends directives via a communication system to the physical domain depending on the evaluated state. PLCs initially take in commands and break them down into several subcommands to control the physical process before sending them on to the physical system.

The rapid improvement of CPS and implementing the technology in day-to-day exists need addressing the security vulnerabilities in this domain. The

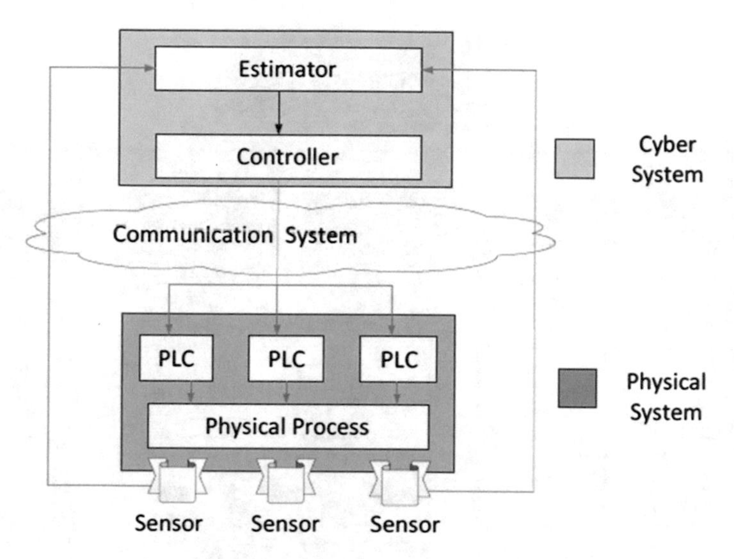

Fig. 1 Prototypical model of a significant cyber-physical system (CPS)

different attacks are discussed in Fig. 2 depicting the attack categorization in the Cyber-physical systems (CPS) field.

3.1 Physical Attacks

Physical assaults refer to network hardware that interferes with the system's ability to function physically.

3.1.1 Reverse Engineering

Sensitive information can be revealed by re-programming and modifying security architecture's algorithm or entire source code [11].

3.1.2 Physical Channel Tapping

Changing Channels Without severing the connection, the invaders utilize the tapping of channel to obtain the signs from the channel. The inconvenience can result from the attackers diverting traffic [12].

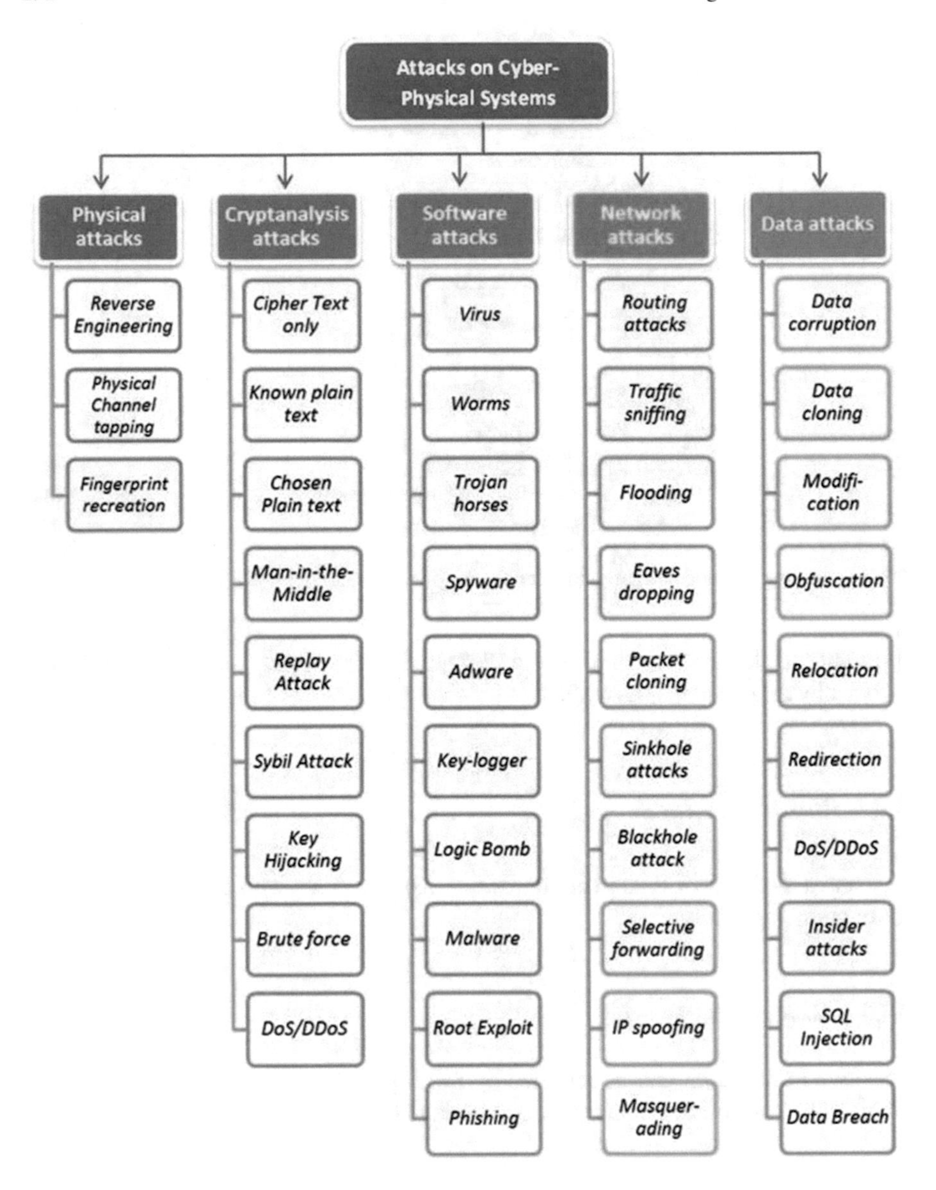

Fig. 2 Taxonomy of attacks in Cyber-Physical Systems (CPS)

3.1.3 Biometric Fingerprint Recreation

Attackers acquire access to the classification and take the biometric template from the device in order to manipulate user data that has been recorded and carry out false acceptance attacks, seriously endangering the system [13].

3.2 Cryptanalysis Attacks

The attacker employs a variety of methods to undermine a network's cryptography. An attack method comparable to this is cryptanalysis, in which the attacker thoroughly examines the cryptographic method to identify the system's weak point. The attacks depend on the algorithm's design and the overall properties of the plaintext.

3.2.1 Cipher Text Only

When the attacker only has interaction to the cipher texts and but not the plain text, but quiet some knowledge of the language or the arithmetical scattering of characters in the plain text are there to utilize this attack model.

3.2.2 Known Plain Text (KPA)

When the cryptanalyst has access to the plain text and the cipher text, the KPA takes place. This type of attack strategy is used to ascertain the secret codes for codebooks and also for encryption [14].

3.2.3 Chosen Plain Text (CPA)

In order to weaken the security of the cryptosystem, the CPA frequently take up that the attacker is accomplished with computing the cipher text for any of the chosen plain text. To obtain that encryption key in this scenario, the rival interrelates with the encryption device, which is represented as a "black box".

3.2.4 The Man in the Middle (MITM)

This type of model has been linked to dynamic listening in and inter-communication among the entities that believe in speaking to one another. Without being noticed, the attacker establishes a personal relationship with the victim and takes control of all communication between them while acting as an impartial third party amongst the entities. To get around two parties' mutual authentication, the attacker interrupts the messages and inserts fresh messages [15].

3.2.5 Replay Attack

This type of attack is also known as attack of playback, is a passive attack in which the data transmission is purposefully or falsely delayed. The assault can be carried out by either party, i.e., the data's creator or an opponent who intercepts it and modifies it before retransmitting it [16].

3.2.6 Sybil Attack

The invader attempts to take over the network by generating several false characteristics. In sensor networks, a fraudulent device node communicating through other nearby nodes using its real identity spreads numerous identities that are illegal. The dependability of the system, data integrity, and security are all compromised by the phony identities [17].

3.2.7 Key Hijacking

When a device on a network is compromised as a result of key hijacking, also known as session hijacking. This method is employed to grant an unidentified user access to the remote server.

3.2.8 Brute Force

The hit and miss method is frequently used to guess simple passwords. The attacker makes several, systematic attempts to estimate the security key generated via the key source, submitting passwords and paraphrases [18].

3.2.9 DoS/DDoS

The DoS attack is detected by hash collisions that take advantage of the worst-case runtime of hash table lookups in the cryptanalysis. Searching takes up system time instead of being used for important tasks [19].

3.3 Software Attacks

Hackers target the devices in addition to trying to access the system and run unauthorized code to carry out a variety of suspicious actions that impede the system's normal operation [20].

3.3.1 Virus

A macro or software called a virus attaches itself to lawful code in order to be executed. The virus programme doesn't run until the conditions are right for its code to run. The system software is damaged by a virus as a result of data corruption [21].

3.3.2 Worms

Attackers utilize independent programmes that spread themselves without the assistance of other host programmes or human users in order to carry out attacks. By using scheme transport services to execute over numerous sensor policies, the device is targeted by the vulnerability attack [22].

3.3.3 Trojan Horses

This is software which appears to be trustworthy and operates through user engagement (Email Attachment or File Download). Attackers destruct the host machine by data stealing, files corruption and aiding the propagation of other malware by tricking users into loading and executing it on their systems [23].

3.3.4 Spyware

These programmes are designed to gather sensitive data about a person or organization while appearing to be real and delivering the data to a third party deprived of the user's awareness. This method enables accessing the device lacking the user's permission [24].

3.3.5 Adware

This programme requires "user clicks" in order to make money from adverts that are displayed on a reliable website. A backdoor is placed on the machine once the user clicks the advertisement, allowing sensitive data to be stolen [25].

3.3.6 Key Logger

The attacker frequently uses a programme that collects keyboard keystrokes to steal the user credentials. Key-logging tools can read passwords and extract all typed information [26].

3.3.7 Logic Bomb

A specific piece of malicious code is injected into the software, and after a specific condition is met, it executes its payload and functionality [27].

3.3.8 Malware

The attackers look for the data about the networks to identify its weak points so they can use malware, or malicious software, to infiltrate the system and steal sensitive data, destroy data, or disable the network. In order to execute and spread malicious software throughout the system and the network, known vulnerabilities are exploited [28].

3.3.9 Attack Using Root Exploit

In order to harm the software and related devices in the network, the root exploit software involves the attacker to obtain rights given by the system administrator [29].

3.3.10 Phishing

Attackers con people by transferring phone messages or spoofed mails to install dangerous software in order to retrieve sensitive data. Attacks are carried out via trustworthy websites and additional information is gathered by tracking user behavior [30].

3.4 Network Attacks

For data sharing and the transmission of other network-related information, the sensor nodes connect with one another through the network. The attackers use a variety of cutting-edge ways to break into a network within an organization or a single system [31]. In an IoT environment, a wide range of potential assaults must be addressed in order to develop protection methods.

3.4.1 Routing Attacks

The attacker identifies definite weaknesses in the network routing software that it attempts to exploit in order to undermine the authentication procedure. Routing attacks include router spoofing, adding fictitious entries to routing tables, and rerouting packets from the router [32].

3.4.2 Traffic Sniffing

Attackers continuously scan the network while using traffic sniffing to capture packets on unexpected targets. In order to learn more about the devices, the attacker examines the packet and obtains the data [33].

3.4.3 Flooding

This method of attack is used to prevent the network from operating normally. The network is overloaded with orders and packets in order to use up all of the network's resources processing them [34].

3.4.4 Eavesdropping

This is a specific type of sniffing attack used to retrieve data being sent between devices linked through an unsecured channel. Because the channel is not secure, information theft happens [35].

3.4.5 Packet Cloning

A packet replication attack, also known as a packet clone attack, is one of the network assaults. One packet from the network traffic can be easily captured by an opponent, and information can be extracted from it; then reprogrammed the same to make a duplicate of a packet that was captured previously. These copies can then be distributed throughout the entire network and accepted as genuine. The receiver will respond to a packet that contained sensitive information and secrets [20].

3.4.6 Sinkhole Attack

By publicizing the alterations to the routing information, the hacked nodes in sinkhole attack draw network traffic. By carrying out related operations such spoofing, modifying router table, selective forwarding and similar attacks, the attackers get an advantage for themselves [36].

3.4.7 Blackhole Attack

When every packet is discarded at a specific black hole node, network connectivity stops. Uncertainty, the sinkhole node is also a black hole node, the issue is even worse. In such a case, there is no communication and the network's traffic is fully stopped.

3.4.8 Selective Forwarding

When the nodes adhere to the "neglect and greed" strategy, this kind of attack takes place. The nodes drop the original traffic that was flowing over the network and deliver their own traffic in its place. The rogue nodes obstruct the network's ability to exchange information, which results in a Denial of Service [36].

3.4.9 IP Spoofing

In a network communication, packets are transferred back and forth between network nodes that include both a sender and a receiver. The source address is contained in the IP header of each packet and is falsified during IP spoofing in order to conceal the identity of a device or to impersonate it in order to launch a DDoS assault against the network substructure [37].

3.4.10 Masquerading

Using valid network identity, this attack is used to improve illegal access to a host system or network in order to obtain the host data. After the organization is on a public network, the attack can be carried out by a member of the organization or a third party [38].

3.5 Data Attacks

Internet of Things (IoT) network and devices encourage people to prioritize convenience over security. Sensitive information has become the target of choice for attackers as a result of the security breach. The attackers employ clever systems to learn about the network and system weaknesses. The system's flaws make the threat to the data potentially vulnerable to a number of major attacks.

3.5.1 Data Corruption

Data corruption can result from any hardware failure, even when an attacker is not involved. When damaged data is input into the system, it produces erroneous outputs that are deemed uncommon for the applications that need them. Data loss could result in system failure if malicious malware modifies the files. The malware may corrupt data as fragment of its development to implement the payload and replace the records through junk code, rendering them useless for use in other activities [29].

3.5.2 Data Cloning

To launch a data cloning attack, a digital copy of the data must first be created. Take credit card theft, for instance, where the card is derivative by the thieves by means of an electronic scanner to carry out false assaults.

3.5.3 Modification

The alteration and interception of the data being transferred constitute an active attack. Hackers (attackers) can manipulate the data to change how the message is perceived and potentially drop off the data at the recipient's location, especially when the network is congested [39].

3.5.4 Obfuscation

An attack enabler that offers information, rules, or strategies to conceal the assault from IDS or additional attack detection mechanisms designed for the purpose, where obfuscation is not a direct attack. Rather, it became an attack facilitator. After strict schema validation (rule validation) is castoff in the network, the DoS attack typically fails.

3.5.5 Relocation

Data are transferred between systems and storage units. Weak security allows data theft as attackers repeatedly sniff the communication route for information. To misappropriate the data for additional attacks, the attackers extract, manipulate, and copy it.

3.5.6 Redirection

A vulnerability called redirection enables an attacker toward persuade application of users to click on a link which takes them to an unreliable outside source or website. Typically, a actual webpage or an application's user interface will include a link to such a site.

3.5.7 DDoS/DoS

With a DoS attack, an attacker attempts to interfere with a network or system's normal operation by flooding a node with excessive data transportation, rendering the server unable to offer the services. The scenario is appropriate for a distributed denial of service (DDoS) attack [40] if numerous hacked mechanisms are used to aim the server by making a demand in order to increase the efficiency of the assault by attacking the network substructure as well.

3.5.8 Insider Attack

A individual known as an insider who has permission to enter the system will carry out this kind of hostile attack. Due to their familiarity with the network, insiders find it easier to carry out the attack than outsiders network substructure [41].

3.5.9 SQL Injection

This kind of attack identifies a web security hole that enables network attackers to inoculate malicious queries and right to use the database. Following the implementation of the query, the attacker obtains right to use to data for which the data is not allowed to access [42].

3.5.10 Data Breach

The goal of attackers is toward recover sensitive information, which may be accomplished by theft, disc loss, hardware theft, etc. Unintentional data breaches can happen as a result of accidental storage media corruption, which exposes private information [43].

4 Potential Threats to CPS Security

4.1 FDI Attack Model

It is define an FDI attack in terms of its objective, the knowledge of the attacker, and its technical prowess. Attacks can have varying degrees of influence depending on their objectives, level of knowledge, and capabilities. attack intent. The state of evaluation S(t) could be different from the genuine state S if misleading data were introduced (t). The objective can be separated into two groups based on various intents: I Add false data to disguise the incorrect state S(t) = sf (sf belongs to F). (ii) Inject data C(t) to make the incorrect commands go out. (S(t) and C(t) is a member of R). The attacker's awareness. Different levels of system knowledge can be possessed by an attacker, including the sensory data (T), (ii) Jacobian matrix (C_{matrix}), and (iii) the system instructions (C). (iv) System A, B, and g parameters as well as (v) states knowledge R and F. It can be imagine as several attack scenarios based on the assumptions made about each of these components. Typically, attacks with flawless and limited knowledge comprise two main settings that are taken into account.

4.1.1 Perfect Understanding

In this scenario, the attacker is presumed to be fully aware of the targeted system. The communication system can be broken apart to acquire components T and C. A hacker may learn the F and C_{matrix} from public knowledge and system theory. R can be obtained by combining historical data and C_{matrix}. A, B, g might be known if the attacker is a system designer.

4.1.2 Limited Knowledge

It is assumed that information may be retrieved from the public dataset and collected by means of breaking the CPS. This indicates that, even though there is a wide range of limited knowledge, the attacker can still access T, C_{matrix}, C, F, and R. A, B, and g can't always be found because the CPS could be complicated.

4.1.3 Attack Potential

Different levels of attack capability are possible for an attacker, including (i) the ability to inoculate any value into a specific sensor, (ii) the ability to concurrently change the sensory data of all sensors, though the inoculated data is constrained, and (iii) the ability to inoculate any sensory data into any sensor.

Situation (i): Many sensor vulnerabilities that make it possible to easily break into some sensors, attack a component of the sensor, and inject bad data [44].

Situation (ii): By means of stated in the study of the authors, it is possible to launch attacks to change time stamps, such as time synchronization assaults. Although all sensor values can be changed at once, the range of injected sensor values is constrained.

Situation (iii): There are many sensors, and they are dispersed throughout the world. It is not practicable to hack into every sensor. A hacker might, however, breach the communication connection and alter the response data sent after the sensors to the state estimator. For instance, the work of the authors of [45], includes that the state estimator receives sensory input from PLCs as feedback. Attackers might potentially alter PLCs to change all of the response data (i.e., attack the firmware website update of PLCs).

4.2 AFDI Attack Model

The AFDI paradigm is first described in terms of attack capability, knowledge of the attacker, and attack aim. The AFDI attack process is then described. The output of inconsistent commands is required for AFDI attacks in order to interfere with the physical system. Continuously disguising the problem status is also required because the established disruption or concert deterioration was the main objective of AFDI assaults.

4.2.1 Attacker's Knowledge of AFDI

The knowledge of T, C_{matrix}, C, F, R, A, B, and g is required in order to be able to remain undiscovered for a prolonged time period for the attackers. C_{matrix}, C, F, T, and R can be obtained by examining public and historical data. In the last section, we suggest a technique through which attackers who are unfamiliar with A, B, and g can seamlessly fabricate states of system. An AFDI attack can be launched with very little expertise.

4.2.2 Attack Capability of AFDI

An AFDI assault necessitates the capacity for attackers to alter all sensory data since attacks can linger for extended period and depths from any sensor may be affected. An attacker must therefore have the capacity to inoculate any data into some sensor in order to conduct AFDI attacks.

5 Securing CPS by Means of AI Algorithms

Cyber Physical Systems (CPS) and Blockchain (BC) technology applications are expanding rapidly. However, because of the intricacy involved, it might be difficult to frame reliable and accurate Smart Contracts (SCs) for these smart applications. SC is updating the dated industrial, technical, and commercial procedures [46]. It is self-executable, self-verifiable, and integrated in the BC, which eliminates requirement for reliable third party systems and as a result, reduces supervision and support outlays. Additionally, it raises system performance and lowers security threats. Although SCs support new technologies of Industry 4.0's advancements, there are quiet numbers of privacy and security issues that need to remain resolved.

In the base of this, a survey taken on SC security flaws in the software code—vulnerabilities that a hostile user may exploit with ease or that could expose the whole BC network, is provided. According to the survey, users from all across the world haven't really explored the problems with SC security and privacy. It has been found from the existing approaches that creating sophisticated SCs cannot resolve its security and privacy problems [47]. In order to safeguard SC privacy, this study explores several Artificial Intelligence (AI) approaches and tools. Then, unresolved problems and difficulties for AI-based SCs are examined. A case study of trade marketing that leverages AI and SCs to maintain its privacy and security is then provided.

5.1 AI Based Security in CPS

Building intelligent devices that can think like people is the goal of artificial intelligence (AI) [48]. John McCarthy, also known as the father of AI was the first to present this idea. AI aims to create expert systems that do cognitive tasks, pick up novel skills, comprehend context, act and provide recommendations just like people [49, 50].

The first straightforward AI application was created and is called "Weather Forecasting". These days, practically all apps (such movie recommendations, early disease detection, Google navigators and others) are AI-based and necessitate a vast quantity of data in order to make appropriate decisions and suggestions. Loss of data security and privacy may result from this. Data-anonymity is one method for safeguarding data privacy. It anonymizes the training data sets by removing any identifying information; however it is not a sure process. The following list of several privacy-preserving AI approaches is provided by [51]:

5.1.1 Federated Learning

Federated Learning is an open-source framework for training artificial intelligence (AI) models using data to which cannot have access [52].

5.1.2 Differential Privacy

Differential privacy is an expedient that makes dataset information publicly available without revealing the individual data of the users [53].

5.1.3 Secure Enclaves

This technique creates a segregated execution environment to guarantee security [54]. It guarantees the data's privacy, accuracy, and veracity.

The two biggest hot technologies mentioned nowadays are BC and AI. Large datasets are predicted and processed by AI, whereas BC offers decentralized data access, immutability, and security [55]. The SC becomes genuinely intelligent when

AI techniques are allowed to be incorporated, and it may be utilized in many different decentralized applications, including autonomous vehicles, recommendation systems, and AI-based model competitions [56]. SC develops self-learning recognitions to AI, and it alters to deviations in the environment. The BC settings can be automated by AI to increase thresholds [57]. Since all data (personal and transactional data) on the BC is accessible to the public, AI is essential for preserving the data's confidentiality and privacy. A BC network is difficult to "hack," as an enemy needs more mining power. AI is helpful in creating an extremely protected BC application that can quickly take the necessary action after identifying the attack (by analyzing attack patterns). In a worst-case scenario, if the attack cannot be prevented, AI may be able to develop isolated environments (secure enclaves) that offer high levels of security. Differential privacy is a machine learning technique that scans users' devices for sensitive information.

The performance of SCs can be utilized by AI, making the analysis's findings highly reliable and without dispute. A survey on AI and BC technology was published by Salah et al. [58] to better recognize how a BC can be useful in AI-based solicitations and vice versa. It is represented about the challenges facing AI today and how the decentralized system can address them. They also highlighted some potential directions for future research and difficulties facing AI- and BC-based systems. Later, [59] put forth the Cortex system, based on BC technology. To take advantage of the BC system's performance, they deployed AI algorithms. Additionally, they added a mechanism for incentives that can both encourage people to submit their own optimized schemes for SC and deprive them of prizes. The system wants to develop its own internal Artificial General Intelligence.

5.2 Mechanisms Based on AI and Blockchain for Reliable CPS

The intricacy of cyber-attacks on Cyber-Physical Systems (CPSs) necessitates the development of a technique that can assess operational behavior and security without impairing live systems' functionality. Digital Twins (DTs) are transforming CPSs in this way. DTs boost the security and integrity of CPSs all the way through the complete product/process lifespan while providing agility to foresee and respond to real-time changes, if the DT data is reliable. While assuming that the DT data is trustworthy, DTs increase the security of CPSs throughout the product lifecycle while enabling swiftness to anticipate and react to real-time variations Prevailing Digital Twins analysis in CPS are restricted by unreliable information distribution amid numerous stakeholders and prompt progress improvement, nevertheless. These constraints highlight the importance of developing reliable de-centralized results through the capacity to produce meaningful real-time perceptions. To achieve this, a system that integrates blockchain with artificial intelligence with a focus on trusted and intelligent DT (AI) should be used. The suitable framework uses a hybrid approach that uses AI models to identify vulnerabilities in securing the sensor data in addition to learning about processes from the CPS specifications. The recent frameworks also incorporate blockchain technology to protect artifact lifespan statistics. We analyze the automotive sector's suitability as a CPS use case for the framework. Finally, we list the unresolved issues that prevent intelligence-driven designs from being used in CPSs.

In this section, the structure for secure CPSs framework is outlined. Four essential elements make up the structure: (i) physical resources and their replications; (ii) data integration and interoperability; (iii) threat intelligence; (iv) blockchain ledger. Prior to entering data into the physical space, the data integration and interoperability is in charge of cleaning up inaccurate, redundant, or missing data, converting disparate data formats into a unified data structure, and combining data to produce a consistent interpretation of a particular object. To continually conduct the digital-physical mapping between the specified system performance parameters received from storage and real-time sensor data from the manufacturing unit, and to check data consistency, the physical asset and its clone counterpart must be linked. Through the storage of data and the recall of events, the blockchain ledger ensures safe data management. To reduce the danger landscape, Thread Intelligence (TI) examines vulnerabilities, threats, and current and potential attack vectors.

Threat hunting and Secured Information and Event Management (SIEM) [60] are two other uses for the TI module. The Digital Twins are constructed through obtaining information of the process from the requirements set by the TI and CPS. We categorize the processes that move data through the system into three categories: (1) The methods for data wrangling and data fusion are provided their starting input; (2) To maintain consistency with the requirements of the underlying application, the continuous process modifies the data or system state depending on a predetermined time period; and (3) Depending on the CPSs' cyber situational awareness and the underlying events, the system-specific process is either scheduled based on those events or triggered. The participating entities (such as sensors, machines, and people) register on the blockchain as authorized entities (step 1). The data wrangler cleans and transforms the multimodal and heterogeneous sensor or actuator data from physical assets (step 2.1) and system-specific input information like engineering and domain knowledge (step 2.2) into a uniform format. Data fusion is used to combine data from many sources, such as supplementary information (step 3.1), data from different instances of the Digital Twins (step 3.2), and incorporated sensory information (step 3.3), before being fed to the physical asset to ensure reliable and thorough data depiction (step 3.4).

The provenance information is recorded on the ledger in order to support track and trace solutions (step 4). Data activities and the whole lineage of the data are recorded in the provenance [61]. The DTs' two security operation modes, simulation and replication, enable the monitoring and playback of CPS events. The physical environment's states or events are continually logged (step 5a) and replayed time to time at DTs in the replication mode (step 5b). This operation's goal is to identify data discrepancies (as shown in Fig. 1). The data sync technique step 6c is used to produce a time-dependent digital-physical mapping in order to achieve this. Any crucial information (such Safety and Security (S&S) regulations) may be accessible by means of the blockchain throughout the process (step 6). S&S guidelines include thresholds and reliability tests to identify assaults, which might help with keeping track of rule infractions.

The conditions or measures are tracked separately from the real-time environment in the simulation mode, and they may be utilized to modernize the twin (step 7) and ultimately the physical asset. The future status of a physical asset need to be predicted for initiating essential measures, such as repairs, in order to prevent subsequent breakdowns that may be brought on by security breaks or distrustful flows. Therefore, to effectively use sensors real-time data streams in DTs, TI obtains data from data sync (via replication mode step 8a) and DT (through simulation mode) (step 8b). When a failure is predicted, ML-based solutions can contact the proper scheduling services (in the physical space) or model calibration services (in the virtual space) to perform the required tasks, such as configuring machine settings based on tool wear data or tuning model parameters to precisely simulate the physical counterpart. The updated models after calibration (step 9b) and S&S rules must be stored and retrieved from the ledger in order to guarantee their dependability (step 9a) and further the argument for integrating blockchain with DTs. We go into great depth about each part of the TI-supported blockchain-based DT system in the sections that follow.

5.3 *Data Interoperability and Integration*

5.3.1 The First Challenge Is How to Prevent Incorrect Data from Being Inserted into the Physical Device and It's Identical

Important preconditions for DT deployment include the Data incorporation and data interoperability from heterogeneous and multi-modal sensors. The data wrangler in the suggested framework is in charge of removing incorrect, redundant, or lost data and transforming various file formats into single structure of sensor data which is entered into the physical assets. By using clever automated methods to complete the gaps in the data and clean it up (such as data de-noising and data de-duplication), AI-enabled data aids in enhancing the data quality [62]. While the blockchain has left behind strict security guarantees, it is equally crucial for vital infrastructures to ensure the reliability of data-generating sources. Ignoring such precautions could lead to the entrance of false data into the system, which would lead to the Garbage-In, Garbage-Out (GIGO) problem [63] since the data would be input to other entities (such DTs, TI), who would then use it to make decisions. The sequence of the tasks followed as: (i) gathering information only from authorized devices, (ii) data from the device is cross-validated against threshold levels, and (iii) Three-fold Integrity Checking Mechanisms (ICMs) (engineering knowledge and domain knowledge) are established to reduce the Age of Information (AoI) based on the amount of time that has passed since the development of the most recent status update received at the destination. We may rely on using Cross-validation of sensor data using tiered trust architecture based on blockchain [64] since it processes do not ensure the reliability of data at the source.

Engineering expertise outlines the network/logic layer design needs of the underlying CPS as follows: (i) components at the system level (specifying logic to control and information about device configuration), (ii) information at the network level (detailing the topology and communication route via logical links and endpoints), and (iii) the connection between the elements (specifying data collection at process-level, limitations and fine-grained procedures). The network configuration of the virtual environment can benefit from such thorough descriptions of the topological, procedural, and control artifacts [65].

Furthermore, implicit security principles can be based on such engineering knowledge. Detecting unfamiliar physical device or unexplained networks, detecting unexpected variations in the control logic, etc. are a few instances of establishing a safer zone based on usual device operation or specific services provided by the method [65]. The sensors deployed in IoT are periodically standardized for problem diagnostics and ageing management and at the beginning of data gathering. The framework comprises domain-specific information from professionals in a variety of domains, including engineers, and reliability specialists, etc., along with basic engineering familiarity. Once created, the domain knowledge may be customized to match the needs of other businesses and utilized as a reference. Following that, DT need to upkeep the association of various occurrences of DT connected using various physical sub-assets in order to enable an abstract perspective of the whole phenomenon. To do this, it is possible to combine numerous occurrences of DTs at computability, where every occurrence represents a physical component or an action/activity. This mimics the overall structure of the actual world. Auxiliary data (data particular to a user or an application) can also be offered. In order to complete the physical-digital mapping through data sync, the combined data is finally inserted into the physical asset. As a result, compared to a single view of data, data integration and interoperability offer a more consistent, thorough, and accurate portrayal.

5.4 *Digital Twins*

The second challenge is how to faithfully simulate CPS behavior throughout the product lifecycle in a remote environment disconnected from real systems. DT is a simulated version of a physical device that uses historical and actual information to evaluate, forecast, and improve processes deprived of compromising lifecycle security or the real environment. DTs can operate in two different ways: replication and simulation, respectively. By reflecting data from the physical surroundings, the imitation manner enables digital tracing of actual occurrences. The virtual duplicate of the physical thing must continually mirror the actual thing via logs, network connections and sensor readings, etc. in order for replication mode to operate. Data collection from the physical area might take place after a specific amount of time or even offline, depending on the needs of the application. The simulation mode operates independently from its actual counterpart in order to assist the trial-and-error method. This method, which also supports the understanding of emergent system behavior, enables conducting examinations recurrently by resetting the prototype across a wide variety of defined circumstances. By running security tests in a virtual environment, the simulation mode helps the security by design methodology by making it easier to analyze process modifications, test devices, or find misconfigurations.

5.5 Using Threat Intelligence and Digital Twins Together

The third challenge is why we need TI when DTs are available for situational awareness and predictive maintenance. While DT security-operation modes provide many benefits, they might also have drawbacks. The time-reliant recording might not produce forthcoming situations for the replication mode. While the balance between budget and fidelity determines the state imitation accurateness, synchronization concerns between the actual item and its many replicas are crucial [66]. Additionally, in order to create the identical stimulus, the input information (events) is necessary beforehand. Similar to the experimental mode, simulation mode relies on user-specified settings and parameters because the system's actual physical state is unknown. Given these drawbacks, we consider how TI may offer supplemental assistance for situational awareness and predictive maintenance. Future scenario creation is difficult for DTs, especially without direct observation data [67]. The predictive potential of AI-based results made existing through the TI module can be extremely important for anomaly identification and risk assessment. As a result, TI may give DTs more assistance in securing CPSs. For accurate forecasts and decision-making, data quality is just as important as data quantity. Given the complexity and constant change of the threat environment as well as the huge volume, velocity, and diversity of big data, this requirement is even more important. In order to prevent/detect abnormalities, forecast system behavior during interruptions, and stop the attack cycle, a TI module which can study beneficial designs from the gathered huge data is recommended.

Additionally, to infer and verify the adherence to security requirements, the organizational security administration framework, such as SIEM, may be connected to the TI unit. TI unit may be linked into the organizational security management framework, such as SIEM [60]. We will examine in what way DT security procedure methods work with TI in the sections that follow, as well as the conditions in which such modes are preferable. Prognostics and Health Management (PHM) may be carried out using the replication mode in conjunction with TI, where DT of the data-supported equipment is used to identify problems or areas in need of repair. The state of a specific asset is forecasted using ML algorithms like Long Short-Term Memory (LSTM) and Support Vector Machine (SVM) based on performance limitations [68]. It schedules the physical asset's maintenance if the asset condition data indicates that failure scenarios might result. Data sync provides data access to TI.

This module's data is accessible since it replicates the sync states of both real and imagined objects, allowing for the analysis of both areas' activity.

6 Cyber-Security/Cyber-Physical Systems Blockchain Models

With the introduction of the cryptocurrency bitcoin in 2009, the term "blockchain" gained widespread recognition. Only lately, however, have discussions about the potential applications of blockchain technology in other fields started. Blockchain is a system that uses computer networks to store and process data in a chain of blocks; it is not specific to any one area. It is possible to describe the potential applications of this technology in production systems since each block in the chain may include any type of data, including manufacturing processes.

6.1 Technology Behind Blockchain

Because of how broadly applicable blockchain knowledge is, most of its applicable areas are currently used across a range of human endeavors. Designing the right blockchain network architecture depending on the activities to be done and choosing the most pertinent tools are essential in order to fully utilize the Industrial Internet of Things (software and hardware) and CPS benefits of blockchain technology in its entirety. When physical processes and computing are combined, the outcome is a cyber-physical system (CPS). Blockchain networks come in two flavors: public and reserved. The former is the utmost sophisticated and frequently used for public or global concerns. Global peer-to-peer (P2P) grids are unsuitable for creating business networks similar to the industrial networks stated above even though they are exceptionally stable owing to their large membership. The biggest drawback is that the cryptocurrency utilized in single or multiple public blockchain is tightly correlated with all data interchange operations. It is challenging to estimate the ownership's cost for the planned CPS since it is nearly impossible to forecast changes in exchange prices in the cryptocurrency market. The currently being created architecture will thus be based on a private blockchain. The CPS should incorporate the following blockchain features:

- The organization of a solitary data space for device-to-device communication within CPS
- Assurance of the CPS's security
- Facilitation of the CPS's simple scaling and restructuring
- Availability of redundant systems and communication pathways
- Establishment of a solitary data storage facility
- Use of smart contracts to put "digital twins" technology into practice
- Using smart contracts to ensure the completion of routine activities for the CPS

The main applications of blockchain technology were to secure notaries, smart contracts, financial transactions, and storage systems. Equally the industry became conscious that it may possibly boost its competence by integrating blockchain, other sectors, including medical, healthcare, transportation, logistics, and cybersecurity, quickly acknowledged its welfares. As a result, this area of study is very busy, with scientists and researchers looking at several applications for this technology. The most often stated applications are those in healthcare, transportation, and cybersecurity.

6.2 Blockchain Categories

We make a distinction between the centralized private strategy and the decentralized public approach. Blockchain technology came into existence as a result of the rise of crypto-currencies, particularly bitcoin. In the peer-to-peer or decentralized bitcoin electronic currency system, cryptography is utilized to verify transactions. Using the decentralized fiduciary mechanism of the protocol, a third party will not be required. Everyone must be able to contribute to the development of the code due to decentralization. An addition to the bitcoin blockchain is made if it has been approved; if not, it is erased. Many cryptographic algorithms are used to preserve transaction security and integrity. The most widely used open blockchain scheme right now is bitcoin. The programme source code states that new bitcoins are generated for each transaction that is processed. The blockchain is created by end-users using their computational influence to validate, preserve, and assured transactions. After being approved, the transaction is time-bounded, published to the blockchain, and made accessible to the beneficiary and the whole system. A public blockchain is the one mentioned above that is utilized to create bitcoins.

If a small, predetermined number of people validate the consensus principle on a private blockchain as opposed to a public one, the blockchain is said to be private. An organization controls both the capability to take part in transactions and the verification duties. The literature describes a reserved network with a set quantity of nodes, similar to a blockchain of places. A private blockchain does not require a cryptography-based method [69]. On a reserved blockchain, there are no PoW, zero miners, and zero financial incentive. Therefore, a blockchain of any kind is an affordable, decentralized, and completely secure data transfer and storage system.

A potent method for identifying diverse cyber-physical threats directed at each CPS tier is the ML technique. However, owing to the peculiarities of the different techniques using ML, such as the demand of enormous attack and significant computing capacity, the vulnerable attacks discovery mechanisms based on ML does not ensure the durability of the CPS always. As a result, there are specific constraints on the usage of ML-based approaches for genuine CPS layers, and required to get around these restrictions in CPS design.

6.3 ML-Based Real-Time Attack Detection

Applications for general machine learning (ML), such obstacle detection, demand high precision and accuracy. As a result, four performance metrics [70] are used to assess generic machine learning applications: accurateness measure, precision value, recall value, and the F1-score measures. Nevertheless, time-related measures are not taken into account by standard evaluation criteria. Due to the real-time nature of physical dynamics in a CPS, critical applications with ML-based security features must be evaluated using time taken to detection measure along with the four conventional metrics. Physical dynamics can theoretically deviate to infinity even if the state of physical systems has a predictable limit in the real-world. Although the state of physical systems in the real world has a finite boundary, theoretically, physical dynamics can diverge to infinity. When the physical system's state crosses the barrier, it is irreparably damaged, which causes disruption. As a result, a cyber-physical attack must be discovered in advanced before the condition of the physical device goes beyond what can be repaired.

System dynamics, the physical system's condition at the time of the attack's launch, and the sort of attack all affect how quickly an attack can be detected. A real-time detection limitation with a familiar physical controller mechanism, the ball-beam control scheme [71], in the context of a DoS attack is illustrated. The ball-beam system is inherently unstable since it inclines a beam to control the location of the ball that would otherwise slide off the beam owing to gravity. Due to the device's inherent instability, the ball rolls off the beam when a DoS attack is performed against a ball-beam control system. We take into account a control scenario in which the reference ball is positioned 0.5 m away from the beam and the beam's range is 0.0–1.0 m. If the ball's location is outside the beam's effective field of vision, the physical system is said to be irreparably broken. In this instance, a DoS attack is launched at 15 s, and at 25.75 s, the ball has moved beyond the beam's range. The ball-beam control mechanism is thereby rendered beyond repair. The cyber-physical assault detection approach must be successful before the physical device crosses the irreversible state border in order to prevent destruction of the physical device. The attack detection deadline is established to be 10.75 s, or the time it takes for the physical system to become irreparably damaged after the DoS assault begins in this scenario.

Although there have been numerous studies on cyber-physical threat detection, there hasn't been enough done on CPS security that takes real-time restrictions into account [72]. A Knowledge-based intrusion discovery technique using KF is suggested in [73] for the physical system layer and provides real-time intrusion discovery from DoS and sensor attacks on a smart grid. A real-time image sensor intrusion discovery approach is provided that uses non-linear physical dynamics knowledge. The suggested technique is authorized for an automobile and can identify a sensor attack in less than 0.175 s. A real-time package manipulation-attack detection approach for CBTC systems is presented for the network layer, where scattered network devices uninterruptedly watch for the abuse of the ARP protocol.

According to [61], a novel framework for CPS offers instantaneous identification against complex sensor attacks across the network, with the actual constraint being set by the physical system's irreparable state condition. For autonomous driving systems, a concurrent anomaly finding method is described in [74], where the ML model quickly distinguishes between legitimate picture inputs and malicious inputs that could lead to harmful circumstances. Actual restrictions on physical systems must be taken into account by ML-based approaches for cyber-physical threat detection. However, a complicated ML model's large computational demands result in a delay in attack detection time, which disrupts the physical system. Minimizing the complication of an ML model is necessary when attaining high levels from traditional ML assessment criteria in order to fulfill present limitations on CPS.

6.4 Design of a Resilient Cyber-Physical System

A cyber-physical attack aims to interfere with the physical system. Particularly, the CPSs for communal structure must be safe and continue to function even in the event of an assault. In order to recover from an attack seamlessly, a resilient CPS design is needed, together with real-time attack detection and appropriate ML-based attack handling algorithms. The physical system becomes irreparably damaged at 25.75 s after the DoS attack begins since the ball's position has substantially moved outside the beam's effective range at 15 s (the red line). On the other hand, if a real-time detection and performance retrieval procedure is used and begins to work at 22.5 s, the ball's location will be timely adjusted to the locus position of 0.75 m with a brief, minimal variation. In this control scenario, the resilience method ensures the stability of the control system against the Denial of Service attacks.

For handling cyber-physical threats and unanticipated system failures, a hardware redundancy method is used in conventional physical system security. When a CPS's computer system exhibits anomalous behavior, the previously set up auxiliary system steps in to take its place. In, redundant controller architecture is used to switch controllers when an assault is identified by the SVM-model indicator in order to lessen the physical influence of a cyber-physical attack that targets biological practices. A redundant system configuration is also taken into consideration by the authors in [75] in order to ensure the permanency of a power plant against cyber-physical attacks. The high assurance control module assumes control of the physical system in the event that the good performing control segment experiences an unanticipated malfunction, such as restarting, providing robustness against system defects. For safety–critical CBTC systems, the simplex design [76] is also used to give CPS resilience against sensor attacks and software errors.

6.5 SPI's Secure Encryption Algorithm

It is suggested to use an incremental update data-based SPI security encryption technique. Prior to encrypting data, our system evaluates user data resources to determine the type of data involved. By encrypting only the data that requires more security and privacy, this successfully addresses the issue of data query delays brought on by conventional approaches in encrypting huge amounts of data. An interference quantification method is used to locate specific data after the data have been encrypted in order to address the issues of poor data recognition rate and effective resource consumption [77]. The results of the experiments demonstrate that the suggested algorithm successfully addresses the drawbacks of conventional approaches and can safeguard users' privacy and information security. We also talk briefly about some possible uses for our research [78]. To our knowledge, the work we've done here is unique in both its application and its outcomes.

The security of user trade secrets and private data has increasingly developed into a research hotspot in this sector with the application and development of network physical systems, and more specifically CPS. The use of information technology (IT) to secure network physical systems is not yet ideal, and the security encryption of private message data suffers from issues including sluggish data acquisition and low recognition rates. In order to solve these issues, this research develops a new SPI efficient encryption technique [79]. The procedure begins by examining user data from public, private, and mixed data sources. It can be inferred from this research which resources need to be encrypted and which data can be freely shared without encryption. The crucial idea is that since not all data fit under the scope of SPI, it is unnecessary to waste computing resources encrypting and decrypting them. Establishing a specific user data subset for sharing and the encryption structure is made easier with the use of data analysis. A method of adaptive data collecting is used to gather user data resources [80]. The analysis of the secure encryption method for SPI is finished by using the interference quantization-based data encryption technique.

The adaptive data collection approach is used in conjunction with the aforementioned definitions of SPI resources to gather SPI. Appropriate data collection tasks and a decision module are chosen to match the requirements of the tasks when instructions are delivered to identify the tasks for data collection. In order to combine the resources, the data processing module is simultaneously gathering privacy information. Assume there are q parameter collection jobs to be completed in order to obtain the data. The purpose of data gathering is to assign the q jobs in a fair and reasonable manner.

7 Experimental Results and Analysis

It is a good idea to conduct research on the encryption techniques that will be utilised on SPI to secure end-users' privacy. This has attracted the attention and focus of professionals and academics to this domain and produced useful exploration findings. Furthermore, like previously said, with SPI possibly working liberally in CPS-based networks, it is imperative to maintain the security of such information. This security may be supplied by some of the existing researches. The recently developed unique encryption algorithm aims to address the subsequent issues with the prevailing procedures:

- Priority to security in CPS
- Delays in data queries using conventional methods
- Slow data gathering rate
- a low rate of data recognition
- Operational resource utilisation

The approach evaluates user data resources to determine the variety of data involved and then encrypts the records in accordance with the analysis results. By encrypting just the data that requires more security and privacy, this successfully resolves the problem of data query delays brought on by traditional approaches in encoding huge bulks of data. After the data have been encrypted, the location of specific data is found using an interference quantification approach, which will be discussed later, to address the issue of poor data recognition rate and efficient resource consumption. The findings of the experiments demonstrate that the algorithms may successfully address the drawbacks of conventional approaches and can safeguard users' privacy and information security. The level of automation in traditional approaches is minimal and requires a lot of user effort while encrypting data. An adaptive data collecting approach is used to gather SPI in conjunction involves the evaluation of user-submitted personal data sources, which may increase the level of automation of information encryption.

7.1 Secure Encryption Algorithm for SPI

The use of information technology (IT) to secure well-connected physical systems is not yet ideal, and the security encryption standard of confidential message data suffers from issues including sluggish data acquisition and poor recognition rates. To solve these issues, the technique creates a novel, SPI-efficient encryption algorithm. The process begins by analysing the user data's public, private, and mixed data resources. This study provides insight into what resources must be encrypted and what information can be freely shared without encryption.

The crucial idea is that since not all data fit under the scope of SPI, it is unnecessary to waste computing resources encrypting and decrypting them. Establishing a specific

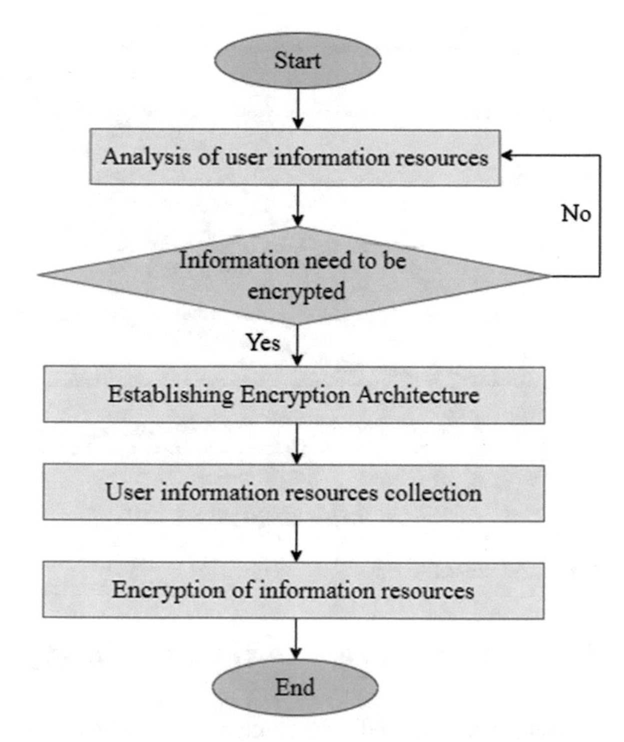

Fig. 3 The detailed Flow chart of Secured Encryption Algorithm

user data subset for distribution and the encryption mechanism is made easier with the use of data analysis. A method of adaptive data collecting is used to gather user data resources. The analysis of the secured encryption standard for SPI is finished by using the interference quantization-based data encryption technique. A detailed description of the overall flow of the process is shown in the below Fig. 3.

The dataset from Google Dataset Search has been chosen as the experimental data source [81]. One-stop dataset shopping is possible with the Google Dataset Search dataset, which includes vast amounts of information of all dimensions and sorted from different sources like NASA and Pro-Publica. The dataset offers a lot of practical usefulness because the data source is extensive.

MATLAB 8.0 was used to build a significant data resource experimental platform for interference quantification, and this platform was employed for data processing. To confirm the efficacy of our strategy, the suggested approach was compared with those of our peers from [82–86] using experimental indicators such as data acquirement period, data resource recognition interval, data inquiry latency, and active resource utilization. We re-ran each of the procedures [83–86] and related them to our approach. All simulations were created using the Matlab R2017b, and the processing and analysis detailed next were applied to them all.

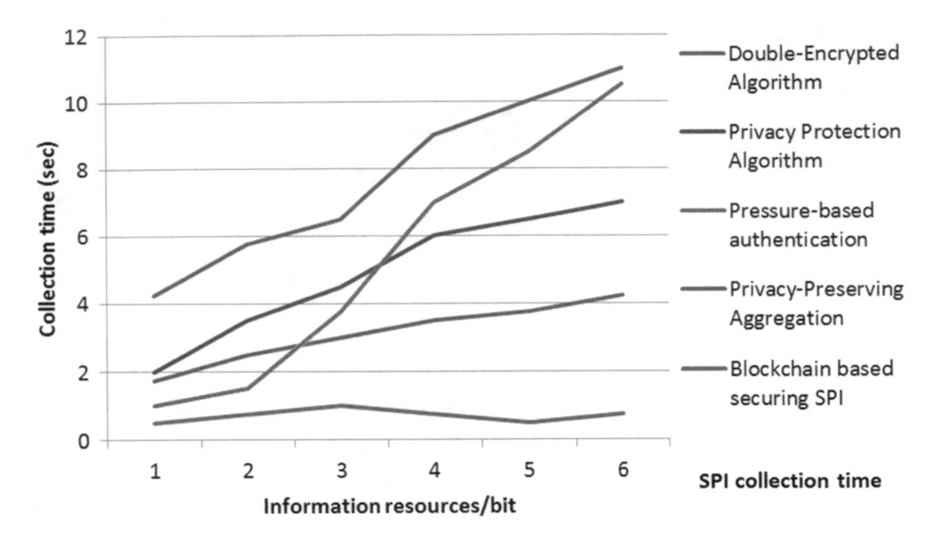

Fig. 4 SPI collection time

7.2 *Examination of Experimental Findings*

The suggested method's data collecting times are contrasted with those of comparable approaches in [83–86] in Fig. 4. We saw faster acquisition times and improved data acquisition efficiency in situations with the same amount of data. As a result, we utilized the time spent collecting data to confirm its effectiveness. Figure 4 displays the specific outcomes. One thing to keep in mind is that whereas most other approaches show a linear rise in collecting time as information resources grow, our method displays more of a steady relationship that remains constant as information resources increase.

The private data gathering times for the five approaches may be seen in Fig. 4 after analysis. The procedures in [83] take between 1.4 and 4.2 s to acquire data, whereas the approach in [84] takes between 2.1 and 6.7 s. The approach described in [85] has an acquisition time that ranges from 1.2 to 10.8 s, and the interval required to obtain private information is considerable. The technique in [86] acquirement times range from 4.5 to 10.8 s. The implemented algorithm's ability is credited to activate the assessment module whereas the data collecting job is being completed, which minimizes a significant amount of time period and satisfies the task criteria, to the production of this legitimate data. We again compared our algorithm to algorithms from [83–86] under various data resource conditions to confirm the approaches' accuracy in data recognition. The outcomes are displayed in Fig. 5.

When the resource capacity is 1×10^3 bits, the data acknowledgement rates for the techniques [83–86] are 69.3, 78.1, 37.8, and 36.6%, respectively, according to the analysis shown in Fig. 5. The blockchain-based methodology recognizes 92.4% of the data. The data identification rates of the algorithms from [83–86] are 59, 80, 62,

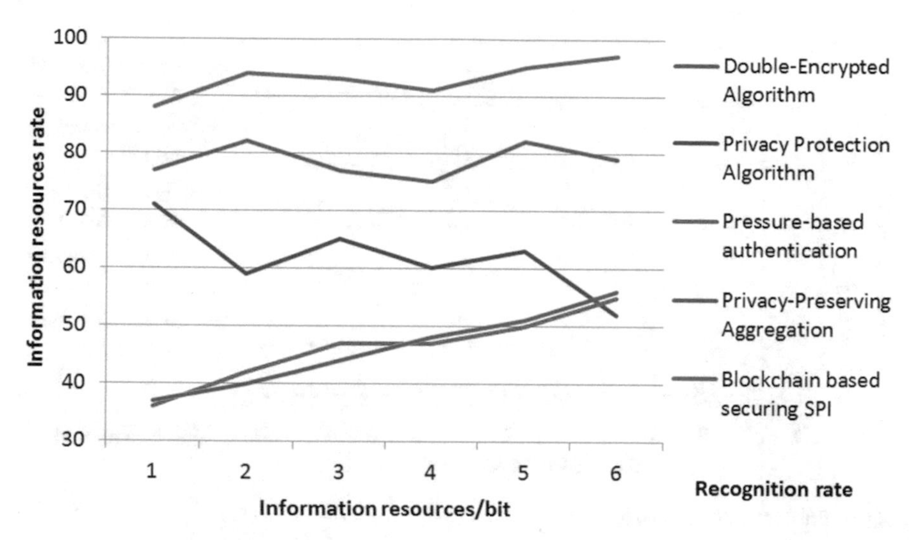

Fig. 5 Recognition rate of information resources

and 64%, respectively, when the resource capacity is 6×10^3 bit. Comparatively, our algorithm's data recognition rate is little over 90%. Figure 5's overall graph shows that the algorithm's data recognition rate is consistently the highest, demonstrating the high data recognition rate and superior recognition performance of our algorithm. This is explained by the requirement for traditional data encryption to resolve difficult non-convex optimization problems. Our approach, referred to as interference encryption, however, solves the issue in a different way. Auxiliary variables are added to the description to increase convenience while lowering interference item influence and increasing effective data recognition rate. One odd feature to take note of is the rate drop at 4×10^3 bit information resource quantities; nevertheless, after this point, as predicted, there is a little rate rise. This unexpected decline may be due to the algorithm's unique behavior with so much data. Figure 6 contrasts the seconds produced by our suggested algorithm's data resource searches with the delays in [83–86].

When the number of data resources rises, Fig. 6's analysis demonstrates that the query delays of the five approaches also increase. The SPI inquiry latencies of [83–86] are 10.75, 8.25, 5.35, and 18.0 s, respectively, when the confidential data utilized in the enquiry are 6×10^3 bit. The delay produced by our approach is also little longer than 1.0 s. The confidentiality data inquiry latencies from [83–86] are 24.75, 16.35, 11.65, and 26.0 s, respectively, when the quantity of confidential data utilized in the query reaches 10×10^3 bit. In contrast, our approach still introduces a latency of about 1.0 s. Even if the volume of confidential data utilized in a query were increased, a user would not perceive a difference in latency. This demonstrates unequivocally that our algorithm's data resource query latency is minimal, its query performance is superior, and it is further practical for big data storing applications. Across the whole range of information resource volumes, our algorithm outperformed all similar reference

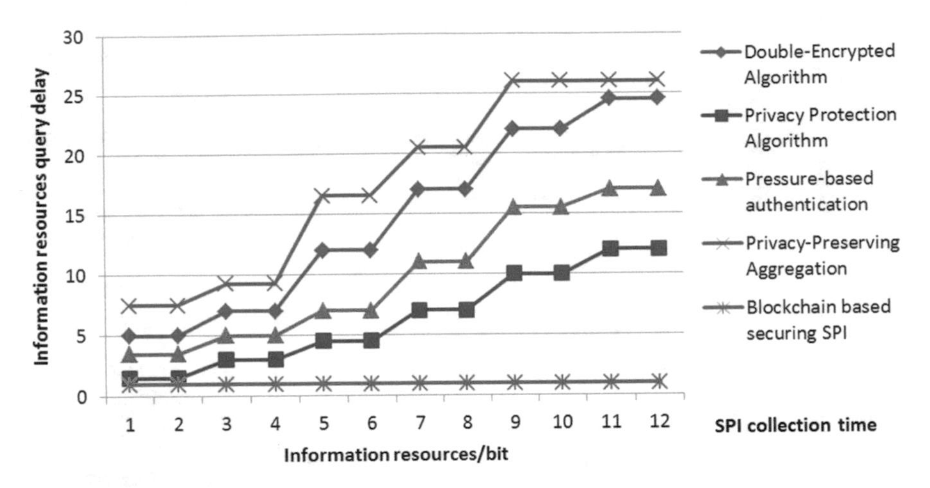

Fig. 6 Privacy information query latency

approaches. Table 1 compares the use rates of information resources (%) for our algorithm versus the approaches from [83–86].

Table 1's analysis reveals that the four approaches use data resources differently depending on the quantity of private data being processed. The resource usage rates of [83–86] are 69, 78, 68, and 85%, respectively, when the private data are 10×10^3 bit. In contrast, our suggested approach uses resources at 90% efficiency. The resource usage rates of the existing [83–86] are 62.1, 80.0, 72.4, and 80.6%, correspondingly, when the private data are 30×10^3 bit. Comparatively, our approach uses 97% of the available resources. Table 1 shows that our algorithm's resource usage rate consistently surpasses 90%, indicating good resource utilization regardless of the volume of private data.

Based on the aforementioned experimental findings, we can conclude that our algorithm may efficiently raise the rate at which data resources are recognized, decrease the delay brought on by inquiries about personal information and increased

Table 1 Utilization rates comparison of information resources

Privacy information ($\times 10^3$ bit)	Methods/algorithms				
	Double-encrypted algorithm	Privacy protection algorithm	Pressure-based authentication	Privacy-preserving aggregation	Blockchain based securing SPI
10	68	77	69	86	89
15	62	84	77	77	91
20	64	79	72	82	94
25	67	83	68	72	92
30	63	81	71	81	96

use of data resources. As a consequence, we can say that Blockchain based encryption method performs better overall than some of the present techniques from [83–86].

8 Future Research Directions

Despite their richness, blockchain promises still have a lot of security problems, as seen by the numerous documented breaches and scams [87]. For a system whose primary promoted features are reliability and inviolability, this is a paradoxical trap to slip into. The majority of computer attacks, on the other hand, target markets because they are susceptible to the well-known weaknesses of centralized-integrated systems, such banking sectors, government sectors, etc., [88]. In common, the problems with blockchain in terms of speed, bandwidth, confidentiality, expandability, operability and adoptability have been demonstrated and widely recorded. Many people disagree on mining concerns, which remain at the core of the PoW that underpins the bitcoin agreement. It is necessary to precisely define the circumstances which will defend against malevolent authentication nodes on a theoretical level [89, 90]. The benchmark level is unquestionably the 51% of computer power owned by a hostile organization. However, there is disagreement about this value among researchers. A crucial topic is the dispersion and number of nodes. Researchers are starting to become more active due to economic concerns [91].

Other blockchain restrictions are also cited, including the conflict between openness and secrecy and between anonymity and stakeholder identification. The registry is extensively circulated, so investors may quickly use the information it contains in plain English. This has advantages when it comes to transaction tracking, but it presents challenges while it arises to corporate secrecy, such as the finance, government, healthcare industries. While participating in events that call for the revelation of any information and the safety of others, the market searches for reliable methods of information concealment [92]. The substantial quantity of energy needed is another restriction. With blockchain requires a lot of self-important cryptography, confirmation, and authentication processes. This method may have large detrimental environmental externalities if it is extensively employed. In spite of the enormous curiosity in blockchain, distributed registries need to be understood to be a limited solution. Blockchain may not be appropriate for the quick processing of massive volumes of records, particularly visual and auditory data, or for application in rapidly changing contexts. Blockchain is the best option for keeping information that changes seldom over the long term and most reliably. The system is therefore potential for collecting client data from financial institutions, healthcare providers, and logistics firms [93]. Current technological capabilities are only a stopgap. Blockchain ongoing development creates opportunities for new sectors to use it. Since blockchain has already surpassed change-resistance, it will continue to advance; any technology must overcome this obstacle in order to advance.

9 Conclusion

A Cyber Physical System is vulnerable to cyber-physical assaults because networks integrate the computational system with the physical system, which might disrupt and result in breakdowns in a real-world system. Additionally, as networks become more connected, CPS will grow in size and complexity. As a result, demonstrating a complex CPS turn out to be challenging and incorrect when compared to a simpler one, decreasing the confidence level of traditional prototype-based cyber-physical encryption techniques. Adopting developing blockchain technology as well as machine learning techniques is important to ensure the security and dependability of big and complex CPSs and to raise the CPS's degree of security. This book chapter provided a thorough analysis of the risks to the CPS as well as attack-detection methods based on ML approaches.

The book chapter discussed cyber-physical assaults for every CPS levels, including intrusion employments, demonstrations, and analyses of various attacks from the viewpoint of physical systems. Additionally, several cyber-physical attack-detection tactics based on ML and blockchain technology as well as security assessment metrics of the ML-based strategies were explored. Commencing from the standpoint of real-time assault discovery and robust CPS scheme, it highlights the future research directions. Despite the limitations of ML and blockchain approaches, these research initiatives raise the security layer of the CPS. With the Internet's quick evolution, information transmission is becoming user-centered. In actuality, safe encryption of confidential data is a prerequisite for all records in the society of information proliferation and resource restructuring. The technologies now in use for privacy encryption of private data does not efficiently improve the security of data and cannot satisfy the various demands of end-users based on unique use trends.

Acknowledgements First and foremost, the authors express their gratitude to the Management, Principal, and Faculty members of Ramco Institute of Technology for supporting our productive work and providing us with the necessary tools. The authors also like to express their gratitude to their loved ones, coworkers, collaborators, and friends for their encouraging remarks during the entire process. The authors also express their sincere gratitude to God for his grace and assistance in making this episode a success.

Key Terms and Definitions

AI Artificial intelligence
BC Blockchain
CPA Chosen plain text
CPS Cyber-physical system
DDoS Distributed denial of service
DT Digital twin
DoS Denial of service

ICM	Integrity checking mechanism
IDS	Intrusion detection system
IoT	Internet of things
KPA	Known plain text
LSTM	Long short-term memory
MITM	Man-in-the-middle
ML	Machine learning
P2P	Peer-to-peer
PoW	Proof of work
SIEM	Security information and event management
SPI	Sensitive private information
SVM	Support vector machine

References

1. Zhou Y, Richard Yu F, Chen J, Kuo Y (2020) Cyber-physical-social systems: a state-of-the-art survey, challenges and opportunities. IEEE Commun Surv Tutor 22(Issue: 1)
2. Zhao C, Zhao S, Zhao M, Chen Z, Gao C-Z, Li H, Tan Y (2019) Secure multi-party computation: theory, practice and applications. Inf Sci 476:357–372
3. Porru S, Pinna A, Marchesi M, Tonelli R (2017) Blockchain-oriented software engineering: challenges and new directions. In: 2017 IEEE/ACM 39th international conference on software engineering companion (ICSE-C), (2017), pp 169–171. https://doi.org/10.1109/ICSE-C.2017.142
4. Teslya N, Ryabchikov I (2017) Blockchain-based platform architecture for industrial IoT. In: 2017 21st conference of open innovations association (FRUCT), pp 321–329. https://doi.org/10.23919/FRUCT.2017.8250199
5. Maleh Y, Shojafar M, Darwish A, Haqiq A (eds) (2019) Cybersecurity and privacy in cyber-physical systems. CRC Press. [Online]. https://www.crcpress.com/Cybersecurity-and-Privacy-in-Cyber-Physical-Systems/Maleh/p/book/9781138346673
6. Olowononi FO, Rawat DB, Liu C (2021) Resilient machine learning for networked cyber physical systems: a survey for machine learning security to securing machine learning for CPS. IEEE Commun Surv Tutor 23:524–552
7. Dorri A, Steger M, Kanhere SS, Jurdak R (2017) BlockChain: a distributed solution to automotive security and privacy. IEEE Commun Mag 55(12):119–125. https://doi.org/10.1109/MCOM.2017.1700879
8. Kim S, Won Y, Park IH, Eun Y, Park KJ (2019) Cyber-physical vulnerability analysis of communication-based train control. IEEE Internet Things J 6:6353–6362
9. Deng R, Zhuang P, Liang H (2017) CCPA: coordinated cyber-physical attacks and countermeasures in smart grid. IEEE Trans Smart Grid 8:2420–2430
10. Xu J, Wei L, Wu W, Wang A, Zhang Y, Zhou F (2018) Privacy-preserving data integrity verification by using lightweight streaming authenticated data structures for healthcare cyber-physical system. Future Gener Comput Syst (in press)
11. Shwartz O, Mathov Y, Bohadana M, Elovici Y, Oren Y (2018) Reverse engineering IoT devices: effective techniques and methods. IEEE Int Things J 5(6):4965–4976
12. Tsague HD, Twala B (2018) Practical techniques for securing the Internet of Things (IoT) against side channel attacks. Internet of things and big data analytics toward next-generation intelligence. Springer, Cham, pp 439–481

13. Yadav P, Feraudo A, Arief B, Shahandashti SF, Vassilakis VG (2020) Position paper: a systematic framework for categorizing IoT device fingerprinting mechanisms. In: Proceedings of the 2nd international workshop on challenges in artificial intelligence and machine learning for internet of things, pp 62–68
14. Banik S, Barooti K, Durak FB, Vaudenay S (2020) Cryptanalysis of LowMC instances using single plaintext/ciphertext pair. IACR Trans Symmetric Cryptol 130–146
15. Mallik A (2019) Man-in-the-middle-attack: understanding in simple words. Cyberspace. Jurnal Pendidikan Teknologi Informasi 2(2):109–134
16. Rughoobur P, Nagowah L (2017) A lightweight replay attack detection framework for battery depended IoT devices designed for healthcare. In: 2017 International conference on Infocom technologies and unmanned systems (trends and future directions) (ICTUS). IEEE, pp 811–817
17. Rajan A, Jithish J, Sankaran S (2017) Sybil attack in IOT: modelling and defenses. In: 2017 International conference on advances in computing, communications and informatics (ICACCI). IEEE, pp 2323–2327
18. Alani MM (2018) IoT lotto: utilizing IoT devices in brute-force attacks. In: Proceedings of the 6th international conference on information technology: IoT and smart city, pp 140–144
19. Kolias C, Kambourakis G, Stavrou A, Voas J (2017) DDoS in the IoT: Mirai and other botnets. Computer 50(7):80–84
20. Gupta A, Christie R, Manjula PR (2017) Scalability in internet of things: features, techniques and research challenges. Int J Comput Intell Res 13(7):1617–1627
21. Haji SH, Ameen SY (2021) Attack and anomaly detection in iot networks using machine learning techniques: a review. Asian J Res Comput Sci 30–46
22. Sahmi I, Mazri T, Hmina N (2018) Security study of different threats in Internet of Things. In: The proceedings of the third international conference on smart city applications. Springer, Cham, pp 785–791
23. Song F, Zhu M, Zhou Y, You I, Zhang H (2019) Smart collaborative tracking for ubiquitous power IoT in edge-cloud interplay domain. IEEE Int Things J 7(7):6046–6055
24. Elmalaki S, Ho BJ, Alzantot M, Shoukry Y, Srivastava M (2019) Spycon: adaptation based spyware in human-in-the-loop iot. In: 2019 IEEE security and privacy workshops (SPW). IEEE, pp 163–168
25. Arul E, Punidha A (2020) Adware attack detection on IoT devices using deep logistic regression SVM (DL-SVM-IoT). In: Congress on intelligent systems. Springer, Singapore, pp 167–176
26. Huseynov H, Kourai K, Saadawi T, Igbe O (2020) Virtual machine introspection for anomaly-based keylogger detection. In: 2020 IEEE 21st international conference on high performance switching and routing (HPSR). IEEE, pp 1–6
27. Magness JM (2020) SLIVer: Simulation-Based Logic Bomb identification/verification for unmanned aerial vehicles
28. Luo Z, Zhao S, Lu Z, Sagduyu YE, Xu J (2020) Adversarial machine learning based partial-model attack in IoT. In: Proceedings of the 2nd ACM workshop on wireless security and machine learning, pp 13–18
29. English KV, Obaidat I, Sridhar M (2019) Exploiting memory corruption vulnerabilities in connman for iot devices. In: 2019 49th annual IEEE/IFIP international conference on dependable systems and networks (DSN). IEEE, pp 247–255
30. Nirmal K, Janet B, Kumar R (2020) Analyzing and eliminating phishing threats in IoT, network and other web applications using iterative intersection. Peer-to-Peer Netw Appl 1–13
31. Kavi Priya S, Vignesh Saravanan K, Vijayalakshmi K (2019) Machine learning techniques to mitigate security attacks in IoT. Secur Priv Issues Sens Netw IoT 65
32. Srivastava G, Parizi RM, Dehghantanha A, Choo KKR (2019) Data sharing and privacy for patient iot devices using blockchain. In: International conference on smart city and informatization. Springer, Singapore, pp 334–348
33. Stiawan D, Idris M, Malik RF, Nurmaini S, Alsharif N, Budiarto R (2019) Investigating brute force attack patterns in IoT network. J Electr Comput Eng
34. Damghani H, Damghani L, Hosseinian H, Sharifi R (2019) Classification of attacks on IoT. In: 4th international conference on combinatorics, cryptography, computer science and computation

35. Kwon S, Park S, Cho H, Park Y, Kim D, Yim K (2021) Towards 5G-based IoT security analysis against Vo5G eavesdropping. Computing 103(3):425–447
36. Pundir S, Wazid M, Singh DP, Das AK, Rodrigues JPC, Park Y (2020) Designing efficient sinkhole attack detection mechanism in edge-based IoT deployment. Sensors 20(5):1300
37. Rajashree S, Soman KS, Shah PG (2018) Security with IP address assignment and spoofing for smart IOT devices. In: 2018 international conference on advances in computing, communications and informatics (ICACCI). IEEE, pp 1914–1918
38. Siddiqui ST, Alam S, Ahmad R, Shuaib M (2020) Security threats, attacks, and possible countermeasures in internet of things. In: Advances in data and information sciences. Springer, Singapore, pp 35–46
39. Bettayeb M, Nasir Q, Talib MA (2019) Firmware update attacks and security for IoT devices: survey. In: Proceedings of the Arab-WIC 6th annual international conference research track, pp 1–6
40. Doshi R, Apthorpe N, Feamster N (2018) Machine learning ddos detection for consumer internet of things devices. In: 2018 IEEE security and privacy workshops (SPW). IEEE, pp 29–35
41. Ahmed A, Latif R, Latif S, Abbas H, Khan FA (2018) Malicious insiders attack in IoT based multi-cloud e-healthcare environment: a systematic literature review. Multimed Tools Appl 77(17):21947–21965
42. Uwagbole SO, Buchanan WJ, Fan L (2017) An applied patterndriven corpus to predictive analytics in mitigating SQL injection attack. In: 2017 seventh international conference on emerging security technologies (EST). IEEE, pp 12–17
43. Vojkovíc G, Milenkovíc M, Katulíc T (2019) IoT and smart home data breach risks from the perspective of Croatian data protection and information security law. In: Proceedings of the ENTRENOVA—ENTerprise Research InNOVAtion Conference (Online), vol 5, no 1, pp 253–263
44. Chung H, Li W, Yuen C, Chung W, Zhang Y, Wen C (2019) Local cyber-physical attack for masking line outage and topology attack in smart grid. IEEE Trans Smart Grid 10:4577–4588
45. Gacia L, Brasser F, Cintuglu M, Sadeghi A (2017) Hey, my malware knows physics attacking PLCs with physical model aware rootkit. In: Proceedings of the network & distributed system security symposium, San Diego, CA, USA, 16–17 February 2017, pp 1–15
46. Tao H, Bhuiyan MZA, Rahman MA, Wang T, Wu J, Salih SQ, Li Y, Hayajneh T (2019) Trustdata: Trustworthy and secured data collection for event detection in industrial cyber-physical system. IEEE Trans Ind Inform 1–1
47. Mistry I, Tanwar S, Tyagi S, Kumar N (2020) Blockchain for 5G-enabled IoT for industrial automation: a systematic review, solutions, and challenges. Mech Syst Signal Process 135:106382
48. Tutorialspoint.com, Artificial intelligence overview
49. Chu R (2018) What is AI? A brief explanation for layman
50. AlayÃşn D, AlayÃşn D (2018) Understanding artificial intelligence
51. Garware B (2019) Privacy-preserving ai (private ai) âĂŞ the rise of federated learning. https://www.persistent.com/blogs/privacy-preserving-aiprivate-ai-the-rise-of-federated-learning/. Accessed 2019
52. McMahan B, Moore E, Ramage D, Hampson S, y Arcas BA (2017) Communication-efficient learning of deep networks from decentralized data. In: Proceedings of the 20th international conference on artificial intelligence and statistics, AISTATS 2017, 20–22 April 2017, Fort Lauderdale, FL, USA, pp 1273–1282
53. Soria-Comas J, Domingo-Ferrer J, Sánchez D, Megías D (2017) Individual differential privacy: a utility-preserving formulation of differential privacy guarantees. IEEE Trans Inf Forens Secur 12:1418–1429
54. Team TO (2018) Towards an open-source secure enclave. https://medium.com/oasislabs/towards-an-open-source-secure-enclave-659ac27b871a. Accessed 2018
55. Marwala T, Xing B (2018) Blockchain and artificial intelligence. arXiv e-prints: arXiv:1802.04451

56. Almasoud AS, Eljazzar MM, Hussain FK (2018) Toward a self-learned smart contracts. CoRR, vol abs/1812.10485
57. Dinh TN, Thai MT (2018) Ai and blockchain: a disruptive integration. Computer 51:48–53
58. Salah K, Rehman MHU, Nizamuddin N, Al-Fuqaha A (2019) Blockchain for ai: review and open research challenges. IEEE Access 7:10127–10149
59. Chen Z, Wang W, Yan X, Tian J Cortex-ai on blockchain
60. Dietz M, Vielberth M, Pernul G (2020) Integrating digital twin security simulations in the security operations center. In: Proceedings of the 15th international conference on availability, reliability and security, ser. ARES'20. Association for Computing Machinery, New York, NY, USA. [Online]. https://doi.org/10.1145/3407023.3407039
61. Kim S, Eun Y, Park KJ (2021) Stealthy sensor attack detection and real-time performance recovery for resilient CPS. IEEE Trans Ind Inform
62. Suhail S, Hussain R, Jurdak R, Oracevic A, Salah K, Matulevičius R, Hong CS (2021) Blockchain-based digital twins: Research trends, issues, and future challenges, 2021, arXiv: 1709.10000. [Online]. https://arxiv.org/abs/2103.11585
63. Suhail S, Hussain R, Jurdak R, Hong CS (2021) Trustworthy digital twins in the industrial internet of things with blockchain. IEEE Internet Comput 1–10. [Online]. https://doi.org/10.1109/MIC.2021.3059320
64. Dedeoglu V, Jurdak R, Putra GD, Dorri A, Kanhere SS (2019) A trust architecture for blockchain in iot. In: Proceedings of the 16th EAI international conference on mobile and ubiquitous systems: computing, networking and services, ser. MobiQuitous'19. Association for Computing Machinery, New York, NY, USA, pp 190–199. [Online]. https://doi.org/10.1145/3360774.3360822
65. Eckhart M, Ekelhart A (2018) Towards security-aware virtual environments for digital twins. In: Proceedings of the 4th ACM workshop on cyber-physical system security, ser. CPSS'18. Association for Computing Machinery, New York, NY, USA, pp 61–72
66. Bitton R, Gluck T, Stan O, Inokuchi M, Ohta Y, Yamada Y, Yagyu T, Elovici Y, Shabtai A (2018) Deriving a cost-effective digital twin of an ICS to facilitate security evaluation. In: European symposium on research in computer security. Springer, Cham, pp 533–554
67. Rasheed A, San O, Kvamsdal T (2020) Digital twin: values, challenges and enablers from a modeling perspective. IEEE Access 8:21980–22012
68. Groshev M, Guimarães C, Martín-Pérez J, de la Oliva A (2021) Toward intelligent cyber-physical systems: digital twin meets artificial intelligence. IEEE Commun Mag 59(8):14–20
69. Daza V, Di Pietro R, Klimek I, Signorini M (2017) CONNECT: CONtextual NamE disCovery for blockchain-based services in the IoT. In: 2017 IEEE international conference on communications (ICC), (2017), pp 1–6. https://doi.org/10.1109/ICC.2017.7996641
70. Gómez ÁLP, Maimó LF, Celdran AH, Clemente FJG, Sarmiento CC, Masa CJDC, Nistal RM (2019) On the generation of anomaly detection datasets in industrial control systems. IEEE Access 7:177460–177473
71. Li J, Xia Y, Qi X, Gao Z (2017) On the necessity, scheme, and basis of the linear–nonlinear switching in active disturbance rejection control. IEEE Trans Ind Electron 64:1425–1435
72. Giraldo J, Urbina D, Cardenas A, Valente J, Faisal M, Ruths J, Tippenhauer NO, Sandberg H, Candell R (2018) A survey of physics-based attack detection in cyber-physical systems. ACM Comput Surv (CSUR) 51:1–36
73. Manandhar K, Cao X, Hu F, Liu Y (2014) Detection of faults and attacks including false data injection attack in smart grid using Kalman filter. IEEE Trans Control Netw Syst 1:370–379
74. Cai F, Koutsoukos X (2020) Real-time out-of-distribution detection in learning-enabled cyber-physical systems. In: Proceedings of the ACM/IEEE international conference on cyber-physical systems (ICCPS), Sydney, NSW, Australia, 21–25 April 2020, pp 174–183
75. Ravikumar G, Govindarasu M (2020) Anomaly detection and mitigation for wide-area damping control using machine learning. IEEE Trans Smart Grid
76. Bak S, Chivukula DK, Adekunle O, Sun M, Caccamo M, Sha L (2009) The system-level simplex architecture for improved realtime embedded system safety. In: Proceedings of the IEEE real-time and embedded technology and applications symposium, San Francisco, CA, USA, 13–16 April 2009, pp 99–107

77. Kashyap S, Bhushan B, Kumar A, Nand P (2022) Quantum blockchain approach for security enhancement in cyberworld. In: Kumar R, Sharma R, Pattnaik PK (eds) Multimedia technologies in the internet of things environment, vol 3. Studies in big data, vol 108. Springer, Singapore. https://doi.org/10.1007/978-981-19-0924-5_1
78. Haque AKMB, Bhushan B, Hasan M, Zihad MM (2022) Revolutionizing the industrial internet of things using blockchain: an unified approach. In: Balas VE, Solanki VK, Kumar R (eds) Recent advances in internet of things and machine learning. Intelligent systems reference library, vol 215. Springer, Cham. https://doi.org/10.1007/978-3-030-90119-6_5
79. Bhushan B, Sinha P, Martin Sagayam K, Andrew J (2021) Untangling blockchain technology: a survey on state of the art, security threats, privacy services, applications and future research directions. Comput Electr Eng 90:106897. ISSN 0045-7906.https://doi.org/10.1016/j.compeleceng.2020.106897
80. Saxena S, Bhushan B, Ahad MA (2021) Blockchain based solutions to secure IoT: background, integration trends and a way forward. J Netw Comput Appl 181:103050. ISSN 1084-8045. https://doi.org/10.1016/j.jnca.2021.103050
81. Google Search Datasets. https://developers.google.com/search/docs/data-types/dataset. Accessed 11 Sep 2019
82. Venkatesh R, Vignesh Saravanan K, Aswin VR, Balaji S, Amudhan K, Rajakarunakaran S (2022) Detection of cracks in surfaces and materials using convolutional neural network. In: Marriwala N, Tripathi C, Jain S, Kumar D (eds) Mobile radio communications and 5G networks. Lecture notes in networks and systems, vol 339. Springer, Singapore. https://doi.org/10.1007/978-981-16-7018-3_18
83. Zhang CL, Xiong L, Lu LC (2018) Simulation of double-encrypted reversible concealment algorithm for real-time network information. Comput Simul 35:201–204+268 (In Chinese)
84. Solomon M, Elias EP (2018) Privacy protection for wireless medical sensor data. Int J Sci Res Sci Technol 4:1439–1440
85. Zhang K, Douros K, Li H, Li H, Wei Y (2015) Systems and methods for pressure-based authentication of an input on a touch screen. U.S. Patent 8,988,191, 24 March 2015
86. Qian J, Qiu F, Wu F (2016) Privacy-preserving selective aggregation of online user behavior data. IEEE Trans Comput 66:326–338
87. Sakhnini J, Karimipour H, Dehghantanha A, Parizi RM, Srivastava G (2019) Security aspects of internet of things aided smart grids: a bibliometric survey. Internet Things 100111
88. Tasatanattakool P, Techapanupreeda C (2018) Blockchain: challenges and applications. In: 2018 international conference on information networking (ICOIN), pp 473–475. https://doi.org/10.1109/ICOIN.2018.8343163
89. Malathy N, Priya SK, Saravanan KV (2022) Pedestrian safety system with crash prediction. Int J Health Sci 6(S2):8707–8717. https://doi.org/10.53730/ijhs.v6nS2.7247
90. Saravanan KV, Priya SK, Thilaga PJ, Vijayalakshmi K, Vikashini S (2022) Mule: multiclass email classification for forensic analysis using deep learning. Telematique 21(1):4670–4686
91. Chang V, Baudier P, Zhang H, Xu Q, Zhang J, Arami M (2020) How blockchain can impact financial services—the overview, challenges and recommendations from expert interviewees. Technol Forecast Soc Change 158:120166. https://doi.org/10.1016/j.techfore.2020.120166
92. HashemiJoo M, Nishikawa Y, Dandapani K (2020) Cryptocurrency, a successful application of blockchain technology. Manag Financ 46(6):715–733. https://doi.org/10.1108/MF-09-2018-0451
93. Kolb J, AbdelBaky M, Katz RH, Culler DE (2020) Core concepts, challenges, and future directions in blockchain: a centralized tutorial. ACM Comput Surv 53(1):1–39. https://doi.org/10.1145/3366370

A Study of Smart Evolution on AI-Based Cyber-Physical System Using Blockchain Techniques

Asmita Biswas, Koustav Kumar Mondal, and Deepsubhra Guha Roy

Abstract The Internet of things emerges in the world's smart evolution through Artificial Intelligence (AI) and Blockchain Technology in Cyber-Physical Systems, which increases the exponential rate of smart manufacturing devices. As more industrial devices are getting connected day by day. Obtaining that massive amount of data transfer easy efficiency and data centralization is much more difficult to process by any human. Blockchain technology has proposed AI-based Cyber-Physical Systems (CPS) for the industry, conferring a secure and efficient financial transaction interoperability, communication security, and processing of a large amount of data. However, there is a wide deployment of industrial compatibility through the Internet of Things (IoT), but this technology will benefit the industrial devices to maintain, predict and awareness on their own. On the one hand, Blockchain Technology advances the device's authentication security enhancement; on the other hand, AI will provide adaptive learning toward cyber attacks during low batteries. Blockchain will also provide load balancing of Edge devices. We explained Blockchain implementation to fabricate a particular AI-based hybrid medical device and Smart manufacturing based on the proposed design. AI authentication will leverage the features of the security systems, and Blockchain Technology will validate the data privacy and Secure data transfer. With Smart monitoring and supervision sub-systems, the CPS can decrease expert technicians' needs and diminish manufacturing limits during implantation. Therefore, CPS integration could also lower the cost of fabrication. This chapter discussed the integration of Blockchain Technology and AI in a cyber-physical

A. Biswas
Institution's Innovation Council, Innovation and Entrepreneur Development Center, Institute of Engineering & Management, Saltlake, Kolkata, West Bengal 700091, India

K. K. Mondal
Inter-Disciplinary Research Division—IoT and applications, Indian Institute of Technology Jodhpur, Jodhpur, Rajasthan, India

D. Guha Roy (✉)
Department of CSE (AIML) CSBS, Institute of Engineering & Management, Saltlake, Kolkata 700091, India
e-mail: roysubhraguha@gmail.com

B. Bhushan et al. (eds.), *AI Models for Blockchain-Based Intelligent Networks in IoT Systems*, Engineering Cyber-Physical Systems and Critical Infrastructures 6,
https://doi.org/10.1007/978-3-031-31952-5_14

system. The integration of CPS in Smart Manufacturing and Medical Devices is well explained. The challenges for CPS are also described in this chapter.

Keywords Internet of Things (IoT) · Cyber-Physical System (CPS) · Artificial Intelligence (AI) · Blockchain Technology

1 Introduction

A Cyber-physical system (CPS) describes as a directed method that relies on some tight synthesis of cyber objects (e.g., connection, computation, and direction) and material gadgets (original and human-made fabrications managed by the rules of physics). Cyber-Physical System is comprehended to obtain the IoT's grown-up generation, developing the IoT interpretation and vision. These are combined into a closed loop, allowing purpose structures and explaining every element of network composite methods robustly linked between users and the Internet and controlled and operated by processor algorithms. Software and hardware systems are interconnected and regularly perform specific time and location-based computations. IoT (by actuators and sensors) authorizes communication between the physical and virtual worlds [1]. The CPS forms a significant quantity of tools, considering one modern model. Its notion includes several interconnected machines relying on actuators, sensors, portable devices, and RFID tags.

The Internet of Things (IoT) proposes creating Smart projects nearby us to transform our idea of living now and executing everyday jobs. Every day, Smart transport, smartphones, and smart watches connect to the Internet, so valuable data can transfer over any interface anywhere and anytime [2]. Traditional wireless technology must satisfy CPS's latency, speed, and performance specifications, including expanding CPS devices. Hence, including the fifth-generation (5G) network enhances availability toward the [3] IoT utilization' increase. Its essential performance standards involve ultra-low latency, security, connectivity, and wireless communication coverage due to many CPS devices' massive contacts. Several technologies are applied to achieve these challenges, such as Blockchain, to satisfy the requirement for abundant wireless service providers for simulating new economic and social development [4]. However, the security risks are becoming ever more moderate [4]. Everything implies that the CPS-IoT has trouble estimating interaction-related vulnerabilities and threats, and further security concerns appear. Digital technologies have usually been verified to improve productivity and more inexpensive costs in production systems. Their appearance is becoming vital in on-demand businesses such as manufacturing and medical devices. Such actions should begin containing Smart systems, including dynamic and integrated digital technologies [5], such as incorporating Blockchain into (CPS) Cyber-Physical Systems, as introduced in this chapter.

CPS builts up cooperating computational elements that communicate among those enclosing the physical world and one, added as an element of the manufacturing industry change [6]. In other words, CPS gets networks with physical and cyber systems composed, measured, managed, and combined by a communicating and computing core [6, 7] the physical context, and humans. Nevertheless, the physical environment does not connect explicitly among different elements (i.e., cyber), inter-communications among the physical and cyber domains, are created possible with actuators' and sensors' guidance [8, 9]. Unlike traditional methods, people have now been acknowledged as an independent and essential component within cyber-physical systems. Numerous physical processes usually demand human interference; such methods regard humans as more identified as human-in-the-loop CPSs [10]. Therefore, there is a requirement to develop those methods to provide standard equipment processes like smart manufacturing devices and medical devices.

A few unresolved concerns, such as security and dependability, which are crucial for cyber-physical systems, have been debated by the design and development of several sectors, including healthcare and smart manufacturing cyber-physical systems. Our proposed approach consists of encryption techniques and Blockchain technology for secure cloud storage-based cyber-physical systems. And for the medical equipment developed by smart manufacturing systems are highly developed CPS with a sub-system that exhibits self-control and awareness properties and may be configured to self-optimize the different real-time manufacturing processes. Malware identification is a critical responsibility for the efficient and smooth operation of cyber-physical systems, and Blockchain integration into those systems helps. Medicines, medical devices, patients, and diseases are just a few healthcare-related topics that might benefit from AI applications. Together, academics, researchers, and healthcare professionals are sifting through vast datasets of patient data to uncover hidden patterns and important information. These sizable datasets are also used to train AI models that categorize patients, aid in the early diagnosis of diseases, and support medical professionals.

2 Cyber Physical System Background

A cyber-physical system (CPS) is a network system in which a device gets observed or controlled via computational algorithms. Dynamic and software elements are intensely tangled, work on various temporal and spatial scales, and manifest different and well-defined interactions among each other in directions that improve, including the context in cyber-physical systems [11]. CPS requires trans-disciplinary strategies, joining the mechatronics theory, cybernetics, form, and method science [13]. Method control is usually assigned to embedded methods. The importance directly impacts the computational components and is more limited to a unique link within the embedded systems' computational and physical aspects. Although CPS performs more eminent coordination and combination within computational and physical components, sharing the related basic design makes CPS comparable to the Internet of Things [14]. Autonomous vehicle systems, smart grids, automated control systems, pharmaceutical monitoring, robotics systems, and mechanical pilot services are all

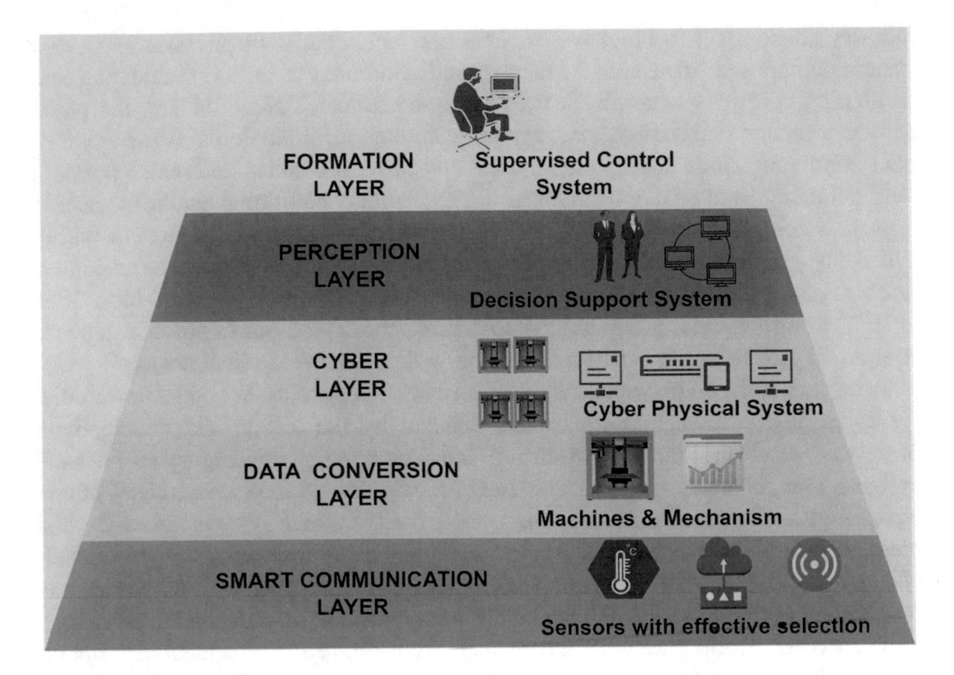

Fig. 1 5 Layer scheme of Cyber-physical systems

included in CPS samples [15]. Ancestors of cyber-physical methods may get located in sectors as varied as manufacturing, automotive, aerospace, chemical processes, energy, civil infrastructure, healthcare, entertainment, transportation, and consumer appliances [15].

The Fig. 1 of the design of the Cyber-Physical System consists of 5 layers of it. Those layers have some attributes. The first layer is the smart communication layer consisting of the sensor network, tether-free communication, and plug-and-play characteristics. The second layer, the information conversion layer, consists of multi-dimension data correlating Smart analytics for machines' components. The third layer is the cyber layer clusters for similar data mining, and this layer is the time machine variation identification and memory. The fourth layer, the perception layer, integrates synthesis, simulation, and remote visualization to make decisions and collaborate diagnostics. The last layer is the formation layer. It has self-optimization, self-adjustment, and self-configuration for disturbance, variation, and flexibility.

2.1 What is Cyber-Physical System

The CPS are systems of Interacting and Computational objects that are in intense link-ing with the encompassing physical environment and its ongoing methods, presenting and applying, by equal time, Data processing and Data-accessing services obtainable proceeding the Internet [16–18]. In other concepts, CPS may describe dynamic systems in which services are regulated, controlled, monitored, and integrated through a communicating and computing core [19]. Communication between the cyber and the physical factors is essential to the crossing, but not the cyber and the physical association. It needs to be more adequate to distinctly recognize the computational parts and the physical elements. We need to understand those intercommunications [21]. CPS is incredibly effective at enhancing all facets of development. Among the ideas created in the past and are now in use are robotic surgery, autonomous cars, intelligent buildings, intelligent power grids, smart manufacturing, and implanted medical devices [18]. Most maximum researchers tend to the roots of Cyber-Physical Systems to im-planted systems [22], described as a computational method inside any automated or electrical operation intended for performing assigned particular purposes by real-time computing limitations.

Robust synthesis and coordination within computation- and environment-based methods represent those implanted operations. In CPS, according to this interpretation, several implanted machines get the interface to monitor, sense, and actuate natural components in this physical world. The analysis lead on cyber-physical systems (CPS) explores new experimental frameworks and technologies to enable speedy and secure expansion and the integration of machine, data-centric physical systems. The lead aims to the user in a different span of material methods that remain incredibly reliable, efficiently manufactured, and proficient in excellent data, communication, computation, and control execution. Regional manipulation and a sense of the dynamic world takes place, and careful, secure, accurate, and authentic adaptation is mounted over a virtual interface. This ability is attributed to Regionally Physical, Globally Virtual [23]. The PCAST (President's Council of Advisors on Science and Technology) emphasizes the absolute magnitude of CPS for manufacturing competitiveness. It is working on a specific estimation of the NITRD (Federal Networking and Information Technology Research and Development). PCAST terminated that the Federal NITRD details need to be confirmed and rebalanced so that the field of cyber-physical systems gets operated as a prime preference problem for federal analysis expenditures [24]. CPS gets into its most eminent priorities and frequently inducts investigation plans by the NSF (National Science Foundation) [25]. The prime applications of CPS, are listed below as follows and the declaration published by the Information Technology Research and Development (NITRD) CPS and federal Networking [25].

- Industry and Smart Manufacturing
- Healthcare,
- Defense,
- Society
- Agriculture,
- Energy,
- Power Response,
- Building Controls, and
- Transportation.

Crosscutting difficulties also signified which are required to succeed in all divisions in the corresponding purpose statement:

- o Social-technical perspectives of CPS.
- o Interoperation
- o Cyber-security,
- o Reliability and Protection,
- o Privacy,
- o Economics.

The (acatech) National Academy of Science and Engineering has performed a foremost part in developing CPS in Germany [16, 17]. CPS capability levels are interpreted following framework basics, formulating clarity, building opinion, promoting choice-making, and self-optimization. CPS, created inside these first layers, while the architectural and organizational conditions perform the four higher layers factor the achievements' development concerning data and experience processing, assistance, and collaboration.

2.2 *Background of Internet of Things Versus Cyber-Physical System*

The terms IoT (Internet of Things) and CPS (Cyber-Physical System) uses reciprocally during the vast, decrepit domain of overlapping. It gets noted that Educational Organizations favor the CPS, whereas Administration Companies and Industries favor the IoT most. Nevertheless, many researchers also apply IoT in Big Data Analytics. The Internet of Things is the interaction network correlating things with Sensing, Naming, and Processing techniques. IoT allows loosely linked Decentralizing operations of interacting Smart Objects, functioning as Smart operators which can interchange data with the users through broadcast media like Wireless Sensor Networks. Interconnections between such Smart Objects (e.g., machines, sensors, actuators, in-stalled processors, and RFID tags) are under every case relied on Standard Interface Protocols (e.g., RFID, Bluetooth, ZigBee, and 6LoWPAN) [26].

On a different note, the Cyber-Physical System is relevant to real-time operations, including classified real-time essential methods that unify communication and computing skills for monitoring and controlling items in a dynamic environment. CPS also illustrates monitoring systems by real-time abilities and categorizes systems with minor human interaction. The Cyber-Physical System is the subsequent generation of Secured Smart Data-Communications technology, collaborative and interdependent. It implements the physical elements methods in different appliances such as manufacturing, healthcare, energy, and transportation with communication and computation. Furthermore, CPS is obtaining real-time information accumulation and decision-making with feedback in various domains, including robotic biopsy, nano-level production, military battle, air-trac controller, and fire-fighting.

3 Blockchain Integration in Cyber-Physical System

Blockchain is a distributed ledger simulated and intercommunicated among Peer-to-Peer (P2P) members. Recently, the Blockchain concept has attracted much attention in distributed technologies, such as IoT, because it increases system fault tolerance, enhances security and privacy, provides faster reconciliation and settlement, constructs a scalable network, and removes intermediaries to help save costs and time.

For example, The transactions of various medical equipment, vaccinations, and patient data may all be stored in such immutable frameworks. Therefore, this technology can have many blockchains for healthcare. The Blockchain of pharmaceuticals, coronavirus patients, and hospital immunizations. By keeping the transactions secure and trackable, the IBM [27] blockchain assists in the transparent distribution of the coronavirus vaccination. With a QR code on the pharmaceutical, QuillTrace [28], a blockchain-based technology, helps monitor medications along the supply chain and spot phony medications.

Numerous sectors have adopted Blockchain Technology, including Machine-to-Machine (M2M) communication, decentralized data sharing in healthcare applications, electrical grid systems, food supply chain traceability, financial services for the banking sector, and decentralized logistics operations. However, more research must be conducted on applying Blockchain Technology to manufacturing systems. A distributed ledger needs certain critical features to be used successfully in IoT-CPS applications, including the type of network access (permitted or unpermitted), the capacity to execute orders (such as Smart Contracts) automatically, energy consumption, and the type of network (public or private). According to a survey, Ethereum, Hyperledger Fabric, and IOTA are the three distributed ledgers best suited for IoT-CPS applications. The following are some crucial features that the structural architecture of blockchain technology can naturally produce:

1. Decentralized and Distributed System: There is no individual decision-making and transfer control from any centralized system, and the system is distributed in various organizations and locations.
2. Transparency: Each system has a manuscript of the ledger in the network. 3. Immutability: Hash functions cannot be changed or modified.
3. Authentication: Every party has real-time, accurate, consistent, and complete data in the transaction.
4. No Intermediates: Self-executable algorithms (e.g., Smart Contract).

4 Architecture

4.1 Physical Layer

The Physical Layer consists of devices and various devices such as RFID sensors. Those are in charge of capturing real-time data (e.g., temperature, vibration) during the formation process. It also maintains the subsequent data processing and model improvement with the sensed information. The primary manufacturing system devices can detect these devices' operating conditions and status, informant by configuration and identify the manufacturing method, and sustain stable composition. For instance, when the device temperature is unusual, the connected temperature sensors transmit a warning message to get the related device support and design arrangement to avoid enormous damage to the organization, explained in Fig. 2.

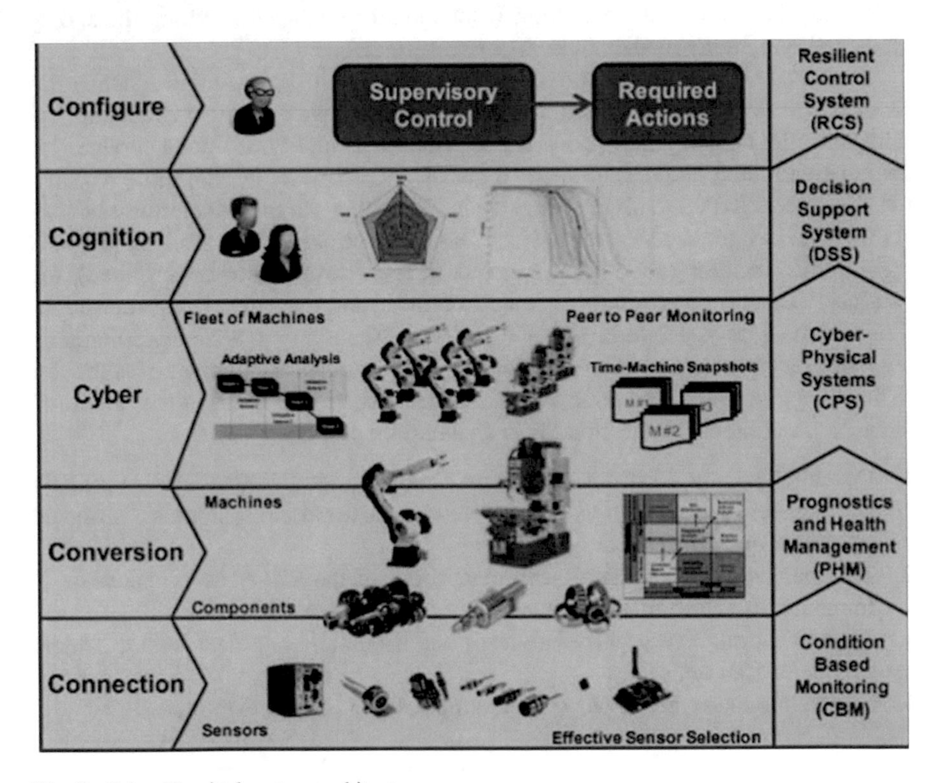

Fig. 2 Cyber-Physical system architecture

4.2 Cyber Layer

This layer is in charge of converting data into meaningful data and overseeing cyber-cyber and cyber-physical relationships to implement fault tolerance, integrity, and solidity. Cyber security, Cloud Computing, Big Data, privacy, network connectivity, and transparency are practical issues at this layer [30]. To enhance system stability, computing speed, and network scalability, as well as efficiently manage idle resources, it is common practice to use various strategies, including Grid and Cloud Computing. By utilizing these strategies, the other computers on the network can share the computing and storage constraints. By enabling distributed data access via a (P2P) peer-to-peer network, providing dispersed data security, and providing dispersed data storage. For the development of distributed computing, a Blockchain system might be essential.

AI tool integration is essential to turn unstructured data into knowledge and give individual Nodes access to intelligence in this layer. For today's manufacturing systems, a network with AI capabilities is necessary [31] and Distributed, and Decentralized A [32] will function better than centralized cloud computing platforms [33]. Distributed learning, retraining, and knowledge coordination are made possible for AI systems by Blockchain Technology. As Distributed and Decentralized AI modules are trained on more reliable and global data, their robustness and reliability increase; their credibility increases due to immediate feedback from operations, and implementation costs are significantly decreased due to peer-to-peer resource sharing and automatic machine learning.

4.3 System Layer

The System Layer introduces an application service assisted through model improvement and data processing. Manufacturing process identification is the foremost concern in this research. It can be executed by analyzing the output predictive of the Blockchain paradigm generated and the identical data saved in the database. Furthermore, any execution developments are observed in the improved manufacturing system identification appearance that can evaluate through three KPIs (Key Performance Indicators), including examining efficiency, training samples, and training time. Consequently, the Smart manufacturing method's proactive controlling method is extra accurate for assisting enterprise systems' execution because more certain data is proposed. Simultaneously, the enterprise database's workload is decreased during the compressed computation produced by more limited training time and several training samples.

5 Implementation

Industries 4.0 reaches a unique structure and maintains entire value-adding methods. The fundamental aspiration is to satisfy particular consumer demands at mass production expenses. It improves every area of system supervision, analysis, improvement, production, transportation to the value, and manufactured goods recycling—the framework for new possibilities in the digital generation, including Cyber-Physical Systems' guidance. Hence complete associated devices are similar mechanics, goods, resources, and schemes combined as self-organized, Smart, real-time, and independently optimized examples. The aimed CPS notion also applies to obtaining the fabrication of medical devices to explain their practicability and efficiency. CPS can utilize production preparation for easy assembling and being a reliable supervision mechanism to experience the supervisors. Implementing the proposed CPS with Blockchain develops a Smart manufacturing system more efficient and secure transaction network in the industry. The technology can recognize from the implementation in CPS. Therefore, the versatility of the CPS is prominently bound to the controlled circumstances. When the algorithms' reconfiguration to provide modifications in the controlled circumstances is available.

5.1 CPS-Based Smart Manufacturing

Technology's rapid development has driven Smart Manufacturing towards the CPS (Cyber-Physical Systems), representing a profoundly interactive method between the physical and the cyber computation layers. As mentioned earlier, the five-layer architecture of CPS presents manufacturers with a guideline to implement CPS [34]. Implemented CPS technology to improve a proactive element approach to solving the lengthy expected time and the useless power consumption during the logistics affected by the restriction of buyers for all machines [35]. Introduced a self-adaptive control CPS mechanism, developing job performance and improving the synchronization among logistics and generation [36]. Introduced a [37] Cloud-based CPS converging on succeeding adaptive schedule and standard condition-based support in the plant, which preferred production stimulating to servitization and digitalization [38].

A CPS-enabled power architecture for intense energy manufacturers to decrease power waste, fast manufacturing during power supervision, and more reliable composition. Newly, Digital Twin has illustrated extensive deliberation, acknowledged as a necessary technology to obtain CPS, and eventually comprise Smart Manufacturing. The Digital Twin [39] also improved the automated internet recommendation structure that has been performed, which explained the interconnection between industrial Internet and the Digital Twin and explained the complete process mechanism and implementation. A Tri-model approach has been introduced to promote digital twin

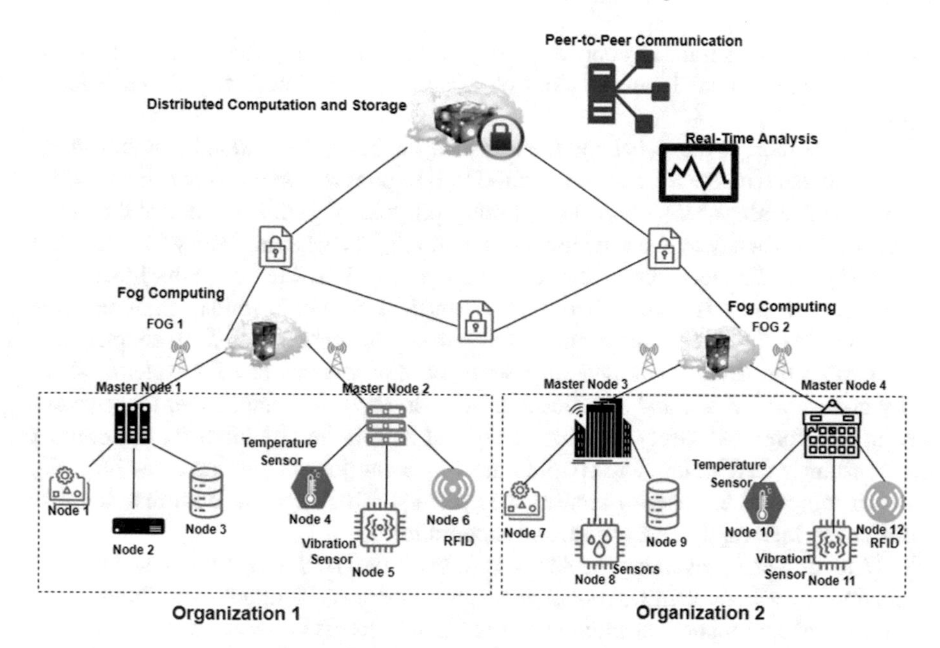

Fig. 3 CPS-based smart manufacturing design

development in the outcome layer, supporting synchronizing the physical goods and virtual design [39].

Smart manufacturing incorporates Distributed Computation and storage and is integrated by CPS in Fig. 3. When moving data from the connection level to the conversion layer and then to the cyber layer, the 5C-CPS architecture prioritizes security, privacy, and capacity. Incorporating"Master Nodes" as a middleman in the first layer of the CPS-based manufacturing structure has been suggested to address this weakness. It will allow their capacities to be shared with other Nodes in the same local network. Major security concerns in cyber-physical interactions [40] include cyber-attacks on sensor communication networks and physical interfaces. The proposed CPS-based manufacturing structure may resolve these issues.

6 Security Discussion of CPS

Control Security and Data Security comprise the two main categories of CPS Security. Control Security entails resolving all control issues within the context of the network and protecting the Control System from multiple attacks on estimating the system and Control algorithms [41]. Data Security comprises ensuring erudition due to the information collection, method, and high-scale distribution in the pervasive loosely-coupled network context. For instance, Data Security converges on information assurance by applying an encryption method, whereas Control Security converges on shielding the Control System dynamic from any Cyber-Attacks [42].

As CPS merges the Cyber and the Physical methods, the system integrates many requirements with the Blockchain method [43] to consider when composing a safety device concerning that system. The primary purpose of CPS is to control the physical process's behavior and actuate moves to modify its behavior. Every CPS element (e.g., Fig. 4. Manufacturing, biometrics, aerography, IoT, Medical, Smart technologies) can classify and centralize. Cyber-Physical System's fundamental features are coupling the environment and heterogeneity in techniques. This chapter also focuses on the Blockchain integrated security requirements of CPS systems, security goals and threats, and significant attacks on CPS. In comparison, few surveys regarding foremost safety concerns carry out in this field. Moreover, the context is constantly developing, and coupled machines can join in various areas [41, 42], which improves the needed security complexity. Challenges in outlining a safety device are interception, exposure, and moderation.

Detecting and preventing the attack is incredibly challenging due to the communication layer connecting physical and cyber systems, including the Cloud. Only some attackers prepare depending upon straight vulnerabilities and further attempts to drive Cross-Layer attacks. The system needs more detection techniques in all CPS layers. The significant difficulty is creating a safety mechanism that decreases the

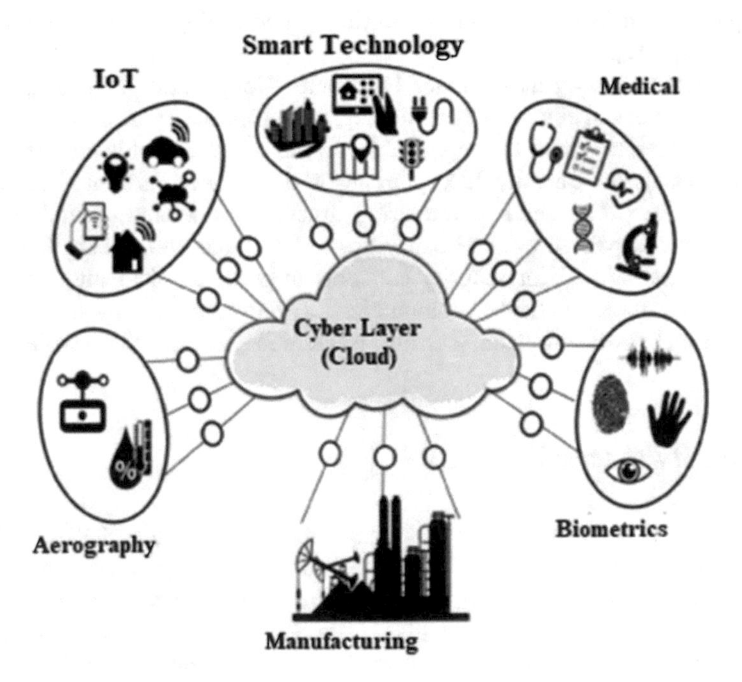

Fig. 4 Connectivity of cyber layer with different technological domains

breach's effects on the system for exceeding the detection and prevention protection aspects.

6.1 Security Requirements

The CPS's security challenges can classify into two categories: The electing challenge of the implemented safety purposes to perform the required protection and the erecting challenge of independent technologies connecting to perform the needed purposes. Because of its comprehensive Internet Connectivity, CPS security design incorporates, for instance, specific security concerns of CPSs and also the approaches in the Internet, Power Networks, WSN, manufacturing, and Medical Devices in Fig. 5. CPSs do not have computational method skills [43] or steady performance to obtain large-security conditions. Therefore, adopting any safety mechanism relying on a switching context is challenging.

6.1.1 Device Access Protection

The first requirement is to protect the device access. If the Authentication does inadequately establish, then the Unauthorized things obtain access and can manipulate the operation. Therefore, neither believing any Underlying Paired Cryptograms nor executing on the Guaranteed Application Layers.

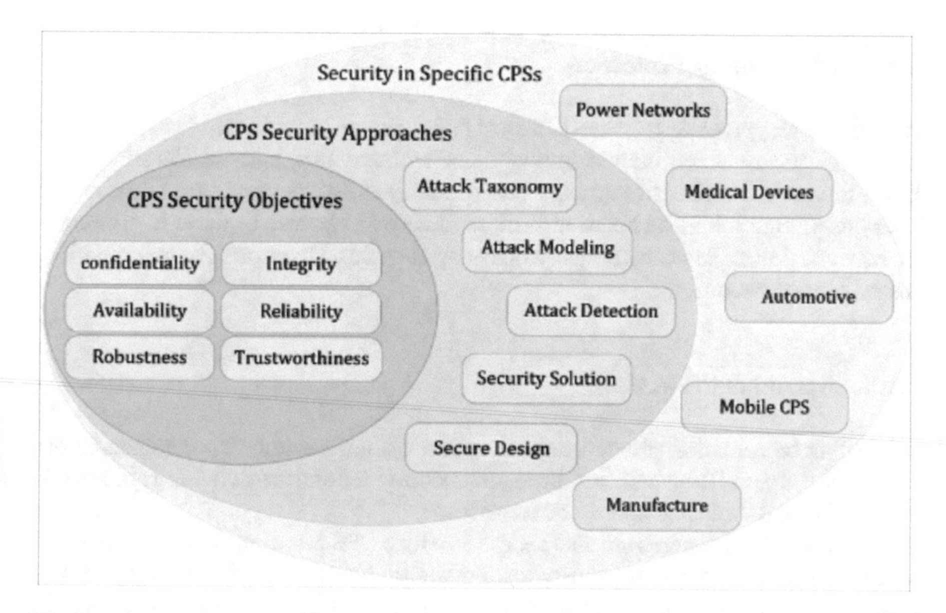

Fig. 5 Security framework of CPS

6.1.2 Data Interface Protection

A data interface is necessary for preventing Unauthorized Access and spotting fraud and vindictive acts in CPS transmission systems. Attackers, for instance, try to stop the energy consumption system's dynamic characteristics and timing activities so they may examine the data supplied and received. Few attackers purposefully disrupt the routing topology or launch DoS assaults to interrupt channels. A few closing instruments need more comprehensive computer systems, significant data processing, conversational capabilities, and enough storage potential. These instruments are also used in an unguarded manner during insertion. The fact that the design process is indefinitely constrained by hardware availability, processing conditions (speed), and energy consumption is another element that immediately points to unprotected skills.

6.1.3 Application Protection

Connecting multiple apps and security issues is the Application Layer. Attackers might utilize specific information provided by users to their advantage in this situation, leading to private data leaking and retirement tragedy. The occurrences and places that the users visited may be included in this data. Position simulation, undifferentiated time, or period encryption are all used in data assurance processes for data stability at this layer. This layer's application areas are social and must be protected as a result.

6.1.4 Data Storage Protection

In CPS devices, protected private data must be protected. Most CPS objects, including sensors, have resource-limited connections and are tiny and wirelessly linked. Even though a number of software-based solutions use encryption algorithms to encrypt information in such systems, more is needed because of these instruments' storage constraints and constrained processing capacity. Therefore, compact stability instruments are required.

6.1.5 Actuation Protection

According to actuation privacy, any actuation operations must be delegated from authorized experts. It ensures that the feedback provided and the switching commands are precise and protected against adversaries.

Internet safety concerns are also necessary since CPS associations use the Internet as a transit layer. As usual, security testing should be done for the entire system as a single end-to-end security design before testing each layer's operational security device. Additionally, massive memory requirements and heavyweight estimations are requirements for any desired security solution. The majority of security-offered solutions make an effort to explore distinct security issues at each tier. Even though these initiatives could aid in securing the system's desirable component, potential customers may nevertheless pass via other system components. The security architecture of CPS is used to maintain security throughout each layer, including information collecting, synchronization, and processing, from the base to the top layer to overcome this issue.

7 Cyber-Physical System Challenges

Introducing the emerging technology according to the CPS framework, some specific challenges have been discussed below:

7.1 Diversity

Diversity originates from many implementations of the Cyber-Physical System and the various problems involved. Such diverse perspectives include variable input size, interaction methods, power consumption, versatility, flexibility, storage space, endurance, and computational complexity.

7.2 Security

Privacy and security appeared in performance in a variety of utilization. There are many reports of cyber-attacks in almost every area that incorporates a cyber element. Data confidentiality and information security privacy need to be appropriately managed to secure a complete IoT ecosystem.

7.3 Trusts

It is primary to the cyber-physical system implementation. IoT security and privacy's fundamental interest is data flow to the web that switches and is applied through different service providers.

7.4 Range

While creating a planned weak CPS architecture, the main idea is to support the implied extension in the application's number and produced data amount. CPS terms are contemplated in Smart Agriculture, Smart Manufacturing, Smart Logistics, Smart Grid, Smart Transport, Smart Environmental Protection, Healthcare, Smart Defense, Smart Home, and others, signifying high-range issues. After keeping this difficulty of manageability in mind, new architectural approaches need to design.

7.5 Latency

After the machine's input port registers the information, a delay occurs, but before, the output port transfers the current data. Simultaneously, the transmission latency and time transpire later in the first end's output port. But before the subsequent input port element holds the data, the port transmits the information.

8 Maintenance Support of CPS

For this instance, the required defense technique emerged on an alliance of different datasets and searching real-time, historical, and planning data. To increase the contestation and globalization of the market, the quality of the product and distribution authenticity have become essential factors in the success of manufacturing industries. The growing features of sustainable creation demand the effectiveness and efficiency of the resources and the production, process, and product system life-cycle. The continued pressure to minimize the price and parallelly increases the customer amusement consequent in an elaborate maintenance process. So, the necessity to decrease the cost and machine parks have built production methods growing vincible to dander as the equipment's breakdown can result in a production loss.

On behalf of constant machine loads-based large-scale production, applied the operable maintenance strategies while in customer orders moved unproductive production methods, usual access of prohibitive maintenance usually fails because they need to report to individual load spectrum. The mentioned case concerns protecting the machine tools to manufacture cars' engines and gearboxes. Opel GmbH has been in Austria and Vienna, providing car engines, including gearboxes, for almost 30 years for automobiles and industrial transportation. Opel's gearbox production methods were taken out with operating 68 tools in two squads where three axes tools for drilling, milling, and thread cutting are identical in production. Two tools are identified: The devices running in a hazardous way and the tools running on side paths. Machine defeats with the complex way of pointing over the whole bearing industry's disruption, whereas the breakdown of devices running on side paths can be recompensed using substitutive tools. The subsistence approach is recently distributed toward machines' policies with the complex way and tools constructing side-paths.

Establishing defensive maintenance proposals is stocked tools in a demanding way, such as the oscillation reports every six months or device monitoring for looking out device corrosion. The additional component control departs the"elements in stock" with the demanding way from the"elements established on-demand" subsequent device crash on side paths. The recently executed sustenance administration consists of various tools and techniques with multiple operation fields, manually updated sustaining files, and applies non-uniform conversation concerning preservation and non-holistic methods for receiving, processing, assessing maintaining data. An innovative sustaining way is generated to predict and determine machine breakdown consequences, applying 6 of the 68 machines as a reference. First, different data compared to position monitoring, the nature of the product, and sustenance records are obtained from various preservation tools. By describing the making equipment on a segment level, a reason for connecting this data is generated. Anticipatory maintenance parameters are identified based on this data from sensor feedback, analytically known corrosion outcomes based on traditional primary data, and a methodical design for preserving essential elements of the making tools.

Similarly, real-time load spectrum data originating from the administration method are obtained. In the subsequent move, the sustenance-related parameters also the currently transpiring load events are harmonized to obtain outcomes concerning load-induced damage and condition variations. Data prospecting techniques facilitate semi-automatic and rule-based data association and concentration. Ultimately, a stability design proposes anticipatory conditions and sustaining propositions by a combined assortment of practices. The outcome of the invention is a sustaining administration center for composition policies. The preservation administration center is based on an anticipatory preservation policy, encouraging the most immeasurable possible product feature, optimized plant availability, and decreased preservation prices. At the corresponding period, it allows immediate acknowledgments to breakdown moments. Therefore, it is probable to think of rule-based anticipative preservation propositions mixed in a stability model, implementing a multivariate optimization of subsistence and rehabilitation propositions.

9 Future Direction

Future works comprise improved CPS testing on human subjects, with both the subjective and objective appraisement methods and differentiating metrics like training time, production time, error rate, and product quality within different conventional CPSs. Also, its future work comprises plans to develop the system used in the CPS to include more critical uncertainty. It can be done by training the network with various data and higher quantities. Another possible resolution is to condiment through Transfer Learning. The security challenges and the requirements also can be solved by integrating the R- Blockchain method. Improving understandability with explainable AI would be crucial in applying Blockchain to CPS to overcome the abovementioned issues. The fundamental issue with existing Blockchain systems is that they need to be more transparent to users, which reduces human trust in the system. Explainable AI will aid in increasing the transparency of ML models and human trust in Blockchain technologies. It may be used to uncover machine learning applications' weaknesses and weak points to exploit them to improve the system's adaptation to new cyber-physical data or threats.

10 Conclusion

Here in this chapter, the collateral improvements in Information and Communication Technologies (ICT) and Computer Science (CS) are driven. Besides, in Manufacturing Science and Technology (MST), they are indicated, highlighting their reciprocal impact. The idea behind CPS-based Blockchain was initiated in short. Here also highlighted a few challenges of CPS. Here explained that, with the outcome of the activities of CPS in past decades, the MST had surfaced way for soft entry within the CPS era. In the end, the CPS has been effectively constructed. It constitutes a subsystem, a Blockchain-based developed human–machine interface, and the device's perspective technology, and the system is more favorable to advanced settings. CPS was employed in the absolute fabrication method of an intelligent manufacturing device, a hybrid medical device, and its effectiveness. With its emerged limitation of the sub-system, a developed CPS gives guidance and quality review and more adaptability and flexibility. Besides, the system's performance, which is -based, can be developed with the amount of input layer and training data variety. Overall, the system with Blockchain in CPSs furthers the potential of its optimization. This paper's novelty is the Blockchain in combination with another sub-system like IoT in AI-based CPS. It increases CPS's ability to guide and control humans in their manual work, pointing to its appropriateness in manual assembly processes. We discussed how CPS was used to create specific Smart manufacturing based on the suggested design. A was used to extract low-dimensional properties in the pre-trained design. It uses a fine-tuning way to focus on a specific manner of perception work. With Smart monitoring and supervision sub-systems, CPS can reduce the demand for specialist personnel and reduce manufacturing restrictions during implantation. As a result, CPS integration might significantly reduce fabrication costs.

References

1. Shih C-S, Chou J-J, Reijers N, Kuo T-W (2016) Designing cps/iot applications for smart buildings and cities. IET Cyber-Phys Syst: Theory & Appl 1(1):3–12
2. Sonar K, Upadhyay H (2014) A survey: Ddos attack on internet of things. Int J Eng Res Dev 10(11):58–63
3. Li S, Da Xu L, Zhao S (2018) 5g internet of things: A survey, Journal of Industrial Information. Integration 10:1–9
4. Hatzivasilis G, Fysarakis K, Soultatos O, Askoxylakis I, Papaefstathiou I, Demetriou G (2018) The industrial internet of things as an enabler for a circular economy hy-lp: A novel iiot protocol, evaluated on a wind park's sdn/nfv-enabled 5g industrial network. Comput Commun 119:127–137
5. Ho N, Wong P-M, Hoang N-S, Koh D-K, Chua MCH, Chui C-K (2021) Cps-based manufacturing workcell for the production of hybrid medical devices, J Ambient Intell Humlzed Comput 1–15
6. Wang L, Torngren M, Onori M (2015) Current status and advancement of cyber-physical systems in manufacturing. J Manuf Syst 37:517–527
7. Liu C, Vengayil H, Zhong RY, Xu X (2018) A systematic development method for cyber-physical machine tools. J Manuf Syst 48:13–24
8. Sowe SK, Simmon E, Zettsu K, de Vaulx F, Bojanova I (2016) Cyber-physical-human systems: Putting people in the loop. IT professional 18(1):10–13
9. Ma M, Lin W, Pan D, Lin Y, Wang P, Zhou Y, Liang X (2018) Data and decision intelligence for human-in-the-loop cyber-physical systems: Reference model, recent progresses and challenges. J Signal Process Syst 90(8):1167–1178
10. Bordel B, Alcarria R, Robles T, Martın D (2017) Cyber–physical systems: Extending pervasive sensing from control theory to the internet of things. Pervasive Mob Comput 40:156–184
11. Sun Y, Song H (2014) Secure and trustworthy transportation cyber-physical systems, Springer, 2017. [12] Suh SC, Tanik UJ, Carbone JN, Eroglu A, Applied cyber-physical systems, Springer
12. Suh SC, Tanik UJ, Carbone JN, Eroglu A (2014) Applied cyber-physical systems, Springer
13. Rad C-R, Hancu O, Takacs I-A, Olteanu G (2015) Smart monitoring of potato crop: a cyber-physical system architec-ture model in the field of precision agriculture. Agric Agric Sci Procedia 6:73–79
14. Khaltan SK, McCalley JD (2014) Design techniques and applications of cyberphysical systems: A survey. IEEE Syst J 9(2):350–365
15. Acatech, acatech-National Academy of Science, Engineering, Cyber-Physical Systems: Driving Force for Inno-vations in Mobility, Health, Energy and Production, Springer Berlin Heidelberg, 2012
16. Geisberger E, Broy M (2012) AgendaCPS: Integrierte forschungsagenda Cyber-Physical systems, 1, Springer-Verlag
17. Wavering A (2013) Foundation for innovation: Strategic r&d opportunities for 21th century cyber-physical systems, Natl Instution Stand Technol, Report January (2013)
18. Rajkumar R, Lee I, Sha L, Stankovic J (2010) Cyber-physical systems: the next computing revolution. In: Design Automation Conference, IEEE, pp 731–736
19. Lee EA, Seshia SA (2012) Introduction to embedded systems: A cyber-physical systems approach, Mit Press, 2017.
20. Park K-J, Zheng R, Liu X, et al. (2012) Cyber-physical systems: Milestones and research challenges
21. Tao F, Qi Q, Wang L, Nee A (2019) Digital twins and cyber–physical systems toward smart manufacturing and industry 4.0: Correlation and comparison, Engineering 5(4) 653–661
22. Marburger JH, Kvamme EF (2007) Leadership under challenge: Information technology R&D in a competitive world: An assessment of the federal networking and information technology R&D Program, NCO/NITRD
23. Monostori L, Kadar B, Bauernhansl T, Kondoh S, Kumara S, Reinhart G, Sauer O, Schuh G, Sihn W, Ueda K (2016) Cyber-physical systems in manufacturing. CIRP Ann 65(2):621–641

24. Roy DG, Mahato B, De D, Buyya R (2018) Application-aware end-to-end delay and message loss estimation in internet of things (iot)—mqtt-sn protocols. Futur Gener Comput Syst 89:300–316
25. Hamze L (2021) Blockchain-based solution for covid-19 vaccine distribution, Ph.D. thesis, Worcester Polytechnic Insti-tute
26. Verma R (2022) Smart city healthcare cyber physical system: characteristics, technologies and challenges. Wireless Pers Commun 122(2):1413–1433
27. Xu X (2012) From cloud computing to cloud manufacturing. Robot Comput-Integr Manuf 28(1):75–86
28. Jay L, Jaskaran S, Azamfar M (2014) Industrial ai: is it manufacturing's guiding light, Manuf Leadersh Counc (2019) 26–36
29. Bond AH, Gasser L (2014) Readings in distributed artificial intelligence, Morgan Kaufmann
30. Montes GA, Goertzel B (2019) Distributed, decentralized, and democratized artificial intelligence, Technological Fore-casting and Social. Change 141:354–358
31. Mahato B, Roy DG, De D (2021) Distributed bandwidth selection approach for cooperative peer to peer multi-cloud platform. Peer-to-Peer Networking and Applications 14(1):177–201
32. Wang W, Zhang Y, Zhong RY (2020) A proactive material handling method for cps enabled shop-floor. Robotics and computer-integrated manufacturing 61:101849
33. Guo Z, Zhang Y, Zhao X, Song X (2020) Cps-based self-adaptive collaborative control for smart production-logistics systems. IEEE transactions on cybernetics 51(1):188–198
34. Mourtzis D, Vlachou E (2018) A cloud-based cyber-physical system for adaptive shop-floor scheduling and condition-based maintenance. J Manuf Syst 47:179–198
35. Roy DG, Ghosh A, Mahato B, De D (2018) Qos-aware task ooading using self-organized distributed cloudlet for mobile cloud computing. In: International Conference on Computational Intelligence, Communications, and Busi-ness Analytics, Springer, pp 410–424
36. Zheng P, Wang Z, Chen C-H, Khoo LP (2019) A survey of smart product-service systems: Key aspects, challenges and future perspectives. Adv Eng Inform 42:100973
37. Bezzo N, Weimer J, Pajic M, Sokolsky O, Pappas GJ, Lee I (2014) Attack resilient state estimation for autonomous robotic systems. In: 2014 IEEE/RSJ International Conference on Intelligent Robots and Systems, IEEE, pp 3692–3698
38. Ashibani Y, Mahmoud QH (2017) Cyber physical systems security: Analysis, challenges and solutions. Comput Secur 68:81–97
39. Lu T, Lin J, Zhao L, Li Y, Peng Y (2015) A security architecture in cyber-physical systems: security theories, analysis, simulation and application fields. Int J Secur Its Appl 9(7):1–16
40. Roy DG, Das P, De D, Buyya R (2019) Qos-aware secure transaction framework for internet of things using blockchain mechanism, J Netw Comput Appl 144 59–78
41. Mahmoud R, Yousuf T, Aloul F, Zualkernan I, Internet of things (iot) security: Current status, challenges and prospective measures, in, (2015) 10th International Conference for Internet Technology and Secured Transactions (ICITST). IEEE 2015:336–341
42. Roy DG, Mahato B, Ghosh A, De D (2019) Service aware resource management into cloudlets for data ooading towards iot. Microsyst Technol, 1–15
43. Roy DG, Mahato B, De D, A competitive hedonic consumption estimation for iot service distribution, in, (2019) URSI Asia-Pacific Radio Science Conference (AP-RASC). IEEE 2019:1–4

Integration of AI, Blockchain, and IoT Technologies for Sustainable and Secured Indian Public Distribution System

S. Kavi Priya, N. Balaganesh, and K. Pon Karthika

Abstract The exponential growth of low-powered IoT devices in smart city applications opens up a challenge to perform on-device analytics of sensor data. Tiny Machine Learning (TinyML) is a new Artificial Intelligence (AI) solution that combines low-power Internet of Things (IoT) systems and Machine Learning (ML) networks to create intelligent IoT devices by fitting ML models onto IoT hardware. By incorporating Blockchain technology into the TinyML-IoT architecture, the IoT network gains an extra degree of security. Blockchain is viable for IoT security since it provides tamper-proof data protection, resistance to unwanted IoT device access, and autonomous shutdown of compromised devices in an IoT network. This work aims to develop a novel framework for end-to-end traceability of the Public Distribution System (PDS) using low-power IoT devices, TinyML, and blockchain technology. It also describes how to integrate TinyML and blockchain technology in PDS comprising low-powered Green IoT components. To efficiently track the food grain quality, an intelligent food quality monitoring system is developed combining ML models and smart contracts. A prototype of the PDS network is created based on the Ethereum platform and NS3 simulator to assess the viability of the proposed system. The accuracy analysis reveals that the proposed framework outperforms the other state-of-the-art approaches. Furthermore, the food grain quality detection module using a TinyML model is simulated and tested using the Edge Impulse platform and the testing results reveal that the system achieves good performance scores in terms of False Positive Rate (FPR), False Negative Rate (FNR), precision, accuracy, and F1-scores.

S. K. Priya (✉) · N. Balaganesh · K. P. Karthika
Department of CSE, Mepco Schlenk Engineering College (Autonomous), Sivakasi-626005, Tamil Nadu, India
e-mail: urskavi@mepcoeng.ac.in

N. Balaganesh
e-mail: balaganesh@mepcoeng.ac.in

K. P. Karthika
e-mail: ponkarthika.k@mepcoeng.ac.in

B. Bhushan et al. (eds.), *AI Models for Blockchain-Based Intelligent Networks in IoT Systems*, Engineering Cyber-Physical Systems and Critical Infrastructures 6,
https://doi.org/10.1007/978-3-031-31952-5_15

Keywords Smart city · Indian PDS · Low-powered IoT · TinyML · Artificial intelligence · Blockchain

1 Introduction

Cities act as the driving force of economic growth for every nation. In the next ten years, Indian urbanization will reach 40% of the population and give rise to 75% of India's Gross domestic product (GDP). With this radical increase in the urban population [12], it is vital to handle the sustainability of social, economic, and environmental resources. The revolutionary initiative 'Smart city' ensures a high quality of life for the residents by providing a firm, neat, and steady environment through Smart services. The Government of India has allocated funding of $6.6 billion for the Smart Cities Mission [13, 14] and the Atal Mission for Rejuvenation and Urban Transformation (AMRUT). The vision of the Smart City initiative is to enhance the living standards of the public and impel the growth of the economy through the development of local areas exploiting transformative technologies leading to intelligent outcomes. The essential kernel components of Smart Cities' mission are:

- Ample and guaranteed supply of Water and Electricity
- Obligatory sewerage inclusive of Solid Waste Management
- Coherent and Cost-effective Urban Transportation
- Housing for all at affordable costs
- Public wireless broadband services at free of cost
- Resilient IT infrastructure for digitalization
- Effective e-Governance administration with the active involvement of citizens
- Viable and balanced environment
- Nurture and nourish consistent agricultural growth
- Eminent technologies for healthcare and education
- Public safety and security enforcement policies, especially for children, women, and senior citizens
- Intelligent systems for street light monitoring, environmental monitoring, and parking facilities

Figure 1 highlights the thrust application areas promoting the wealthy, safe, and secured Smart City ecosystem. The exhaustive evolution of Smart City solutions will upgrade the lifestyle of citizens, generate new employment opportunities, and increase the overall income which in turn will enrich the economic status of cities and eradicate poverty. The underlying technologies to make the Smart City Vision [4, 5] into reality include the Internet of Things (IoT) [6], Wireless networking technologies, Machine learning (ML), Artificial Intelligence (AI), Big Data, and Blockchain. The integration of all these pioneering technologies brings up new research challenges and opportunities to build optimized and efficient Smart City applications.

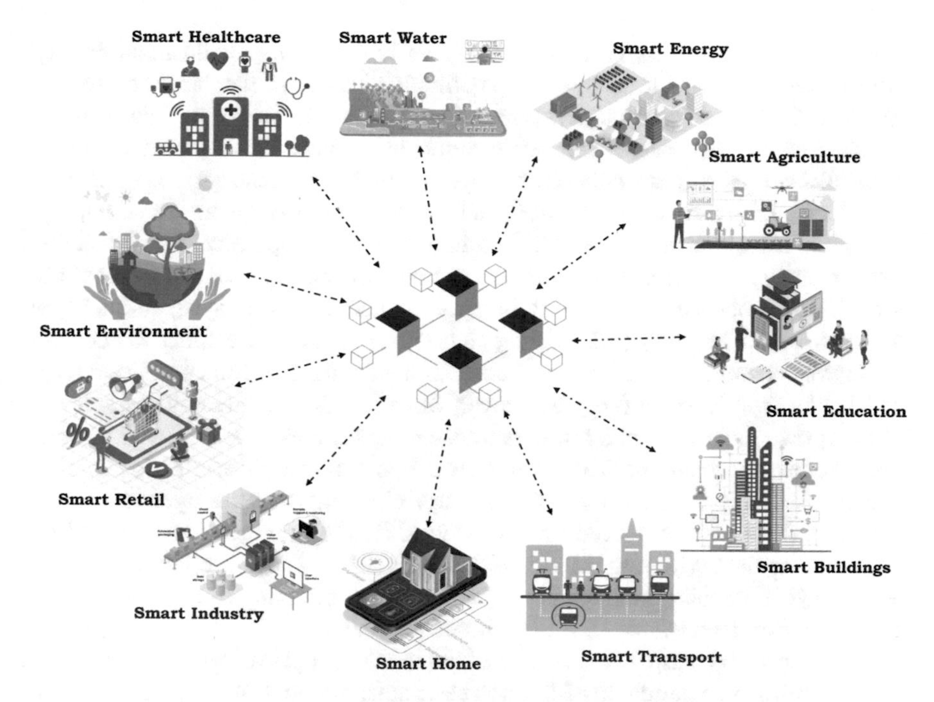

Fig. 1 An illustration of Blockchain and IoT-enabled smart city ecosystem

A secure and extensive storage environment is required to administer and manage the intricate smart city infrastructure and systems [15], which generates a significant volume of sensitive data. The encryption, distributed ledger, and encryption features of blockchain can serve as a cornerstone for several social and business interactive applications with improved security, efficiency, and privacy to alleviate cyberattacks and provide guaranteed data acquisition, data storage, and data retrieval. Blockchain technology has advantages including improved reliability, indelibility, and fault resilience that make it a viable solution for addressing authentication issues and preserving information integrity. Moreover, the incorporation of blockchain in smart cities can ensure real-time, reliable data transfer between smart city technologies without the requirement for a centralized system to monitor and manage all the data packets. The fusion of blockchain with IoT offers a more reliable and safe infrastructure for smart devices and sensors to function together smoothly in a smart city ecosystem. The tremendous growth of IoT has made intelligent automation and ubiquity of connectivity possible, which has aided in the creation of smart cities. Blockchain has the potential to significantly improve IoT networks, optimize resource usage, and boost data processing performance to progress IoT-enabled smart cities. Transparency, easy data integration, and improved security are made possible by the synthesis of blockchain with IoT. Blockchain technology can allow automated processes, safe device connections, and trustless transactions in IoT.

Artificial Intelligence (AI) plays a dual role in smart city applications. Firstly, AI techniques are used to analyze the enormous quantity of data acquired by IoT gadgets [16, 17] and generate new types of information in the form of correlations and patterns that are used to enhance governance decision-making processes and, eventually, to establish new policies. Along with information, energy is a key component of smart cities since many processes that contribute to economic and employment growth rely on it. Therefore, it is imperative to emphasize the connection between energy and algorithms to predict and plan the energy usage of smart cities. Secondly, AI models enable edge intelligence and decision-making capabilities [18] through Machine Learning (ML) models embedded in IoT devices. Since IoT devices often have constrained resources, it is challenging for them to run complicated ML models like deep learning (DL). Moreover, when IoT devices are connected to the cloud for data processing and raw data transfer, the primary risks involved are data privacy exposure, response latency, increased communication overhead, etc. Tiny Machine Learning (TinyML) is a pioneering technology that has opened the way to address these concerns faced by IoT devices. TinyML permits local and instant processing of data within the device rather than sending it to the cloud. Furthermore, TinyML enables the inference of ML models pertaining to DL models on a resource-constrained device such as a Microcontroller.

Food security is a basic human right that has to be ensured to obtain nutritious food for every citizen in a nation. In addition to basic nutrition, food security is associated with enhanced global security and stability, poverty alleviation, better healthcare, and economic stability. Food supply chains are highly complex networks [24] due to a lack of transparency and integrity. Leveraging modern approaches to validate and identify the provenance, quality standards, and transfer and storage of information is crucial. This chapter presents a comprehensive strategy for enhancing food supply chain security in Indian Public Distribution System (PDS) and elevating consumer confidence in the PDS sector. By safeguarding each component that makes up the system, it attempts to increase the openness of food supply chain tracking systems. Public Distribution System (PDS) is one of the largest and obligatory Smart city applications in India. The PDS of India ensures the Food and Nutritional Security of the Nation. The PDS system facilitates the availability of food grains to needy people at an affordable price to eradicate the poverty and malnutrition problem. The existing PDS supply is prone to a wide variety of malpractices such as food grains leakages, black marketing, procurement issues, storage issues, and unauthorized access to the food grains management system. Due to these issues, there is an imbalance in the PDS demand–supply chain. Also, food grains wastage and expiration rates increase because of improper storage management. The PDS authorities need to take care that the food grains have been stocked appropriately throughout the entire supply chain at a specific temperature and other environmental factors. To alleviate these issues and to automate the entire system with reduced manpower involvement, this work aims to integrate Blockchain, AI, and IoT technology for effective tracking and traceability of food grains in the PDS supply chain independent of third parties and a centralized system. Blockchain is a promising and tamper-proof technology that can enhance the PDS supply chain by providing faster and cost-effective product delivery, substantially reducing communication and transaction data errors. The convergence of Blockchain and AI towards a Green IoT network for PDS enables the system to assess the quality of food grains and improves the visibility and security of data to a great extent. The major contributions of the proposed work are stated as follows:

1. Provides a basic introduction to the rudimentary technologies in developing Smart City applications.
2. Designed a smart framework for the end-to-end traceability of food grains in PDS leveraging the key concepts of blockchain, AI, and IoT technologies.
3. Proposed an intelligent solution for the food grain quality monitoring system using low powered IoT device embedded with a TinyML model to track the mobility and quality of food grains at various intermediate storage and transportation environments. Thereby, an alert message will be sent to the higher authorities if there is any food grain quality impairment.

The remainder content of the chapter is composed as follows. Section 2 provides a brief literature review. Section 3 introduces the proposed PDS framework combining IoT, blockchain, and big data technology. The performance analysis and the practical implications of the proposed PDS architecture are discussed in Sect. 4 and Sect. 5 respectively. Finally, Sect. 6 concludes the chapter with a neat summarization of the proposed work and its limitation and future prospects.

2 Related Work

The basic concepts of Blockchain, Big data, AI, and Green IoT solutions are outlined in the subsequent subsections. This section also discusses the issues prevailing in the current PDS system and the importance of the proposed framework to build a secure and sustainable PDS network in Smart cities.

2.1 Fusion of Blockchain and Big Data in IoT Applications

Blockchain is one of the trending immutable information recording systems of distributed digital ledger networks. Blockchain is an interdisciplinary system of cryptography, software engineering, digital infrastructure, cyber security, and distributed computing. The Blockchain ledger is a continually growing database of information recordings, known as blocks, that are linked to one another and encrypted by cryptographic techniques. Information about the version, timestamp, parent block hash, nonce (which starts at 0 and rises each time after a hash calculation), transaction count, and the sum of all transaction hashes are included in a general block. Each node within the network will participate in the block authentication process whenever a new block is generated. The block will automatically be added with a reference pointing to its preceding or parent block after it has been checked and authorized.

As a result, unrecognized or unauthorized transactions may be easily detected since the hash value of a modified or faked block will differ significantly from that of an unmodified block. Based on the application domains, blockchain frameworks can be grouped into three clusters namely public, private, and hybrid blockchain. Public Blockchain in which the network is transparent to all participating entities and does not have sole authority. Bitcoin is the best example of a public blockchain. A private Blockchain is a network that provides limited control for the participating entities to access the blockchain. Hybrid Blockchain is a network that provides complete blockchain access only to a specific group of entities and is a semi-decentralized platform. The below-mentioned predominant characteristics of blockchain make the system more unique and popular.

- Immutability—Blockchain is tamper-resistant due to the persistent and immutable network.
- Decentralization—There is no single point of control in blockchain, which enables blockchain access and replication through the web using the private keys of the participating node.
- Security—Blockchain ensures high security by means of complex cryptographic algorithms and hashing.
- Consensus—It is a unique feature of blockchain imparted to determine which nodes are active in the network while transaction validation.
- Rapid financial agreement—Accelerating the speed of transaction processing and money transfer.

Big Data is defined in terms of 4 V's such as Volume, Velocity, Variety, and Veracity. The concise description is given as follows:

- Volume—Big data is known for its voluminous quantity of data. Big data processing experiences problems related to its voluminous nature such as the curse of dimensionality, modularity, class imbalance, data non-linearity, etc.
- Variety—Big data handles a variety of data in structured, semi-structured, and unstructured formats from heterogenous sources and multiple modes. The main problems related to the variety are messy and noisy data, data locality, heterogeneity, etc.
- Velocity—It represents the rapidity i.e., the rate at which the data is generated concerning the application criteria.
- Veracity—It denotes the aspect of big data quality.

The unprecedented growth of big data is associated with inevitable issues such as privacy and security, erroneous data, data reliability, data sharing, etc. The distinctive features of Blockchain technology are used to tackle the challenges addressed by big data [7]. The need to fuse the blockchain and big data is depicted in Table 1. A detailed survey on blockchain for big data is provided in [8]. The authors briefed an overview of blockchain for big data along with the motivation for integration. The survey about various services in blockchain for big data such as blockchain for secure big data acquisition, data storage, data analytics, and data privacy preservation was discussed. At last, the latest studies which use the blockchain in domains such as smart city,

Table 1 Solutions provided by blockchain for problems in big data

Big data challenges	Blockchain solutions
Security and Privacy Issues in the cloud	Decentralized and one-way encryption of data
Data integrity	Reliable and tamper-resistant solution by means of immutable nature
Fraudulent transactions	The transparent nature of blockchain allows the visibility of data to all entities participating in the network, which in turn prevents fraud
Lack of real-time data analytics	Enables real-time data analytics by storing all transactions in a blockchain
Low-quality data	Enhances the big data quality using structured and complete data format
Multiple data sharing and redundant processing	Reduces data leakage and eliminates redundancy in data analysis by storing it in the blockchain
Indirect access to data	Streaming the data online to all the participating nodes

smart healthcare, smart transportation, and smart grid were reviewed. The authors of [10] proposed the design and implementation of layer-based distributed data storage in blockchain-enabled large-scale IoT systems. This system has been developed to mitigate the trust issues caused by the dependencies in centralized server solutions by using Hyperledger Fabric (HLF) platform for the distributed ledger. The transaction verification and audit records are performed by using HLF peers in the big data system with the help of blockchain technology. The porotype was implemented for deploying the proposed solution in IoT edge computing and the big data ecosystem. The proposed work has been evaluated based on the metrics such as throughput, latency, communication, and computation costs.

The key issues in Big Data, such as data-driven policies, data governance, identification, trust, and decentralization, have been dramatically addressed by blockchain. Blockchain technology may be used to process, organize, and store Big Data when there isn't an ideal means to do so. With the aid of Blockchain technology, the authors [20] suggested an architecture to solve the issues faced by Big Data. The authors of [19] presented a thorough analysis of the security enhancements made possible by blockchain technology in IoT systems as well as the difficulties that arise during their integration. The article presented by [22] aims to provide a concise overview of recent developments in Supply Chain Management and extensively demonstrates several approaches for incorporating blockchain technology into Supply Chain Management and Logistics in line with various use cases. The review conducted by [21] discusses techniques for implementing blockchain into quantum cryptosystems as well as quantum cryptography. The paper examines attacks that compromise quantum security, as well as the strategies provided to counter the assaults. The authors of [23] presented a thorough examination of blockchain technology to shed light on its security dangers, highlight the need for privacy in modern applications, and define the problems that it faces as well as how they can be addressed.

2.2 *AI in IoT Applications*

IoT applications have numerous issues, including big data analytics, privacy and security, energy efficiency, and traffic congestion. Many AI techniques for IoT applications incorporate designs like machine learning, deep learning, and neural networks to address these problems. The problem of energy efficiency in IoT is notably resolved by the application of deep learning methods. The analysis of the data association and its forecast for energy management enables designers to choose the most important settings for regulating energy usage. Machine learning-based IoT authentication offers a security solution that includes techniques like supervised, unsupervised, reinforcement learning, access control, and safe offloading for the preservation of data privacy. Decisions for new ecosystems are made based on identifying patterns provided by the analytics capabilities of AI for the collection of data in IoT applications. The present utilizes TinyML for on-device computation. Tiny ML is one of the trending and emerging embedded computing technologies for edge devices. It is a special variant of Machine Learning that aims to design, develop and deploy ML algorithms on low-powered microcontroller devices. The fundamental goal of incorporating TinyML into an IoT framework is to achieve energy efficiency, optimal bandwidth usage, data confidentiality and privacy, low latency, and make it economical. The advantages of microcontroller devices designed for the TinyML paradigm are:

- Tiny and Energy efficient devices
- More economical and powerful equivalent to computers and servers
- More convenient, comfortable, and easy to use
- Perfect fit for large-scale applications

TinyML enables the edge import of pre-trained ML models for offering ML-as-a-Service (MLaaS) [27] to IoT systems. The research outlined a TinyMLaaS (TMLaaS) framework for potential Internet of Things applications. In terms of resource consumption, security, confidentiality, and congestion, the TMLaaS architecture naturally poses various architectural trade-offs for large-scale IoT applications. A TinyML-based method for detecting catastrophic gas leaks is demonstrated in [25]. With the use of BLE technology, the system can be trained to identify anomalies and alert the users via an application notification and a message sent to their mobile. With their high CO_2 emissions and other harmful chemicals, vehicles are the primary contributor to air pollution in smart cities. To offer predicted results of CO_2 emission levels, the authors [26] developed a soft-sensor approach based on an embedded platform that is intended to receive data from automobiles using their OBD-II interface. To improve the overall accuracy of the soft sensor without interacting with cloud-based servers, an unsupervised TinyML technique is also developed to eliminate outliers from the generated data stream.

2.3 *Green IoT Solutions*

The growth of IoT technology and its incorporation into smart cities [7] has transformed our way of life and improved society. The authors of [11] delivered a review on an overview of IoT-enabled smart cities, detailed and highlighting the impacts of embedding IoT with smart city services. IoT technologies do, however, create numerous difficulties, including increases in energy usage, toxic pollutants, and the production of E-waste in smart cities. In order to create a sustainable green smart world, the authors [29] conducted research that provides a comprehensive overview of Green IoT (GIoT) and numerous integrated technologies as well as GIoT constraints. The empirical investigations and outstanding challenges in GIoT have also been highlighted. Based on the findings, it is proven that GIoT has the potential to provide several benefits, including environmental conservation and protection, end-user happiness in various IoT sectors, and the reduction of adverse ecological effects and human life.

GIoT [1] is necessary since applications for smart cities must be eco-friendly. Eco-friendly environments are more sustainable for smart cities as a result of green IoT. The approaches and strategies for lowering the risks associated with pollution, traffic waste, resource utilization, and energy consumption, as well as for ensuring public safety, a high standard of living, and environmental sustainability, must all be addressed. There are several IoT applications where efficient energy usage is needed to facilitate a sustainable society. The authors [28] presented the core concepts of Green IoT (G-IoT) and sustainability, and G-IoT facilitated sustainability initiatives for smart cities. The authors of [9] summarized the IoT energy consumption methodologies and approaches to designing energy-efficient IoT devices. The proposed work focuses on developing a low-powered Green IoT network for PDS application in the Smart City initiative for making cities smarter, sustainable, and eco-friendly.

2.4 *Public Distribution System*

Since Independence, the eradication of poverty and reduction of malnutrition has been the central focus of Indian public policy. Alleviating Malnutrition plays a vital role in the Sustainable Development of the Nation. Malnutrition among children is firmly linked with persistent poverty. Malnutrition will cause many short-term and long-term adverse effects including poor physical health, cognitive disorders, infant mortality in the short term, and the probability of spreading long-term non-communicable diseases. The food security system of India was launched by the Indian government under the Ministry of Consumer Affairs, Food and Public Distribution to deliver food and non-food essential commodities to deprived households in India at subsidized rates. The main commodities distribution includes food grains such as rice, wheat, sugar, and essential fuel like kerosene, through a large network of Fair Price Shops (FPS)/ration shops located in several states all over the country. The Food Corporation of India (FCI), a corporation owned by the government is responsible for the procurement and maintenance of the Public Distribution System (PDS) [2, 3]. The PDS system of India is the largest ever welfare scheme established to reduce poverty and malnutrition. The Indian Government launched the Targeted Public Distribution System (TDPS) in June 1997, to primarily focus on the poor people. The targeted beneficiaries are divided into two subcategories under TDPS: Below Poverty Line (BPL) households and Above Poverty Line (APL) households. In September 2013, the Parliament approved the National Food Security Act (NFSA) to distribute food grains to the beneficiaries of TDPS as legal entitlements. The Indian government spends around Rs. 750 billion for grain storage. The state government is responsible for the distribution of food grains to poor households. As of 2018, 5.27 lakhs of fair-price shops are functioning across the nation. Though the PDS distribution network is a complex and important food security network with high coverage and public expenditure, it has some major concerns about the efficiency of the distribution process. Some of the major problems associated with the existing PDS system in India are:

- Beneficiary Identification: Due to urban bias, the TDPS system fails to serve the poor population effectively.
- Food grains leakage: TDPS suffers from the transportation leakage of food grains and black marketing by the FPS owners. By the TDPS evaluation conducted by the Planning commission, 36% leakage of wheat and rice is found at the national level.
- Procurement and Storage Issues: Issues raised due to the open-ended procurement and the shortage of storage capacity.
- Time consumption: The existing PDS process is a time-consuming process that will take around 5–15 days for approval at each level from various parties.

These issues led to corruption and tampering with the food grains while transporting them to the FPS. This will prevent the proper delivery of an allocated quantity of food grains to the poor people and will create malnutrition impacts. The proposed intelligent PDS system will be useful for the Indian government to overcome these issues and guarantee the delivery of food grains to the appropriate households with the allotted quantity and quality. The proposed system is resistant to leakage, tampering, corruption, and black marketing of food grains in the existing PDS system. This system will highly contribute to the Integrated Management of Public Distribution System (IM-PDS), a new scheme launched by the Indian government to implement the One Nation One Ration Card initiative. The proposed work focuses on developing a cost-effective framework to provide end-to-end traceability and monitoring of all transactions and interactions between all entities in the PDS food grains supply chain via smart contracts. It will speed up the process of the PDS supply chain and can substantially reduce the cost of transactions and the cost of operating an effective logistics system. The framework uses RFID technology to track the movement of food grains. The RFID tags attached to the food grain bags allow the system to track the mobility of food grains in transit and stored in warehouses. RFID tags avoid the problem of manual counting of grain bags since multiple objects can be identified simultaneously using radio waves. The system aims to design low-cost and low-powered IoT devices to monitor the food grain quality in terms of temperature, humidity, and moisture content. The quality of the food grains will be assessed based on the parameter values obtained from IoT sensors. Upon any quality deterioration, further distribution of food grains is detained as well as an alert report is sent to corresponding authorities to replace the food grains or take other steps to ensure that distribution continues. The system also includes a customer satisfaction feedback module to provide updated information on the quality of food grains and to register complaints if there is any issue.

3 Proposed PDS Framework

The present invention introduces a blockchain framework to interlink the major entities of the PDS system including the Ministry of Food and PDS, FCI, rented warehouses of central and state governments, FPSs all over India, and the target beneficiaries. The overall architecture of the novel traceability system designed for digital tracking of food grains in the PDS supply chain from procurement level to end customer level is illustrated in Fig. 2. The framework follows a four-layered architecture i.e., data layer, traceability layer, blockchain layer, and storage layer.

a. Data Layer: The data layer handles the communication between the entities of the PDS food grains supply chain. The interactions involve the distribution of food grains along with auditable delivery proof. While transferring data from one node to another, several concerns and threats to data privacy exist in the device layer. AI uses blockchain technology to solve these problems, with Bitcoin and Ethereum used to carry out transactions from one node to another. The communication level, which serves as an intermediary for transmitting data from one node to another, receives the information gathered in real-time. A consensus method is used in the blockchain and AI integration for IoT-based PDS to achieve security and scalability.
b. Traceability Layer: The traceability layer is responsible for tracking all transaction data and IoT sensor data at every level of food grains distribution. The data transfer towards the blockchain layer is in charge of data management and establishing infrastructure criteria among networks for the application layer. Blockchain and AI integration enable cryptographic hashing, digital identification, and encryption codes. Finally, the data is passed to the blockchain application layer, which is in charge of global management. The combination of AI and blockchain creates a deep learning data analytics tool to ensure network data privacy and security.

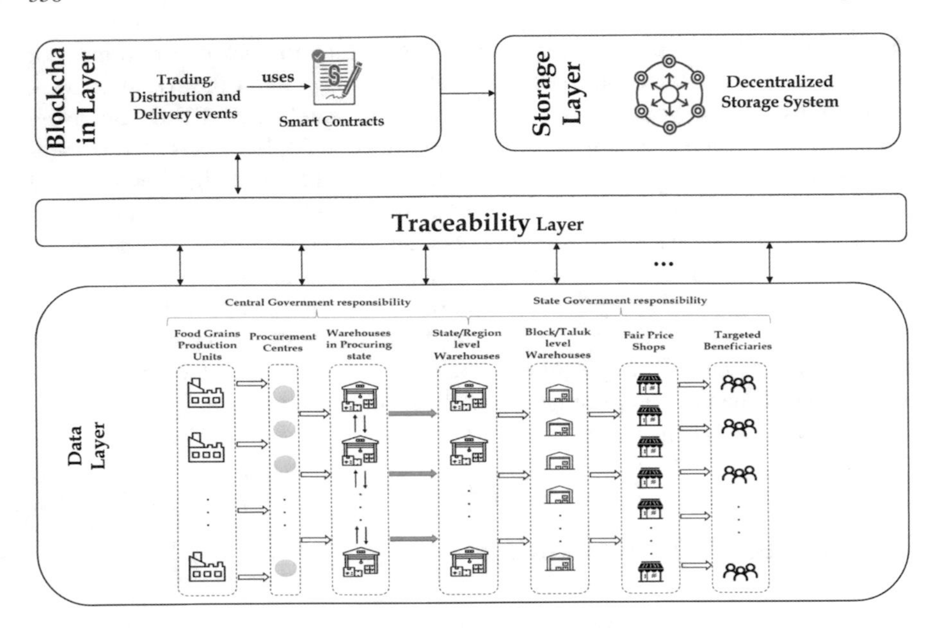

Fig. 2 The overall layered architecture of the proposed PDS system

c. Blockchain Layer: The blockchain layer handles the transactional data of the distribution and delivery events that occurred between the entities involved. To prevent unauthorized access to the unauthorized to the data, this layer imposes strict control access strategies. In this layer, the integration of AI in blockchain makes use of a distributed platform of cloud, intelligent storing, and smart contracts to ensure secure validation.
d. Storage Layer: Decentralized storage system is used to store all kinds of blockchain transaction data. The decentralized storage system enhances the performance of the proposed system with scalability, high throughput, and low latency. Figure 3 shows the storage of data blocks from all entity nodes in the blockchain storage servers.

3.1 Modular Description

Figure 4 depicts the modular diagram of the proposed Blockchain-driven IoT framework. IoT management module is used to develop an environmental condition monitoring application with multiple Traceability Units for the dispersion of food grains in the PDS supply chain. The collected data are stored in the decentralized storage system.

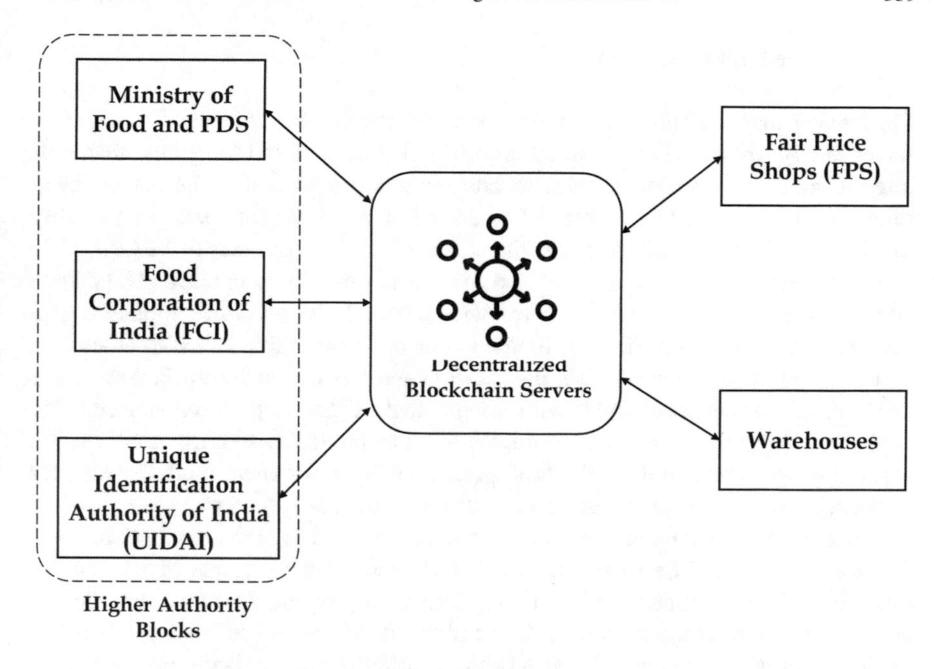

Fig. 3 The data storage of all entities in decentralized blockchain servers

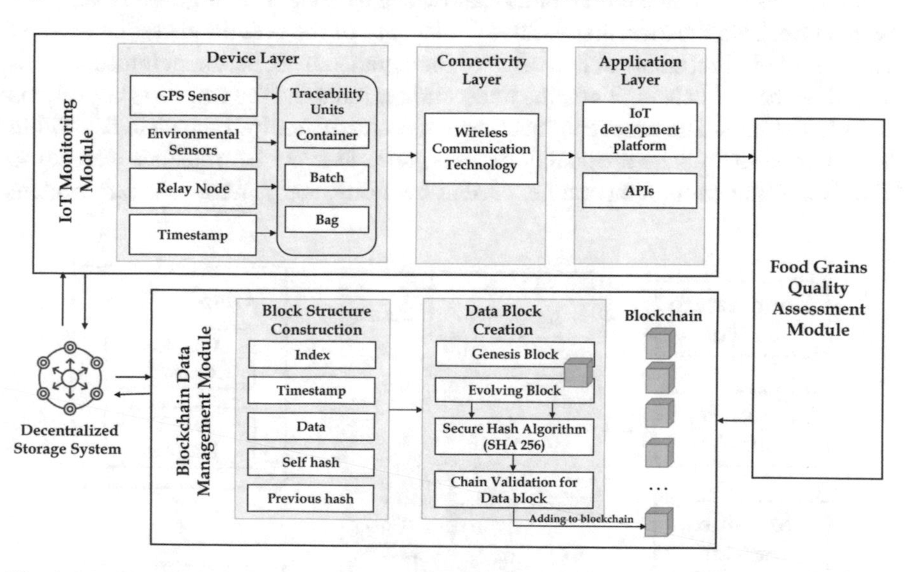

Fig. 4 Modular diagram of the proposed PDS system

3.1.1 IoT Monitoring Module

The three primary components of the IoT network are the Device Layer, Connectivity Layer, and Application Layer. The environmental conditions of food grains are monitored using the environmental sensors and relay nodes present in the device layer. Figure 5 presents the low-powered IoT device architecture for food grain quality monitoring. The device developed in the proposed work contributes to the Green IoT solution since it is energy efficient and facilitates on-chip computing. The device consists of a low-powered MCU fitted with temperature, humidity, moisture, and gas sensors. An on-chip TinyML model is embedded into the device to assess the quality grain of food grains. Also, the location and movement are tracked using the GPS sensor. The collected sensor data along with the timestamp are transferred to the application layer using wireless communication technologies like Bluetooth, WiFi, etc. Using the collected data, the food grains quality assessment module evaluates the quality in terms of temperature, humidity, and moisture content.

Table 2 provides the temperature, relative humidity, and moisture content levels for rice and wheat. The food grain quality is monitored for levels of traceability units, including container, batch, and bag levels. As a result, data recording begins at the stage of food grain storage. To implement this, the *dataStorage()* function in the smart contract shown in Algorithm 1 automatically collects pertinent data from devices and stores it on a blockchain when a batch of food grains is kept in a warehouse. This function first verifies the legality of the present operating account and the registration status of this batch of food grains. If so, pertinent information is stored on the blockchain. If not, the transaction expires and the contract state returns to its initial state. The smart contract function *evaluateQuality()* shown in Algorithm 2 is invoked to evaluate the quality of food grains at every intermediate warehouse and FPS. The current temperature, current humidity, and current moisture content

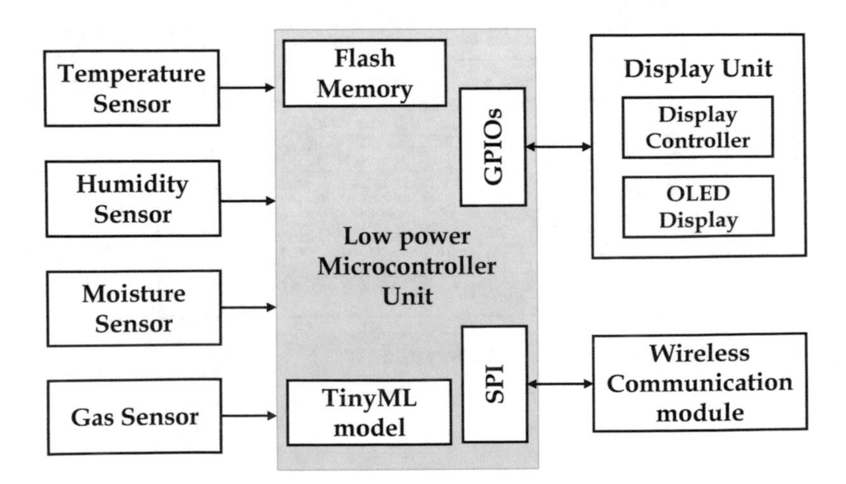

Fig. 5 The low-powered IoT device architecture for food quality monitoring

Table 2 Temperature, relative humidity, and moisture content levels for rice and wheat

Food grain	Temperature in °C	Relative humidity $(\%)gm/cm^3$	Moisture content (%)
Rice	22–32	50–75	12.2–13.8
Wheat	15–37	20–45	12.2–13.9

readings are taken from sensors and it is checked with the corresponding threshold values. When there is any deterioration in the quality of food grains is detected, the system immediately restricts the food grains from further distribution. Further, it generates a report to the respective authorities and sends an intimation to replace the food grains or take necessary actions for further distribution. Moreover, the collected data can be stored in storage servers for further querying.

Algorithm 1. ***dataStorage()***

Input:

Batch number (*batch.num*), food grain ID (*FGID*), sender ID (*sender.id*), warehouse address (*warehouse.addr*), storage temperature (*str.temp*), storage humidity (*str.humid*), storage moisture (*str.moist*), storage duration (*str.dur*), storage location (*str.loc*), authorization list (*AL*), current time (*curr.time*)

If *sender.id* ∈ *AL* then

  If *batchNumber* not existent then

    The data *batchNumber, FGID, warehouse.addr, str.temp, str.humid, str.moist, str.dur, str.loc, curr.time*

  Else

    Revert the contract state and display an error message

Else

  Revert the contract state and display an error message

End

Algorithm 2. ***evaluateQuality()***

Input:

Batch number (*batch.num*), current temperature (*curr.temp*), current humidity (*curr.humid*), current moisture (*curr.moist*), authorization list (*AL*), temperature threshold (*temp.threshold*), humidity threshold (*humid.threshold*), moisture threshold (*moist.threshold*), sender ID (*sender.id*)

If *sender.id* ∈ *AL* then

If *curr.temp, curr.humid, curr.moist* lies between *temp.threshold, humid.threshold, moist.threshold* then

Generate a notification that this batch of samples complies with the standard

Else

Stop the distribution of food grains and send a notification to the higher authority

Else

Revert the contract state and display an error message

End

3.1.2 Blockchain Data Management Module

This module is used to construct the blockchain structure and create data blocks. The blockchain structure contains a block index, timestamp, data, self-hash value, and previous hash value. The genesis block represents the first block in the blockchain and Evolving blocks are the intermediate blocks. The SHA-256 algorithm is used to create a unique fingerprint for blocks by encryption. Once the data block is validated, the block will be added to the blockchain.

3.2 *Food Grains Traceability*

The PDS supply chain contains a large number of entities involved in the process of production and transportation of food grains from procurement to the targeted beneficiaries. It is a challenging task to track and traces the entire PDS supply chain process. To achieve traceability of the complete system, the transactions are recorded from the initial stage by adding the unique RFID attached to the food grain bags and batch number for each succeeding transaction. The records are hashed and added to the hash chain. A batch is a group of food grain bags to be distributed in the warehouse and the batch number is the unique identifier. The transaction data are stored in storage servers to maintain the hash chain. To ensure the privacy and confidentiality of the data, an access control strategy is enforced to write or access the blockchain data. Only authorized and registered users are allowed to access the data and perform the transaction.

Three smart contracts are developed for the traceability process i.e., Registration contract, Batch Contract, and Transaction Contract. A registration contract is used to register the distribution entities involved in the PDS supply chain and the food grains available with the distributors. A batch Contract is used to add food grain details of the particular batch in the warehouse. Transaction Contract is used to add transaction events and approve the transaction based on the terms and conditions of the parties involved. To maintain the transaction traceability chain, all the contracts are deployed to get the addresses of respective contracts.

3.3 Distribution and Delivery Mechanism

For the distribution of food grains between each level of central and state governments, the Logistic Companies (LC) are used to transfer the food grains between two distributing parties. To carry out the distribution process, the distribution entities and LC is registered using the Registration Contract. Then the contract between the two entities will be initiated. Then the details of the transaction are uploaded to storage servers, which return the hash value. This hash value proves the authenticity of the food grains being transferred. Once the transaction is confirmed, the receiving entity pays the amount to the preceding distribution entity. Once the transaction is completed, the smart contract between the current distributing entity and LC is initiated. This contract handles the transportation of food grains from the current location to the next destination. To avoid tampering/leakage of food grains while transportation, the contract also collects transportation security from both entities. To assure the authenticity of food grains, the storage hash is used to access the food grains' details and match them with the original data.

3.4 Blockchain Transactions of PDS Supply Chain

The Ministry of Food and PDS, FCI, and UIDAI are the higher authority entities of the PDS system. Warehouses and FPS are the other two significant entities, responsible for the storage, transportation, and delivery of food grains to the target beneficiaries. First, all the required information of all the entity officials, managers, and owners should get stored in the decentralized blockchain servers and as depicted in Fig. 2. As illustrated in the diagram, the mandatory information including the biometrics of the warehouse managers and officers, FPS owners and officers, target beneficiaries, and the Ministry and FCI officials are stored in the servers and provided by UIDAI. At any point in time, the stored information can be accessed and retrieved by authenticated officials. To initiate the transaction, the PDS department creates the genesis block of

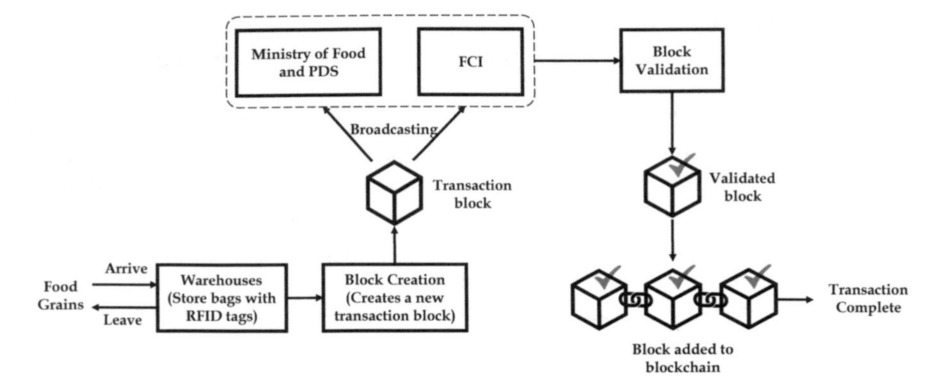

Fig. 6 Warehouse transaction scenario

the blockchain in favor of the Ministry of Food and PDS. The department appoints a dedicated team to handle the PDS blockchain network. Once the genesis block is created, the preceding blocks can be created for every transaction that occurred in the PDS blockchain network.

3.4.1 Warehouse Transaction Scenario

As illustrated in Fig. 6, when the procured food grains reach the respective central or state warehouses, where the RFID tags will be attached to the food grain bags. Then the RFID tag information of all the grain bags is recorded in newly created blocks and is secured with a hash (warehouse manager's fingerprint or digital signature). The blocks are then broadcasted to all the entity nodes (Ministry and FCI) for further approval and validation. Once the warehouse blocks are validated, they are added to the blockchain.

3.4.2 FPS Transaction Scenario

Using FCI apps, the FPSs raise their demand for the supply of food grains. The FCI application records the demand raised by the FPS and connects it with the nearest warehouse to reduce the cost of transportation incurred. Then, the warehouse sends the required grain bags to the FPS after recording the respective FPS information and RFID tag information of the specific grain bags. As depicted in Fig. 7, once the food grains reached the FPS, the RFID tags are verified and the data is uploaded to the FPS block. The data block is secured by the hash of the FPS owner. The block is then broadcasted to the Ministry, FCI, and the warehouses from where the food grains are obtained. The entities verify and validate the data block, which in turn will be added to the blockchain on successful validation. Now, the obtained food grains are ready to be distributed to the beneficiaries.

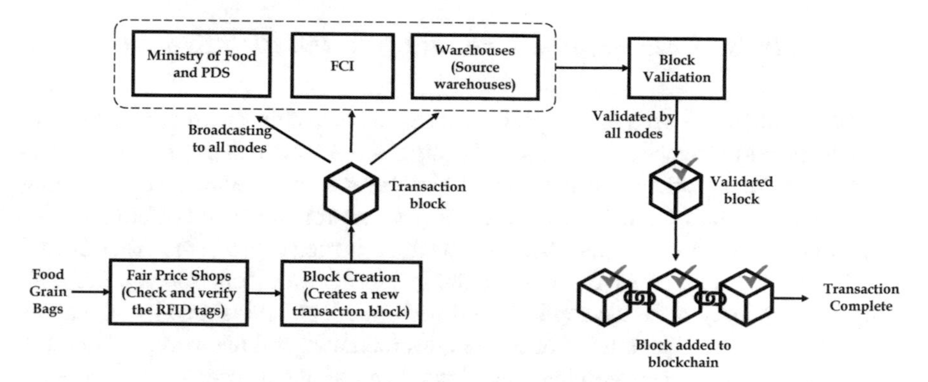

Fig. 7 FPS transaction scenario

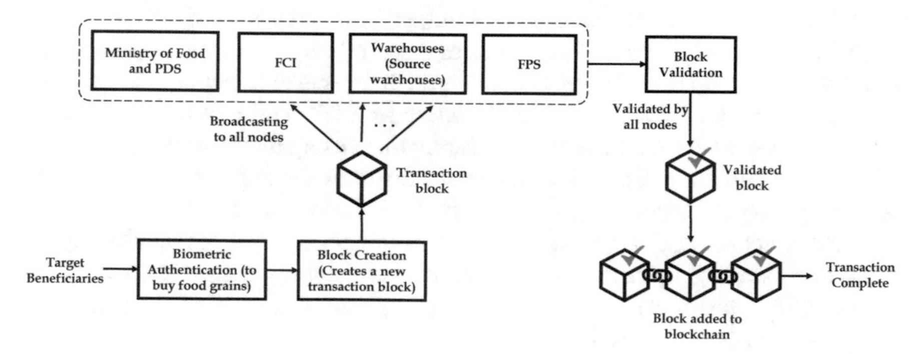

Fig. 8 Transaction scenario of food grains delivery

3.4.3 Transaction Scenario of Food Grains Delivery

As shown in Fig. 8, the beneficiaries are first authenticated by the biometric information using the PoS machine in the FPS shop. The authenticated beneficiaries are allowed to get the allocated food grains from the FPS. The transaction is closed after the food grains are delivered to the customer. The delivery transaction block is secured by the customer hash (biometric of the customer) and is validated by the Ministry, FCI, and the FPS shop from where the food grains are purchased.

4 Performance Analysis

This section presents the experimental simulation of the proposed PDS framework. An open-source Ethereum blockchain platform is used for the simulation of the proposed network. When compared to the other existing methodologies, the proposed framework differs significantly in terms of its core functions.

4.1 Analysis of Blockchain Technology in the PDS System

The NS3 simulator is used for the performance analysis of the smart city network to evaluate the performance of the proposed design. It is advantageous to compute low-resource device performance using the NS3 simulator. Another advantage of utilizing this simulator is that it improves overlay network performance and offers effective peer-to-peer network evaluation. Accuracy analysis is used to gauge the experimental viability of the proposed AI-BC architecture in a low-powered microcontroller unit. The accuracy analysis for the application is performed, and it is found that the accuracy increases when the number of edge nodes increases and the increased number of instances used for deep learning operations. The training dataset grows as a result of the contributions of more nodes, which also improves the accuracy. In the IoT PDS network, the investigation of security and privacy is done by calculating the object's Euclidian distance and similarity index. The similarity index value drops as Euclidian distance increases. As a result, as the similarity index value declines, the security and privacy of IoT gadgets rise. Figure 9 depicts the accuracy comparisons of the proposed unified AI-BC-IoT architecture for PDS. The maximum percentage accuracy observed for intelligent device deployment using blockchain technology is 85%. The maximum percentage accuracy for distributed computing and decentralized computing using blockchain technology is 88% and 92%, respectively. Without blockchain technology, the greatest percentage accuracy at intelligent device deployment, distributed computing, and decentralized computing is reported to be 57%, 64%, and 78%, respectively. As a result, adopting blockchain for IoT applications is always noticed as having high accuracy.

AI is essential for processing data or automating processes so that the information can be used in blockchain technology. Numerous trials have shown that AI can

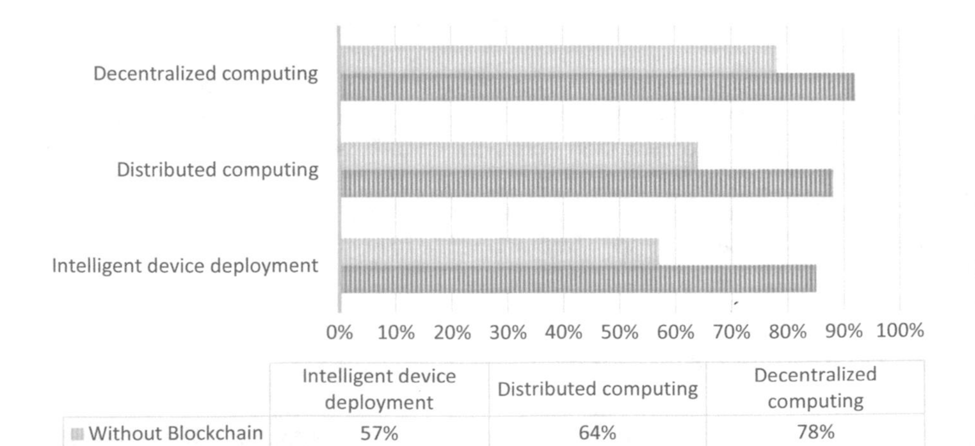

	Intelligent device deployment	Distributed computing	Decentralized computing
Without Blockchain	57%	64%	78%
With Blockchain	85%	88%	92%

Fig. 9 Accuracy comparison graph

perform tasks as well as humans, and it is suggested that AI is superior since it can operate around the clock without experiencing failure or being prone to human error. It is explored how artificial intelligence will be integrated with blockchain technology for IoT-based PDS. The intricacy, network scale, and high transaction costs of blockchain technology are only a few of its drawbacks. Artificial intelligence is being combined with blockchain to alleviate these constraints. AI resolves the implications caused by hardware, security and privacy, scalability, efficiency, and energy consumption through continuous monitoring of networks, knowledge acquisition, data integrity, authenticity, privacy, path discovery, aggregation, association, classification, scalability, efficiency, etc. By incorporating blockchain technology, AI issues for the proposed system are addressed. Blockchain addresses five distinct areas of AI challenges such as data sharing, security and privacy, explainable AI, effectiveness, and trust.

4.2 *TinyML-Based Food Quality Monitoring Mechanism*

A TinyML-based food quality monitoring system is provided in the succeeding paragraphs. The scenario is meant to demonstrate the viability of designing hardware capable of maintaining the integrity of a system, but the device, algorithms, and methods are all subject to modification in light of various use cases. A device that is appropriate for this use must satisfy a number of requirements. A TinyML-compliant development board is the first and most important component. The board then has to be set up to accommodate several sensors or contain pre-installed sensors. The device must also be able to transmit data to the blockchain system. Therefore, the Arduino Nano 33 BLE Sense board that is completely TinyML compliant is used in this work. It is equipped with all required sensors and a communication chip that can transfer information to the blockchain. It is required to construct a dataset of typical data instances. To do this, the device must be mounted in food containers and continually checked using the equipment. A dataset is created for the test case using lab simulations. For model training and inference, the Edge Impulse framework is used. A shallow customized Convolutional Neural Network (CNN) model is developed to detect the quality of food grains.

The False Positive (FPR) and False Negative (FNR) rates are evaluated for the model's assessment in order to determine how frequently the system failed to identify an anomaly and how frequently it mistakenly perceived a normal result as an anomalous one. Moreover, the number of true anomalies is determined using precision, and the entire model performance is calculated using the F1 score and accuracy metrics. The models are tested for identifying outliers in temperature, humidity, and moisture content. The test trials for the temperature yield 0.1415 FPR and 0.1053 FNR. The model achieved a precision measure of 0.85, an accuracy of 0.8756, and an F1-Score of 0.8718. For humidity, the model obtains 0.1058 FPR and 0.0918 FNR, a precision score of 0.89, an accuracy score of 0.9010, and an F1-Score of 0.8990. Similarly, for the moisture content evaluation, the model provides 0.0971 FPR and 0.1212

Table 3 Results of simulation testing

Trial instances	FNR	FPR	Precision	Accuracy	F1-Score
Temperature	0.1415	0.1053	0.8500	0.8756	0.8718
Humidity	0.1058	0.0918	0.8900	0.9010	0.8990
Moisture	0.0971	0.1212	0.8969	0.8911	0.8878

FNR, a precision score of 0.8969, an accuracy score of 0.8911, and an F1-Score of 0.8878. The experimental findings for the developed machine learning model to detect anomalies in temperature, humidity, and moisture are presented in Table 3. The results illustrate that it is possible to construct a device that can remotely execute machine learning models and anomaly detection methods into monitoring devices to enhance their security. In the future, it is planned to evaluate the overall model performance under real-world conditions including food grain trucks and containers with RFID technology.

5 Practical Inferences

The Ministry of Food and PDS (Central Government), FCI, and state governments that regulate the PDS in India can implement the intelligent and smart PDS system using the proposed conceptual framework. The framework will maintain transaction transparency because all recorded transactions are immutable and can only be executed by those who have been permitted to do so because the blocks are safeguarded with each user's unique digital hash (fingerprint). Since the transaction would only be added after his or her approval, if the person engages in malpractice, they will be immediately accountable (digital hash–signature). The proposed PDS combines smart contracts and machine learning technologies for procurement and tracking data recording and food quality monitoring that occur during the storage and distribution process. If the sample quality at the storage and distribution process is designated as unqualified, further distribution is not carried out to avoid extra losses. By merging off-chain models with on-chain data, our solution can leverage blockchain traceability and smart contract auto-execution to achieve reliable and efficient quality monitoring. The weight of the product at the time of packing, the time and date of packaging, and any other specifications required would all be recorded on the RFID tag. The addition of RFID tags to the bag will prevent it from being altered or mishandled while in transportation. RFID tags on bags are registered and locked in the block when they leave the warehouse. The identical RFID must connect to the chosen FPS (with the same RFID Tags). The FPS block will close once these RFID tags have been counted (thus ensuring RFID tags are untampered while transportation). The grains will be delivered to the consumer with biometric confirmation to guarantee that they are the intended recipient. Consumers will use their biometric digital signature to close their block after getting grains from FPS. The freedom for customers to close their block will guarantee that they get grains. The customer will not close the block if there is a discrepancy and may file a formal complaint with higher authorities. Higher authorities inspecting blockchain transactions will notice that this specific consumer's block is unclosed or unadded and can investigate further.

Deploying Blockchain, AI, and IoT-driven PDS is a cost and time-effective solution to the smart city initiative. The investment for the deployment of the proposed system will be a one-time investment in the first year and is estimated roughly Rs. 23 billion for the permissioned blockchain which can execute 500,000 transactions per day. The cost will be reduced gradually in the subsequent years. Initial investment cost includes the software and hardware ideation, development, implementation, cloud service and storage costs, maintenance, review, and monitoring costs. Also, the remittance of the technical experts and non-technical human resources involved in the development and deployment of the proposed solution. The components required to build the hardware for food quality monitoring are also available in the market at affordable prices. Even if the number of transactions increased exponentially, the cost would be less than the amount lost due to corruption. Furthermore, the use of blockchain will ensure that the grains reach the intended recipients, thus achieving the goal.

6 Conclusion, Limitations, and Future Work

The proposed AI-Blockchain-IoT framework was designed for PDS supply chain management in Smart Cities, in which food is distributed properly to the intended recipient. The framework can guarantee integration, security, shared databases, accountability, traceability, and transparency. It could potentially be extended with appropriate modifications for use in the other supply chain management, which similarly faces transparency and corruption concerns.

The framework is predicated on specific assumptions. To enable blockchain use, there should be reliable internet connectivity, enough infrastructure availability, and technological skills. To use blockchain technology, warehouse officers and FPS owners must have adequate training. Even after applying the framework, there are still risks that dishonest officials will replace the grains without letting the RFID tags linked to bags be affected. Also, the framework is only applicable from the warehouse level to the consumer level. The future extension could create a framework that includes farmers and other intermediaries. This will allow tracking of where grains are procured as well as testing grain quality. Future research can look into the perspectives of PDS stakeholders and whether they will embrace such a framework if it is adopted.

References

1. Almalki FA, Alsamhi SH, Sahal R et al (2021) Green IoT for eco-friendly and sustainable smart cities: future directions and opportunities. Mobile Netw Appl
2. Kumar A (2020) Improvement of public distribution system efficiency applying blockchain technology during the pandemic outbreak (COVID-19). J Human Log Supply Chain Manag
3. Mishra H, Maheshwari P (2021) Blockchain in Indian public distribution system: a conceptual framework to prevent leakage of the supplies and its enablers and disablers. J Global Operat Strat Sourc
4. Sharma A, Podoplelova E, Shapovalov G, et al (2021) Sustainable smart cities: convergence of artificial intelligence and blockchain. Sustainability 13:13076
5. Singh S, Sharma P, Yoon B et al (2020) Convergence of blockchain and artificial intelligence in IoT network for the sustainable smart city. Sustain Cities Soc 63:102364
6. Bellini P, Nesi P, Pantaleo G (2022) IoT-enabled smart cities: a review of concepts, frameworks and key technologies. MDPI in Appl Sci 12
7. Sandeep BK, Shajimon KJ (2020) Blockchain integration with low-power internet of things devices. Handbook of Research on Blockchain Technology, Elsevier, pp 183–211
8. Deepa N, Pham QV, Dinh CN et al (2022) A survey on blockchain for big data: approaches, opportunities, and future directions, future generation computer. System 131:209–226
9. Henkel J, Pagani S, Amrouch H, et al (2017) Ultra-low power and dependability for IoT devices, design, automation & test in Europe conference & exhibition (DATE)
10. Pajooh HH, Rashid MA, Alam F et al (2021) IoT big data provenance scheme using blockchain on Hadoop ecosystem. J Big Data 8:114
11. Bauer M, Sanchez L, Song JS (2021) IoT-enabled smart cities: evolution and outlook. Sensors 21(13):4511
12. Jawaid MF, Khan AF (2020) The smart city mission in India and prospects of improvement in the urban environment. IOP Conf Ser: Mater Sci Eng 955(1):012001
13. Prasad D, Alizadeh T (2020) What makes Indian cities smart? A policy analysis of smart cities mission. Telemat Inf 55:101466
14. Russell MS, Pathak PA, Agrawal G (2019) India's "smart" cities mission: a preliminary examination into India's newest urban development policy. J Urban Aff 41(4):518–534
15. Rejeb A, Rejeb K, Simske SJ et al (2022) Blockchain technology in the smart city: a bibliometric review. Qual Quant 56:2875–2906
16. Kabir MH, Hasan KF, Hasan MK, Ansari K (2022) Explainable artificial intelligence for smart city application: a secure and trusted platform. In: Ahmed M, Islam SR, Anwar A, Moustafa N, Pathan ASK (eds) Explainable artificial intelligence for cyber security. Studies in computational intelligence, vol 1025. Springer, Cham
17. Zamponi ME, Barbierato E (2022) The dual role of artificial intelligence in developing smart cities. Smart Cities 5:728–755
18. Alajlan NN, Ibrahim DM (2022) TinyML: enabling of inference deep learning models on ultra-low-power IoT edge devices for AI applications. Micromachines 13:851
19. Shivam S, Bharat B, Mohd AA (2021) Blockchain based solutions to secure IoT: background, integration trends and a way forward. J Netw Comput Appl 181:103050
20. Kummar S, Bhushan B, Bhatia S (2022) Blockchain based big data solutions for internet of things (IoT) and smart cities. In: Sharma R, Sharma D (eds) New trends and applications in internet of things (IoT) and big data analytics. Intelligent systems reference library, p 221
21. Kashyap S, Bhushan B, Kumar A, Nand P (2022) Quantum blockchain approach for security enhancement in cyberworld. In: Kumar R, Sharma R, Pattnaik PK (eds) Multimedia technologies in the internet of things environment, vol 3. Studies in Big Data, p 108
22. Bhushan B, Kadam K, Parashar R, Kumar S, Thakur AK (2022) Leveraging blockchain technology in sustainable supply chain management and logistics. In: Muthu SS (eds) Blockchain technologies for sustainability. Environmental footprints and eco-design of products and processes. Springer, Singapore

23. Bharat B, Preeti Sinha K, Martin S, Andrew J (2021) Untangling blockchain technology: a survey on state of the art, security threats, privacy services, applications and future research directions. Comput Electr Eng 90
24. Tsoukas V, Gkogkidis A, Kampa A, Spathoulas G, Kakarountas A (2022) Enhancing food supply chain security through the use of blockchain and TinyML. Information 13(5):213
25. Gkogkidis A, Tsoukas V, Papafotikas S, Boumpa E, Kakarountas A (2022) A TinyML-based system for gas leakage detection. In: 2022 11th international conference on modern circuits and systems technologies (MOCAST), pp 1–5
26. Andrade P, Silva I, Silva M, Flores T, Cassiano J, Costa DG (2022) A TinyML soft-sensor approach for low-cost detection and monitoring of vehicular emissions. Sensors 22:3838
27. Zaidi SAR, Hayajneh AM, Hafeez M, Ahmed QZ (2022) Unlocking edge intelligence through tiny machine learning (TinyML). IEEE Access 10:100867–100877
28. Sinha M, Chacko E, Makhija P, Pramanik S (2021) Energy-efficient smart cities with green internet of things. In: Chakraborty C (eds) Green technological innovation for sustainable smart societies. Springer, Cham
29. Thilakarathne NN, Kagita MK, Priyashan WDM (2022) Green internet of things: The next generation energy efficient internet of things. In: Iyer B, Ghosh D, Balas VE (eds) Applied information processing systems, advances in intelligent systems and computing, p 1354

Artificial Intelligence (AI) and Internet of Things (IoT): Application in Detecting and Containing the Spread of COVID-19

Mohd Anas Wajid, Aasim Zafar, Bharat Bhushan, Akib Mohi Ud Din Khanday, and Mohammad Saif Wajid

Abstract The goal of this work is to provide a high-level overview of the ways in which the Internet of Things and Artificial Intelligence are being used to control COVID-19. The current work begins with an overview of IoT, AI in healthcare, and the origins of the COVID-19 virus. The work has gathered prior research from scholarly journals to see how AIoT has been employed in healthcare, and has also stated the AIoT technologies that were used to contain the infection. We have discussed the applications of these technologies in the pharmaceutical industry. We also outlined some of the most noteworthy IoT and AI technologies and use cases that aim to control COVID-19. Challenges of introducing AIoT technologies into healthcare environments are discussed. This work has been carried out so that researchers just beginning their careers in this discipline would gain a comprehensive overview of the field. This will encourage them as well as other stakeholders to utilise technology to prevent future pandemics.

M. A. Wajid (✉) · A. M. U. D. Khanday
Department of Computer Science & Applications, Sharda University, Greater Noida, India
e-mail: mohd.wajid1@sharda.ac.in

A. Zafar
Department of Computer Science, Aligarh Muslim University, Aligarh 202002, India

M. S. Wajid
School of Engineering and Sciences, Tecnológico de Monterrey, Monterrey, Mexico

B. Bhushan
School of Engineering and Technology (SET), Sharda University, Greater Noida, India

A. M. U. D. Khanday
United Arab Emirates University, Al Ain, UAE

B. Bhushan et al. (eds.), *AI Models for Blockchain-Based Intelligent Networks in IoT Systems*, Engineering Cyber-Physical Systems and Critical Infrastructures 6,
https://doi.org/10.1007/978-3-031-31952-5_16

1 Introduction to IoT and AI in Healthcare

The inception of COVID-19 in the year 2019 is marked as a pandemic in history that has touched the lives of people all over the globe. As is the spread of disease around the world, so is the research which has been undertaken to contain the pandemic. Researchers around the globe are actively searching the ways to contain and treat this virus using present technology. When we talk about present technology in the twenty-first century, we cannot abstain from mentioning the use of IoT and AI in healthcare and monitoring. The sudden hit by the pandemic has left the government and other stakeholders awestruck since the available pharmaceutical means were not adequate to identify and contain the pandemic in the first instance. The dedicated research in the field of IoT and AI has provided ways to identify, monitor, and treat such pandemics at the first instance by making use of the real-time network for better monitoring purposes. IoT-backed technologies with ubiquitous sensing ability and seamless connectivity are well efficient in gathering real-time data such as vital signs and symptoms to detect COVID-19. This can be inferred from Fig. 1, where an IoT-enabled drone is controlled using a virtual reality device to collect real-time data. This data is transmitted to the connected devices, where this data is treated using machine learning algorithms and AI to predict the infection or health state of a person.

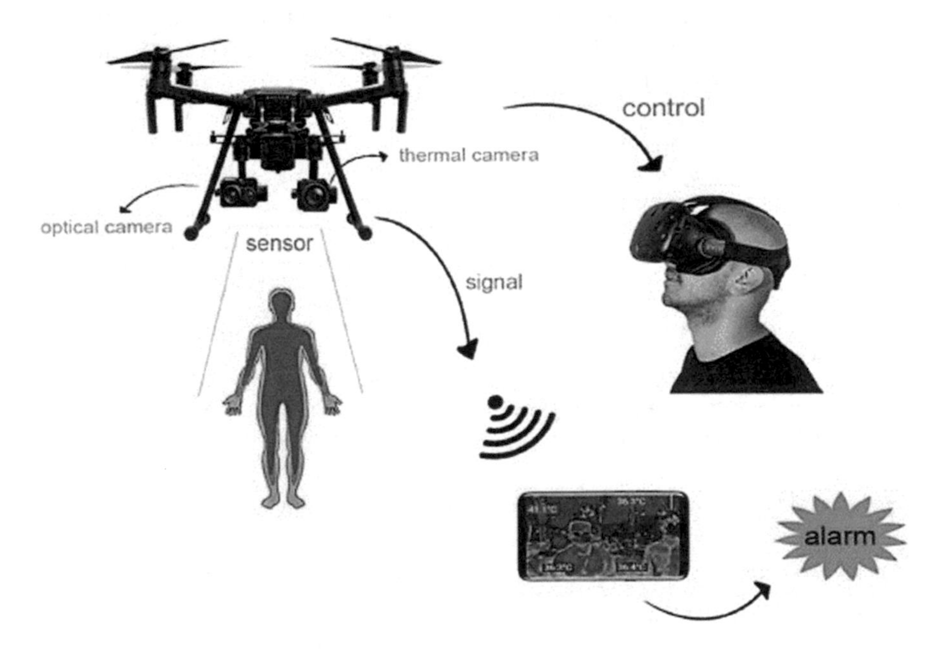

Fig. 1 A smartphone magnetometer-based diagnostic [1]

Therefore it has become significant to utilize AI-assisted IoT technology for better healthcare monitoring purposes in a real-time environment. The AIoT-assisted technologies will not only aid healthcare workers in treating such pandemics but would also enable the general public to monitor their health status by the use of technology such as wearable devices and installed sensors at home [2]. These devices would aid in analyzing various aspects of day-to-day activities such as temperature, food habits, heart rate, respiration rate, sleeping patterns, and various other physiological values to identify and make a judgment at the first stance on their own. The use of IoT in the frontline can help in the development of modern intelligent healthcare, intelligent diagnosis, treatment, and prevention of such pandemics in the future [3]. The applications of IoT in containing COVID-19 can be inferred from Fig. 2.

To understand how AIoT helps in treating such pandemics in the first instance, let us consider the case of the present pandemic, i.e., COVID-19. The major problem which was faced during COVID-19 was screening and testing to identify the infection. Intelligent remote screening using IoT has done away with the problem of overcrowded emergency rooms where people were more exposed to the virus. The intelligent diagnosis methods using IoT can help in removing existing slow speed and less accurate manual reading of the CT Scans, which was a major problem when the pandemic broke. IoT-enabled devices help in gathering real-time data using faster 5G/6G networks and aid doctors in isolating a person at the earliest point possible so that he does not need to visit a doctor and keep himself/herself isolated. This will lead to slow transmission of the infection.

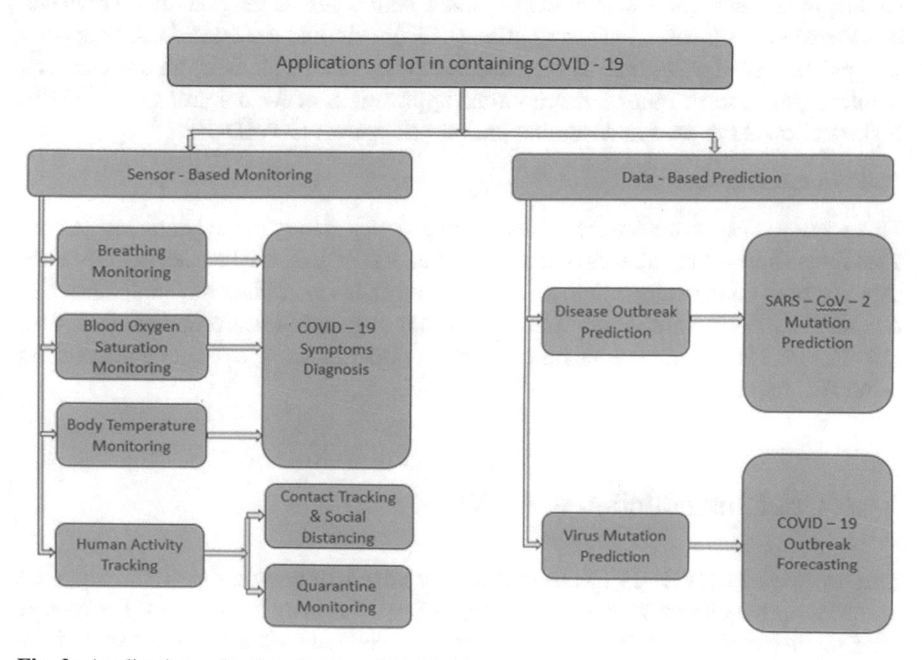

Fig. 2 Applications of IoT in fighting COVID-19

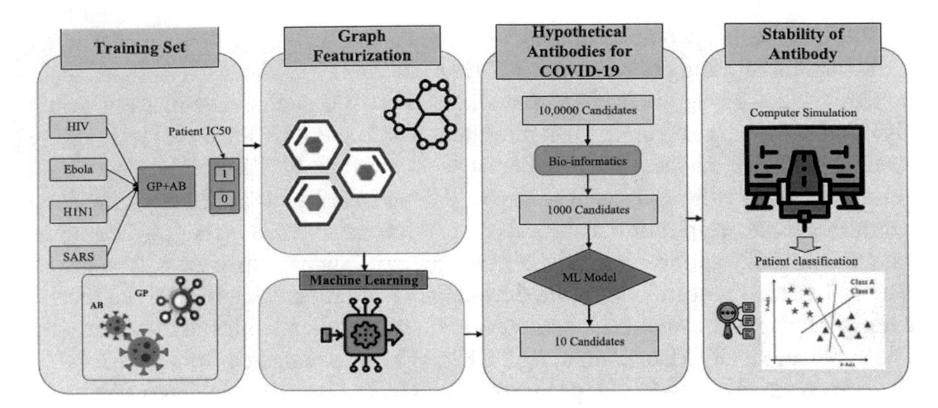

Fig. 3 Antibodies against COVID-19 designed using AI [5]

Not only IoT but AI has a significant role to play in this regard. Now let's discuss them briefly by explaining scenarios where it has helped in containing the current pandemic, i.e., COVID-19. When we monitor any epidemic, the task of identifying the trajectory of virus infection is of utmost importance for controlling the diseases. AI has the potential to learn from huge available datasets where these datasets are based on many known factors such as demography, deaths, environmental conditions, etc. [4]. Based on the analysis of this data, AI can predict future pandemics and can create prediction models with much accuracy. AI has shown the potential which can be utilized in healthcare monitoring, disease diagnosis, and the creation of drugs that will provide great assistance to the frontline workers as well as to the masses as a whole. Figure 3 shows one of the important applications of AI in fighting COVID-19. It shows how AI is used in designing antibodies against COVID-19.

Take-home Message

This section presents before the reader the use of technology in the healthcare sector. This shows how the most recent technology, i.e., AI and IoT, can be used in the healthcare sector to contain any infection. This also explains to readers the applications of AI and IoT, which stakeholders are using to fight against the current pandemic, i.e., COVID-19. These applications range from disease diagnosis to drug manufacturing using AIoT technology.

2 A Brief Introduction to COVID-19

Coronavirus or COVID-19 has caused deadly havoc all around the globe. Since its inception in the city of Wuhan, China, COVID-19 has taken many lives globally. A brief history of COVID-19 will enable us to understand what this disease is, how it is identified, what symptoms people develop at first when they get infected with the virus, and how the disease spreads. The understanding of all these concepts will enable us to explore the applications of IoT and AI in identifying and containing the spread of the infections like COVID-19.

The virus behind the spread of COVID-19 is SARS-CoV-2 [6]. This virus causes respiratory tract infection and other mild symptoms which could easily be identified using IoT-enabled devices. It has its effects on both: the upper respiratory tract and the lower respiratory tract, and corresponding symptoms are ongoing chest pain or pressure, bluish lips or face, trouble breathing or shortness of breath, can't wake up fully, and a state of confusion. The researchers have demonstrated that these symptoms are among the most prominent in patients infected with COVID-19. Their studies show the presence of these symptoms as follows:

- Mucus/phlegm 27%
- Shortness of breath 31%
- Body aches 35%
- Lack of appetite 40%
- Cough 59%
- Fatigue 70%
- Fever 99%

AIoT has a much wider role to play in this regard. IoT devices will easily recognize these symptoms and could be used for modeling AI algorithms for identifying and containing infections like COVID-19. Like in Fig. 4, a new approach is proposed for measuring blood SpO2 in a noncontact way using the camera-recorded video of a person. Similarly, in Fig. 5, we can see how IoT-enabled drones can be used to identify the above-mentioned symptoms and can easily detect a person infected with COVID-19.

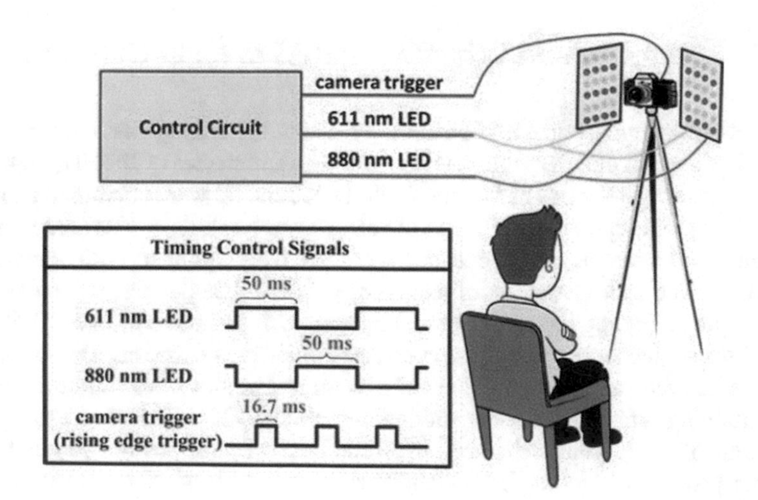

Fig. 4 IoT: SpO_2 measured using IoT-enabled camera [7]

Fig. 5 IoT: Use of IoT-enabled drones for detecting COVID-19 using infrared thermography imaging [8]

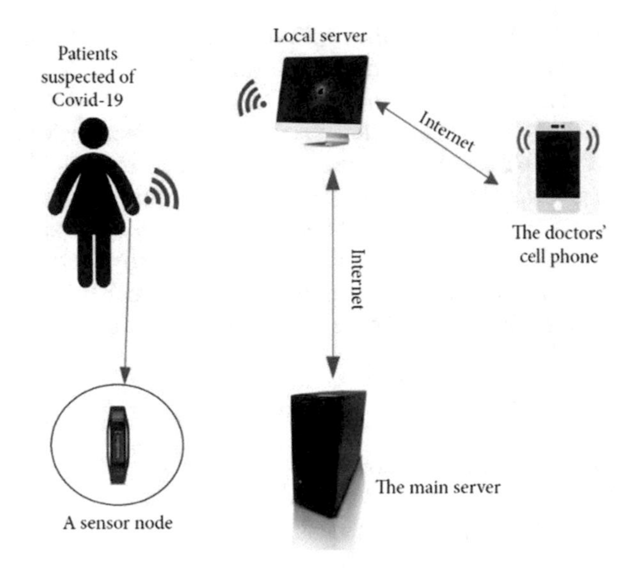

Take-home Message

This section presents a brief history of COVID-19, its symptoms and how AIoT technologies can utilize the real-time data to contain such infections in the future. The section also presents two recent examples of AIoT-assisted technology which has been used to contain COVID-19.

3 Motivations Behind Using IoT and AI to Fight COVID-19

The development of science and technology marks the twenty-first century. Technology has entered our lives and has become part and parcel of life. Though there have been many pandemics before the COVID-19, and all were contained using the conventional techniques where the use of science and technology was near minimal. This happened because earlier decades were not well equipped with science and technology. The widespread use of technology like mobile phones, wearable smart devices, and smart appliances at home, together with fast internet like 5G/6G, has enabled researchers to explore its applications in various domains. The domain of medicine has seen an exponential use of current technology while treating patients and containing infections, but the sudden outbreak of COVID-19 has exposed the limitations of the present technology. The sudden outbreak was not handled using available pharmaceutical methods also. There was a lack of surveillance systems, and decision-making was based on the information collected from various sources that were unreliable most of the time. Also, the available conventional methods were not accurate and were time-consuming. The spread of infection was much faster than its diagnosis and treatment. Therefore there was a need for faster ways that are accurate and efficient in terms of identifying infection at an initial stage. IoT and AI have come along as a boon during this pandemic. The pandemic at its initial stage needs to be identified using various symptoms among people, which could easily be done using a surveillance system equipped with IoT devices and fast internet. The monitoring can be done in various ways. This includes monitoring breathing patterns, quarantine, blood oxygen saturation, body temperature, and contact tracing.

The collected data using AIoT technologies will help us in forecasting the outbreak of any infection. The AI techniques like neural networks assisted with LSTM (Long Short Term Memory) are capable of predicting any illness using collected data and also taking data from social media platforms. The capability of AI in detecting the virus using various techniques could be utilized in knowing how the virus will spread. This prediction will be done using data from previous health records, conditions prevailing in environment, and how it is transmitted from one person to another. The IoT and AI have not only provided help in detecting and containing COVID-19, but they played an important role in maintaining the mental health of quarantined patients. An example of AIoT-enabled technology for this task is Paro (Fig. 6) which is a social robot. This robot is equipped with a technology called the Sefucy IoT button, which was earlier used to track missing children. This button enables the robot to collect data about the patient, and in case of an emergency, it can inform the doctors and family members of a confirmed COVID-19 patient.

AIoT technology has also helped in maintaining hygiene in hospitals and quarantine centers. This technology has helped the frontline workers remotely to clean these hospitals and quarantine centers. An example of such an AIoT-enabled machine is XDBOT (Fig. 7), a human-operated robot used to sanitize the contaminated areas.

So, overall we can see the motivation behind using IoT and AI to contain COVID-19. In further sections, we will have in-depth knowledge of materials and methods utilized for this task and the IoT-AI architectures proposed to fight COVID-19.

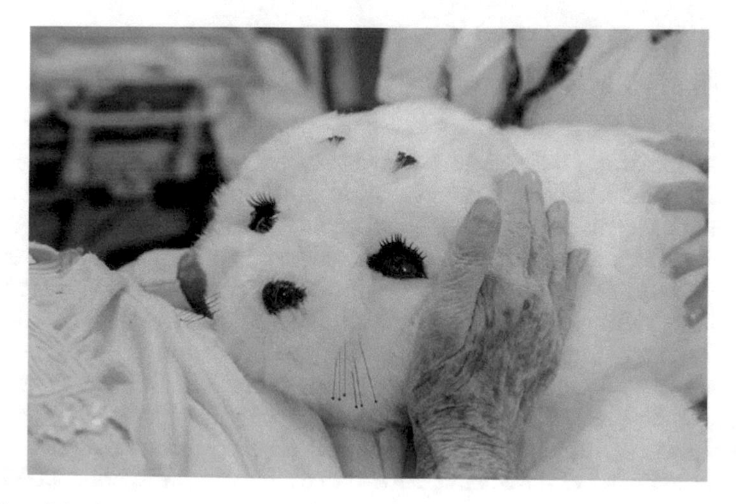

Fig. 6 A social robot Paro

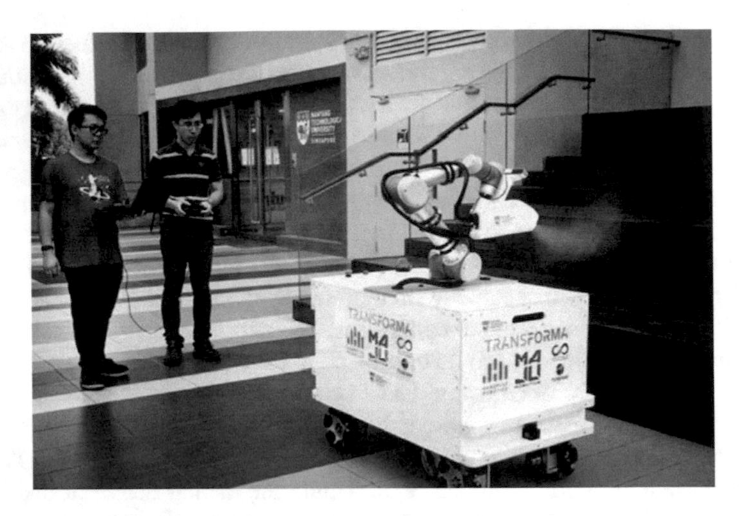

Fig. 7 An AI-based XDBOT for sanitizing the contaminated areas

Take-home Message

This section explains the motivation behind using AI and IoT technologies to contain the infection, i.e., COVID-19. The section first highlights the limitations of the prevailing healthcare systems and later in the section presents how these limitations of present the healthcare system can be removed using AIoT technologies. We have presented two recent technologies from the real world, i.e., Paro and XDBOT, and highlighted how these are utilized in the healthcare system to contain COVID-19. After going through this section, the reader will understand the motivation behind using these technologies.

4 Methods and Materials

This section provides the research overview of IoT and AI technologies used to contain COVID-19 and explains how various researchers adopt these technologies to fight this deadly disease or any other infection. For this purpose, we have collected several scientific papers which contain the tools and techniques of AIoT and map them with strategies used to contain the spread of COVID-19. COVID-19, IoT, AI, deep learning, and machine learning are used as search strings. This is done to collect scientific research from top-tier & peer-reviewed journals, symposiums, books, and conferences. The databases used for the purpose are ACM digital library, IEEE Xplore, ScienceDirect, and Springer Link. After going through the collected papers, the contribution of IoT and AI in fighting COVID-19 is identified. Tables 1 and 2 present the summary of the existing AIoT-based solutions for fighting COVID-19.

Table 1 Summary of AI-based solutions for containing and controlling COVID-19

Author	Title	AI technology
Vinay Chamola et al. (2019) [9]	"A Comprehensive Review of the COVID-19 Pandemic and the Role of IoT, Drones, AI, Blockchain, and 5G in Managing Its Impact"	Mitigating the impact of COVID-19 using AI
Daniel Shu Wei Ting et al. (2020) [10]	"Digital technology and COVID-19"	Describes the role of Deep learning in containing the virus, i.e., COVID-19
Adedoyin Ahmed Hussain et al. (2020) [11]	"AI Techniques for COVID-19"	Provides a study of AI technologies that are used for detecting and diagnosing COVID-19. This also predicts the transmission
Abu Sufian et al. (2020) [12]	"A Survey on Deep Transfer Learning to Edge Computing for Mitigating the COVID-19 Pandemic"	Describes the application of Deep learning, deep transfer learning, and edge computing for containing COVID-19
Agam Bansal et al. (2020) [13]	"Utility of Artificial Intelligence Amidst the COVID 19 Pandemic: A Review"	Applications of AI in screening and predicting COVID-19
Sweta Bhattacharya et al. (2020) [14]	"Deep learning and medical image processing for coronavirus (COVID-19) pandemic: A survey"	COVID-19 data analytics using AI algorithms
Ghouali S. et al. (2020) [15]	"Artificial Intelligence-Based Teleopthalmology Application for Diagnosis of Diabetics Retinopathy"	Describes the use of AI algorithms in data analytics and drug discovery
Thanh Thi Nguyen et al. (2021) [16]	"Genomic mutations and changes in protein secondary structure and solvent accessibility of SARS-CoV-2 (COVID-19 virus)"	Studies Protein structure of virus using AI techniques which helps in drug discovery to contain COVID-19
Dinh C. Nguyen et al. (2021) [17]	"Enabling AI in Future Wireless Networks: A Data Life Cycle Perspective"	Proposes a wireless AI architecture for COVID-19 tracking using ML and Deep learning
Kantilal P Rane (2020) [18]	"Design and Development of Low-Cost Humanoid Robot with Thermal Temperature Scanner for COVID-19 Virus Preliminary Identification"	Proposes AI techniques based on temperature monitoring using smart helmets to identify infected people in public places
Aasim Zafar et al. (2020) [6]	"A Mathematical Model to Analyze the Role of Uncertain and Indeterminate Factors in the Spread of Pandemics like COVID-19 Using Neutrosophy: A Case Study of India"	Proposes a mathematical model based on Neutrosophic Logic for the identification of factors responsible for the spread of COVID-19

(continued)

Table 1 (continued)

Author	Title	AI technology
P. K. Gupta et al. (2021) [19]	"COVID-WideNet—A capsule network for COVID-19 detection"	For detecting COVID-19 from X-ray images authors have proposed a capsule network based on deep learning

Take-Home Message

This section presents a brief review of the research work from peer-reviewed journals related to fighting COVID-19. These are the AI and IoT technologies that have been used to contain the spread of COVID-19. This section clearly explains what AI and IoT technologies are used for this task, how they are used, and the principles behind these technologies. This will enable the reader to better understand ongoing research in the area and its applications in the real world.

5 AI-IoT Architecture for Controlling and Containing COVID-19

In this section, we present an AIoT architecture that has been used to fight COVID-19. The presented architecture integrates AI and IoT to fight COVID-19 and is organized into four different layers. These layers include coronavirus data source, IoT concepts & functions, AI concepts & functions, and the stakeholders. To better understand the architecture, we explain the workflow in Fig. 8.

The first step in this regard is the collection of data from various sources such as IoT-enabled devices, clinical data, data from social media, etc. This collected data becomes what we refer to big data. This data could be X-rays, CT scans, cases of historic infections, location of outbreak areas, information about the structure of the virus, and data collected from the World Health Organization. IoT plays a vital role in collecting data later processed using AI. IoT devices are extensively used to collect data on breathing, blood oxygen saturation, and body temperature monitoring. IoT devices can also be used for quarantine monitoring, contact tracing, and SARS-CoV-2 mutation prediction. This can be inferred from Fig. 9, which shows social distancing and contact tracing being done using IoT-enabled devices.

The big data collected using the above-mentioned IoT applications is utilized in predicting COVID-19 and assists AI in fighting the infection at the root level. Having known the importance of data related to COVID-19, many countries have made big datasets, and some have shared them over GitHub and Kaggle for research usage. AI uses this data and aids in fighting the infection. AI-assisted technologies such as neural networks, regression, and classification aids in coronavirus detection and analytics. These technologies also aid in vaccine/drug development and prediction of any future outbreak. At the top of the architecture come the stakeholders,

Table 2 Summary of IoT-based solutions for containing and controlling COVID-19

Author	Title	IoT Technologies
Wang et al. (2020) [20]	"Abnormal respiratory patterns classifier may contribute to a largescale screening of people infected with COVID-19 accurately and unobtrusively"	Introduced depth cameras using deep learning for monitoring and identifying six types of COVID-19 breathing patterns
Miad Faezipour et al. (2020) [21]	"Smartphone-Based Self-Testing of COVID-19 Using Breathing Sounds"	Distinguishes healthy and unhealthy users based on data collected from the microphone
Guochao Wang et al. (2014) [22]	"Application of Linear-Frequency-Modulated Continuous-Wave (LFMCW) Radars for Tracking of Vital Signs"	Radar-based breathing monitoring for health monitoring
Heba Abdelnasser et al. (2016) [23]	"UbiBreathe: A Ubiquitous non-Invasive WiFi-based Breathing Estimator"	Proposes a wifi-based breathing sensor UbiBreathe for monitoring breathing patterns indoor and outdoor
Dangdang Shao et al. (2017) [24]	"Noncontact Monitoring of Blood Oxygen Saturation Using Camera and Dual-Wavelength Imaging System"	Introduces remote IoT-based cameras for blood SpO2 measurement
Mohammed et al. (2020) [8]	"Toward a Novel Design for Coronavirus Detection and Diagnosis System Using IoT Based Drone Technology"	Introduces IoT-based infrared thermography imaging for identifying persons infected with COVID-19
S–H. Hsiao et al. (2020) [25]	"Measurement of body temperature to prevent pandemic COVID-19 in hospitals in Taiwan: repeated measurement is necessary"	IoT-based isolation technique in hospitals for containing the spread of COVID-19
Nor Aini Zakaria et al. (2019) [26]	"IoT (Internet of Things) Based Infant Body Temperature Monitoring"	Proposes IoT-based temperature monitoring for remote health care
Peng Hu (2021) [27]	"IoT-based Contact Tracing Systems for Infectious Diseases: Architecture and Analysis"	Introduces IoT-based technology for contact tracing during COVID-19
Sandeep K. Sood et al. (2017) [28]	"Wearable IoT sensor-based healthcare system for identifying and controlling chikungunya virus"	Proposes IoT, fog computing, and cloud computing-based solutions to contain the infection

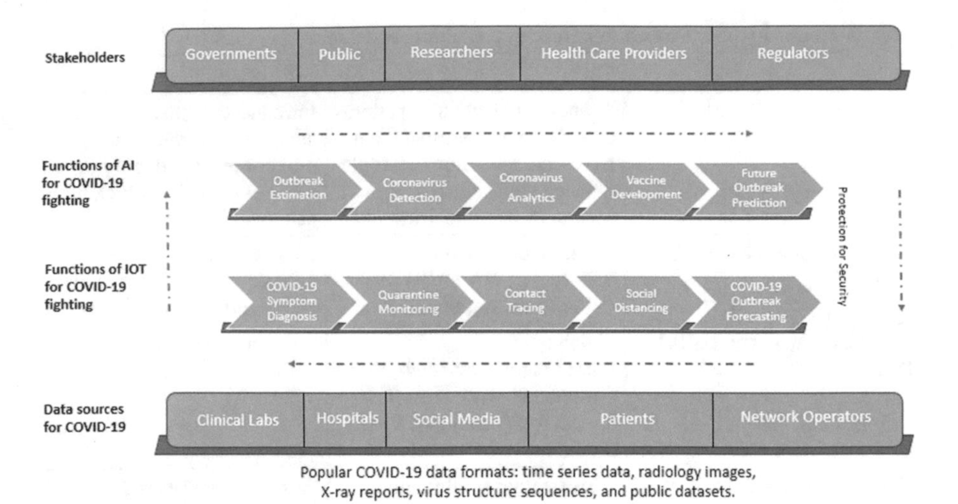

Fig. 8 AI and IoT for coronavirus fighting

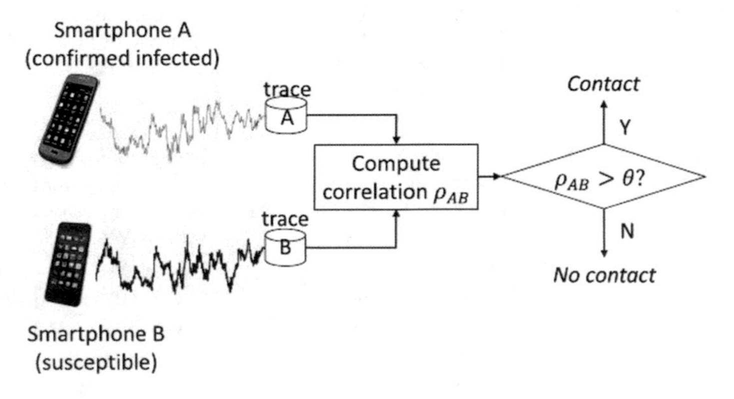

Fig. 9 Contact tracing and social distancing using magnetometer-based smart phones [29]

which include frontline workers, researchers, regulators, and governments. These stakeholders benefit from the AIoT-based solutions. IoT helps the stakeholders by providing them with real-time data about the patient activities and allows remote monitoring of patients, which was the need during COVID-19. IoT-assisted wearable devices help to collect data that plays an essential role in health care monitoring and can reduce any possibility of community spread of the infection. To summarize the AIoT architecture being discussed above, we now present a flow diagram (Fig. 10) to further enhance our understanding of the AIoT architecture.

Take-Home Message

This section provides a detailed architecture that has been used to fight COVID-19. The architecture comprises four layers, and each has its significance. After going

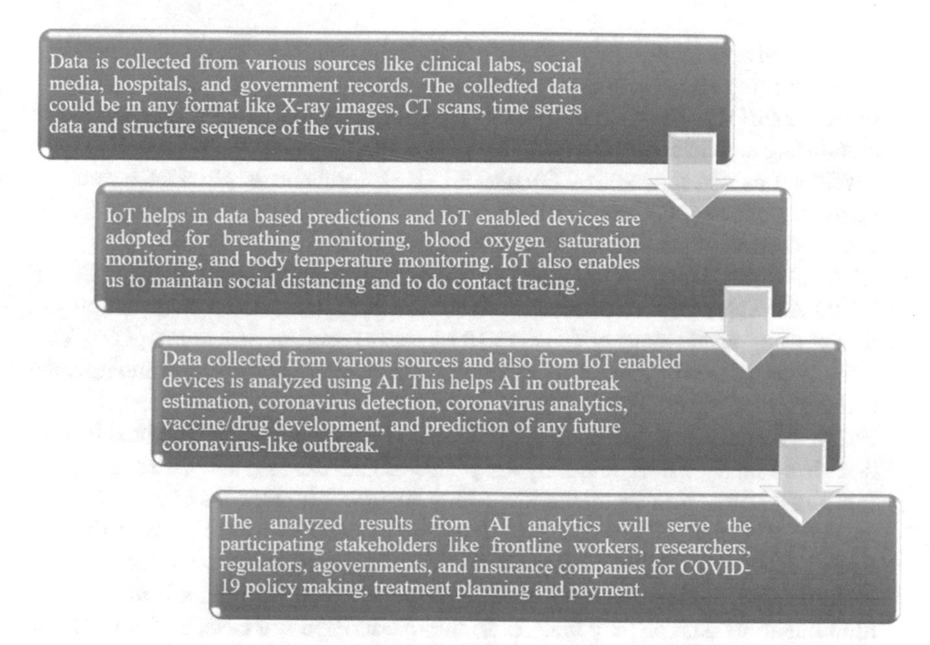

Fig. 10 The flowchart of AIoT architecture for fighting COVID-19

through this section, the reader will have a clear understanding of AIoT architecture that has been used to fight COVID-19.

6 AI and IoT Use Cases for Containing COVID-19

6.1 AI Use Cases

The applications of AI have always been explored in healthcare by companies and academia. Companies in the recent past have worked for efficient detection and diagnosis of infections like COVID-19 using AI. Some of the projects and use cases on AI for containing COVID-19 are highlighted below:

- **BLUEDOT**: This project was undertaken by a Canadian start-up to detect COVID-19 using AI-based techniques. The project uses AI algorithms and is the first to detect the spread of COVID-19 in the city of Wuhan hours before the authority could proclaim its spread officially. This project uses data available over social media, data from healthcare agencies, and government documents to predict the disease using machine learning algorithms and NLP.
- **GOOGLE DeepMind, AlphaFold SYSTEM**: This project of google DeepMind has very recently introduced an AlphaFold system, an AI tool for predicting the structure of proteins of SARS-CoV-2 [30]. The protein structure of a virus provides

information about a possible outbreak. This tool is built using AI techniques like Deep Learning and Machine Learning.

- **INFERVISION**: This is one of the recent AI use cases which aids doctors in monitoring and detecting a particular disease among patients. This has also been developed as a response to COVID-19 [31]. Infervision is efficient because it helps in analyzing CT scans at a faster pace which is highly important for the early detection of pneumonia among the patients.
- **NVIDIA's AI**: This is a very recent innovation being done in Wuhan, China. NVIDIA's AI is a GPU-accelerated AI use case that is being extensively employed for detecting visual signs of COVID-19 by identifying cancers in lung CTs. This AI software is used by doctors for detecting, diagnosing, and early treating infected patients.
- **FEDERATED AI**: A federated AI project has been developed by medical institutions from Japan, China, and Italy [25]. This project uses FL and CT scan images for COVID-19 region segmentation. Many AI-based approaches have performed exceptionally well when it comes to detecting COVID-19 by collecting sufficient data to implement intelligent algorithms, yet there remain privacy concerns as data is shared with data centers for data analytics. Federated AI project is in the limelight due to its data privacy feature. In this AI-assisted use case, each institution trains their Deep learning models using their local repositories of data, i.e., X-ray images and CT scans. Later, only model parameters like gradient are exchanged, preserving the sensitive and actual information.

6.2 IoT Use Cases

Applications of IoT can be seen everywhere, but many of its applications have been explored in healthcare during COVID-19. IoT-enabled devices have become part and parcel of our lifestyle after COVID-19. Below we mention some of the use cases of IoT.

- **CAMERA**: The camera is a crucial sensor in IoT-enabled devices. The camera is used to capture images and videos, which can be analyzed for different purposes such as noncontact monitoring, contact tracing, human behaviour recognition, and health state [32].
- **INERTIAL SENSOR**: This sensor is deployed in mobile devices and other IoT-enabled wearable devices. These sensors are based on the principle of inertia and relevant measurement, including an accelerometer and gyroscope [2]. Both accelerometer and gyroscope work along 3-axis. Where an accelerometer derives 3-axis acceleration measurement for detecting dynamic forces of the device like gravity, movement, and vibration, a gyroscope, on the other hand, is generally a spinning wheel with 3-axis rotation. Both these sensors provide data for identifying human behavior attributes that have extensively been utilized to study COVID-19.

- **mmWave RADAR**: mmWave RADAR is mostly used for non-contacting and non-invasive sensing. mmWave RADAR uses reflecting signals to capture the movements of the objects using its capability to modulate transmission signals to sweep across a specific frequency band [33]. mmWave RADAR has resistance against any obstruction, such as light and sound, therefore is extensively used for higher accuracy activity recognition using COVID-19 [34].
- **COMMODITY WiFi**: COMMODITY WiFi has been used during COVID-19 as a sensor. The Fine-grained PHY layer channel state information (CSI) of WiFi signals is extensively used for sensing purposes because it is stable and is not affected by environmental conditions like its earlier counterpart received signal strength indicator (RSSI) [35].
- **MAGNETOMETER**: One of the critical sensors in most IoT devices is a magnetometer. This sensor works on principles of magnetic field and is used to detect the orientation of the devices. It can detect magnetic fields in three perpendicular directions. This capability of the magnetometer is used to identify the distance between the devices since the magnetic field contains spatial information [36]. Therefore magnetometer is used extensively as a proximity sensing which is usually used for contact tracing during COVID-19.

Take-Home Message

This section presents various projects and uses cases on IoT and AI. These use cases are from the real world and are used extensively to fight against COVID-19. After going through these use cases and projects, the reader will come to know the real-life applications of AI and IoT in fighting any infection. Some of the use cases are so common that the reader can visualize them, enhancing their understanding of these technologies.

7 Challenges and Future Directions

IoT and AI in healthcare are promising, especially when dealing with an epidemic like COVID-19. Though the technologies of AIoT have proven benefits during COVID-19, there are several open issues and critical challenges. These challenges must be considered carefully before deploying AIoT-enabled technologies in the healthcare domain. To move forward with our point, we first highlight some of the research challenges that are faced by these technologies, then we move on to describe the potential briefly in this domain together with future directions.

7.1 Challenges

There have been various challenges associated with IoT and AI, i.e., AIoT, but we have confined our study to the main four challenges. These are:

- **Regulatory Challenge**: Though AIoT has been used extensively in healthcare, its use must adhere to regulatory laws in the country. Also, there has been a lack of a regulatory framework to support the growing use of IoT in the healthcare industry. We also consider legal issues related to content and personal information running on IoT and AI platforms. These issues are related to the copyright infringement and defamation. Laws must be made that govern AI and IoT in the healthcare sector [37].
- **Privacy Challenge**: In the scenario of COVID-19, it has become essential to protect people's privacy. The applications of AIoT like contact tracing have put people's privacy at stake. The data that is being collected using IoT devices must be protected from third-party usage where it can be utilized in the wrong manner. Various applications powered by AI use location-oriented data of people, which must be regulated by the government [38].
- **Challenge of Unified Database**: One of the significant challenges in fighting COVID-19 is the lack of a unified database such as cases infections, history of their medical record, location, and medical supply status. The sources of data in most of the cases are individual and confined, like data coming from individual medical records and data coming from social media accounts [39]. This type of decentralized data is not sufficient for large-scale AI to work and puts restrictions on fighting COVID-19 [40].
- **Implementation challenges**: Implementing AIoT-based solutions to fight COVID-19 infections is becoming increasingly challenging [41]. This is because the data needed by AI algorithms and collected by IoT is enormous and therefore requires strong hardware support and storage. Another challenge is the deployment of IoT-enabled devices due absence of standards or SOPs. This is the reason behind low quality and cheaply designed AIoT devices, which may have undesirable consequences for the users.
- **Security Challenge**: The AIoT-based solutions to fight COVID-19 have increased demand for IoT devices. This increasing demand for devices has shifted the developer's concerns from quality to quantity. This has resulted in the mass manufacturing of poor quality devices, making the security of such devices weak and vulnerable [42]. Further, the homogeneity in the design of such devices has increased the chances of a single security weakness to all the devices having the same features.

7.2 Future Directions

After discussing the potential challenges of AIoT-led solutions to fight COVID-19 infection, let us discuss some future directions of research on AIoT for fighting such pandemics.

- **Improving the performance of IoT-enabled devices**: The performances of IoT-enabled devices must be optimized to achieve high efficiency, scalability, and throughput. The aim of performance optimization must be to make IoT an ideal choice for emergency use in healthcare applications [43]. Further research must be targeted on how IoT devices can aid isolated quarantined patients efficiently manage their daily chores. How IoT-enabled technology is incorporated into business and the marketplace to cover both the safety and efficiency of both must be a matter of concern.
- **Security issues of IoT**: One of the main concerns of using IoT-enabled devices during the current pandemic of COVID-19 was personal data sharing, where patients were asked to share their data [42]. This is a big concern for every person. Therefore there is a need to define secure channels for communications by utilizing efficient encryption techniques before sharing private information. These could be possible research areas.
- **Improving AI algorithms**: Intelligent data analytics needs intelligent AI technologies. This will surely increase the efficiency of intelligent data analytics, which is a key to its applications in healthcare. Intelligent healthcare analytics will include virus structure analysis and drug manufacturing. There is a need to develop adaptive AI models for containing infections like COVID-19 [44]. This will aid in better monitoring, detection & prediction of such infections.
- **Incorporation with other technologies**: IoT and AI-enabled technologies must be incorporated with other technologies such as cloud computing, fog computing and blockchain (BIoT). Incorporating technologies will build a comprehensive and efficient healthcare system [45]. For instance, we can see that Alibaba has integrated cloud computing with AI. This combination has yielded many wonderful results while carrying out AI-enabled data analytics for fighting infections like COVID-19. Likewise, all three technologies must be combined for better outcomes in healthcare sectors and in smart cities [46–48].

Take-Home Message

This section discusses the potential challenges AI and IoT technologies face in their deployment in the real world. This section also highlights the future work needed to meet these challenges. After reading this section, readers who wish to start their careers in this domain will get a clear understanding of the domain. This will motivate them and the stakeholders to put extra efforts into using the potential of these technologies to contain future pandemics.

8 Conclusion

This chapter introduces AI and IoT technologies to the readers. After reading this chapter, the reader will establish a deep understanding of AI and IoT technologies that are used in the healthcare sector. The main focus of the chapter is to introduce to the readers the AIoT technologies used to contain the spread of COVID-19. We have mentioned many applications of AIoT from real life and have also mentioned the research work that is carried out for the enhancement of present technologies. We have also mentioned many use cases on AI and IoT adaption in response the present pandemic, i.e., COVID-19. We have discussed how these use cases have helped the stakeholder to contain the spread of the current pandemic, i.e., COVID-19. Further in the chapter, we have discussed various challenges which need to be taken care of to fully exploit the potential of AIoT technologies in the healthcare sector. In the end, we have mentioned the future directions in order to meet the challenges being faced by AIoT in the healthcare sector.

References

1. Gupta PK, Siddiqui MK, Huang X, Morales-Menendez R, Pawar H, Terashima-Marin H, Wajid MS (2022) COVID-WideNet—a capsule network for COVID-19 detection. Appl Soft Comput 122:108780
2. Kharya S, Onyema EM, Zafar A, Wajid MA, Afriyie RK, Swarnkar T, Soni S (2022) Weighted Bayesian belief network: a computational intelligence approach for predictive modeling in clinical datasets. Comput Intell Neurosci
3. Nasajpour M, Pouriyeh S, Parizi RM, Dorodchi M, Valero M, Arabnia HR (2020) Internet of things for current COVID-19 and future pandemics: an exploratory study. J Healthc Informatics Res 4(4):325–364. https://doi.org/10.1007/s41666-020-00080-6
4. Chen SW, Gu XW, Wang JJ, Zhu HS (2021) AIoT used for COVID-19 pandemic prevention and control. Contrast Media Mol Imaging. https://doi.org/10.1155/2021/3257035
5. Edeh MO, Dalal S, Obagbuwa IC, Prasad BS, Ninoria SZ, Wajid MA, Adesina AO (2022) Bootstrapping random forest and CHAID for prediction of white spot disease among shrimp farmers. Sci Rep 12(1):20876
6. Zafar A, Wajid MA (2020) A mathematical model to analyze the role of uncertain and indeterminate factors in the spread of pandemics like COVID-19 using neutrosophy: a case study of India. Neutrosophic Sets Syst 38:214–226. https://doi.org/10.5281/zenodo.4300498
7. Shao D et al (2016) Noncontact monitoring of blood oxygen saturation using camera and dual-wavelength imaging system. IEEE Trans Biomed Eng 63(6):1091–1098. https://doi.org/10.1109/TBME.2015.2481896
8. Istiqomah NA, Al-zubaidi SS (2020) Toward a novel design for coronavirus detection and diagnosis toward a novel design for coronavirus detection and diagnosis system using IoT based drone technology. https://doi.org/10.37200/IJPR/V24I7/PR270220
9. Chamola V, Hassija V, Gupta V, Guizani M (2020) A comprehensive review of the COVID-19 pandemic and the role of IoT, Drones, AI, Blockchain, and 5G in managing its impact, pp 90225–90265
10. Shu D, Ting W, Carin L, Dzau V, Wong TY (2020) Digital technology and COVID-19, vol 26, pp 2019–2021. https://doi.org/10.1038/s41591-020-0823-6
11. Hussain AA, Bouachir O (2020) AI techniques for COVID-19, vol 8

12. Sufian A, Ghosh A, Safaa A, Smarandache F (2020) Since January 2020 Elsevier has created a COVID-19 resource centre with free information in English and Mandarin on the novel coronavirus COVID- 19. The COVID-19 resource centre is hosted on Elsevier Connect, the company's public news and information
13. Bansal A, Padappayil RP, Garg C, Singal A, Gupta M, Klein A (2020) Utility of artificial intelligence amidst the COVID 19 pandemic
14. Bhattacharya S, Kumar P, Maddikunta R, Pham Q (2020) Since January 2020 Elsevier has created a COVID-19 resource centre with free information in English and Mandarin on the novel coronavirus COVID-19. The COVID-19 resource centre is hosted on Elsevier Connect, the company's public news and information
15. Ghouali S, Onyema EM, Guellil MS, Wajid MA, Clare O, Cherifi W, Feham M (2022) Artificial intelligence-based teleopthalmology application for diagnosis of diabetics retinopathy. IEEE Open J Eng Med Biol 3:124–133
16. Nguyen TT, Pathirana PN, Nguyen T, Viet Q, Nguyen H (2021) Genomic mutations and changes in protein secondary structure and solvent accessibility of SARS—CoV-2 (COVID-19 virus). Sci Rep 1–16. https://doi.org/10.1038/s41598-021-83105-3
17. Nguyen DC et al (2020) Enabling AI in future wireless networks: a data life cycle perspective, pp 1–42
18. Rane KP (2020) Design and development of low cost humanoid robot with thermal temperature scanner for COVID-19 virus preliminary identification. Int J Adv Trends Comput Sci Eng 9(3):3485–3493. https://doi.org/10.30534/ijatcse/2020/153932020
19. Gupta PK, Khubeb M, Huang X, Morales-menendez R (2022) COVID-WideNet—a capsule network for COVID-19 detection. Appl Soft Comput 122:108780. https://doi.org/10.1016/j.asoc.2022.108780
20. Wang Y, Hu M, Li Q, Zhang X-P, Zhai G, Yao N (2020) Abnormal respiratory patterns classifier may contribute to large-scale screening of people infected with COVID-19 in an accurate and unobtrusive manner. IEEE Internet Things J 7(9):8559–8571. 2002.05534v2
21. Faezipour M, Abuzneid A, Science C (2020) Smartphone-based self-testing of COVID-19 using breathing sounds, pp 1202–1205. https://doi.org/10.1089/tmj.2020.0114
22. Wang G, Member S, Gu C (2014) Application of linear-frequency-modulated continuous-wave (LFMCW) radars. https://doi.org/10.1109/TMTT.2014.2320464
23. Abdelnasser H (2015) UbiBreathe: a ubiquitous non-invasive WiFi-based breathing estimator
24. Shao D et al (2015) Noncontact monitoring of blood oxygen saturation using camera and dual—wavelength imaging system
25. Hsiao S-H, Chen T-C, Chien H-C, Yang C-J, Chen Y-H (2020) Measurement of body temperature to prevent pandemic COVID-19 in hospitals in Taiwan: repeated measurement is necessary. J Hosp Infect 105(2):360–361. https://doi.org/10.1016/j.jhin.2020.04.004
26. Zakaria NA, Nadia F, Mohd B, Azhar M, Razak A (2018) IoT (Internet of Things) based infant body temperature monitoring. In: 2018 2nd international conference on biosignal analysis, processing and systems, pp 148–153. https://doi.org/10.1109/ICBAPS.2018.8527408
27. Hu P (2020) IoT-based contact tracing systems for infectious diseases: architecture and analysis, pp 1–6. https://doi.org/10.1109/GLOBECOM42002.2020.9347957.P
28. Wajid MS, Terashima-Marin H, Paul Rad PN, Wajid MA (2022) Violence detection approach based on cloud data and neutrosophic cognitive maps. J Cloud Comput 11(1):1–18
29. Jeong S, Kuk S, Kim H (2019) A smartphone magnetometer-based diagnostic test for automatic contact tracing in infectious disease epidemics. IEEE Access 7:20734–20747. https://doi.org/10.1109/ACCESS.2019.2895075
30. Dananjayan S, Raj GM (2020) Artificial intelligence during a pandemic: the COVID-19 example. Int J Health Plann Manage 35(5):1260–1262. https://doi.org/10.1002/hpm.2987
31. Huang J et al (2020) Care for the psychological status of frontline medical staff fighting against coronavirus disease 2019 (COVID-19). Clin Infect Dis 71(12):3268–3269. https://doi.org/10.1093/cid/ciaa385
32. Vishwakarma S, Agrawal A (2013) A survey on activity recognition and behavior understanding in video surveillance. Vis Comput 29(10):983–1009. https://doi.org/10.1007/s00371-012-0752-6

33. Mitomo T, Ono N, Hoshino H, Yoshihara Y, Watanabe O, Seto I (2010) A 77 GHz 90 nm CMOS transceiver for FMCW radar applications. IEEE J Solid-State Circuits 45(4):928–937. https://doi.org/10.1109/JSSC.2010.2040234
34. Wang G, Gu C, Inoue T, Li C (2014) A hybrid FMCW-interferometry radar for indoor precise positioning and versatile life activity monitoring. IEEE Trans Microw Theory Tech 62(11):2812–2822. https://doi.org/10.1109/TMTT.2014.2358572
35. Wang W, Liu AX, Shahzad M, Ling K, Lu S (2015) Understanding and modeling of WiFi signal based human activity recognition. In: Proceedings of the annual international conference on mobile computing and networking, MOBICOM, vol 2015, pp 65–76. https://doi.org/10.1145/2789168.2790093
36. Pasku V et al (2017) Magnetic field-based positioning systems. IEEE Commun Surv Tutorials 19(3):2003–2017. https://doi.org/10.1109/COMST.2017.2684087
37. Bahalul Haque AKM, Bhushan B, Nawar A, Talha KR, Ayesha SJ (2022) Attacks and countermeasures in IoT based smart healthcare applications. In: Balas VE, Solanki VK, Kumar R (eds) Recent advances in internet of things and machine learning. Intelligent systems reference library, vol 215. Springer, Cham. https://doi.org/10.1007/978-3-030-90119-6_6
38. Ahmad N, Chauhan P (2020) State of data privacy during COVID-19. Computer (Long Beach Calif) 53(10): 119–122. https://doi.org/10.1109/MC.2020.3010549
39. Chen E, Lerman K, Ferrara E (2020) Tracking social media discourse about the COVID-19 pandemic: development of a public coronavirus Twitter data set. JMIR Public Heal Surveill 6(2). https://doi.org/10.2196/19273
40. Cohen JP, Morrison P, Dao L (2020) COVID-19 image data collection. Available: http://arxiv.org/abs/2003.11597
41. Sethi R, Bhushan B, Sharma N, Kumar R, Kaushik I (2020) Applicability of industrial IoT in diversified sectors: evolution, applications and challenges. Stud Big Data Multim Technol Internet Things Environ. https://doi.org/10.1007/978-981-15-7965-3_4
42. Liu Q, Liu T, Liu Z, Wang Y, Jin Y, Wen W (2018) Security analysis and enhancement of model compressed deep learning systems under adversarial attacks. In: Proceedings of the Asia South Pacific design automation conference ASP-DAC, vol 2018, pp 721–726. https://doi.org/10.1109/ASPDAC.2018.8297407
43. Malik A, Gautam S, Abidin S, Bhushan B (2019) Blockchain technology-future of IoT: including structure, limitations and various possible attacks. In: 2019 2nd international conference on intelligent computing, instrumentation and control technologies (ICICICT). https://doi.org/10.1109/icicict46008.2019.8993144
44. Chee ML et al (2021) Artificial intelligence applications for covid-19 in intensive care and emergency settings: a systematic review. Int J Environ Res Public Health 18(9). https://doi.org/10.3390/ijerph18094749
45. Wang Y et al (2020) COVID-19 data visualization public welfare activity. Vis Informatics 4(3):51–54. https://doi.org/10.1016/j.visinf.2020.09.003
46. Onyema EM, Dalal S, Romero CAT, Seth B, Young P, Wajid MA (2022) Design of intrusion detection system based on cyborg intelligence for security of cloud network traffic of smart cities. J Cloud Comput 11(1):1–20
47. Wajid MA, Zafar A (2021) Pestel analysis to identify key barriers to smart cities development in India. Neutrosophic Sets Syst 42:39–48
48. Bhushan B, Sahoo C, Sinha P, Khamparia A (2020) Unification of Blockchain and Internet of Things (BIoT): requirements, working model, challenges and future directions. Wireless Netw. https://doi.org/10.1007/s11276-020-02445-6

'rinted in the United States
ʳ Baker & Taylor Publisher Services